The
MEASURE
OF ALL
THINGS

Jean-Baptiste-Joseph Delambre, *aged fifty-two, as painted in 1803 by Per Eberhard Gogell. Delambre wears the uniform of the Academy of Sciences.*

(Swedish Royal Academy of Sciences. Photo: Georgios Athanasiadis)

Pierre-François-André Méchain, *in the uniform of the Academy of Sciences, painted posthumously in 1824 by Narcisse Garnier, based on etchings taking during Méchain's lifetime.*

(Musée de Laon)

The
MEASURE
OF ALL
THINGS

*The Seven-Year Odyssey that
Transformed the World*

Ken Alder

LITTLE, BROWN

A *Little, Brown* Book

First published in the United States of America by the Free Press in 2002
First published in Great Britain in 2002 by Little, Brown

Copyright © Ken Alder 2002

The moral right of the author has been asserted.

A CIP catalogue record for this book is available from the British Library

ISBN 0 316 85989 3

Cover art: These oil portraits of Delambre and Méchain currently hang in the
Paris Observatory. Both were commissioned in the late nineteenth century,
long after their subjects were dead. Delambre's portrait was based on a bust
of the astronomer made toward the end of his life. The portrait of Méchain,
by the painter Hielre, is a 1882 copy of a posthumous portrait commissioned
in 1824 by Méchain's son. The Borda repeating circle pictured here –
numbered Lenoir No. IIII – was one of the two circles used by Delambre and
is the only instrument still surviving from the expedition of 1792–9. It was
very recently rediscovered among the holdings of the Marseille Observatory.
(Observatoire de Paris for portraits. Observatoire de Marseille and the
Ministère de Culture et Ministère de Recherche for the Borda circle.)

Maps by Chris Robinson

Typeset in Cochin by M Rules
Printed and bound in Great Britain by Clays Ltd, St Ives plc

Little, Brown
An imprint of
Time Warner Books UK
Brettenham House
Lancaster Place
London WC2E 7EN

www.TimeWarnerBooks.co.uk

For Bronwyn and Madeleine

It is the star to every wandering bark,
Whose worth's unknown, although his height be taken.

CONTENTS

DRAMATIS PERSONAE

The leading players

Jean-Baptiste-Joseph Delambre (1749–1822). Astronomer who led the northern portion of the meridian expedition in 1792–9. Delambre finished his career as Permanent Secretary of the Paris Academy of Sciences.

Pierre-François-André Méchain (1744–1804). Astronomer who led the southern portion of the meridian expedition in 1792–9, with the assistance of the cartography engineer, **Jean-Joseph Tranchot**. Méchain married **Barbe-Thérèse Marjou** in 1777; their oldest son **Jérôme-Isaac** served on Napoleon's Egyptian expedition, and their younger son **Augustin** assisted Méchain on his second mission to Spain, from which he never returned.

Joseph-Jérôme Lalande (1732–1807). Astronomer and *philosophe* in the high Enlightenment tradition. An avid atheist, a friend of Voltaire and the self-styled 'most famous astronomer in the universe', he was the *maître* of both Delambre and Méchain.

The supporting cast

Jean-Charles de Borda (1733–99). Veteran naval commander and France's leading experimental physicist. He invented the repeating circle, the scientific instrument used by Delambre and Méchain.

Jean-Dominique de Cassini, known as Cassini IV (1748–1845). The fourth member of his family in succession to direct the Royal Observatory of Paris under the *ancien régime*. Appointed to lead the meridian expedition, he withdrew as a protest against the Revolution.

Marie-Jean-Antoine-Nicolas Caritat de Condorcet (1743–94). History's great optimist for human progress. He served as Permanent Secretary of the Academy of Sciences under the *ancien régime*. An ardent revolutionary, he touted the political and egalitarian virtues of the metric system. He committed suicide in 1794 to avoid execution at the hands of the Revolutionary police.

Pierre-Simon Laplace (1749–1827). The leading mathematical physicist of his age. His crowning achievement, the *System of the World*, represented the eighteenth-century culmination of Newtonian physics. A crucial part of Laplace's theory concerned the shape of the earth. He was a leading proponent of the metric system.

Antoine-Laurent Lavoisier (1743–94). One of the principal founders of modern chemistry and – thanks to his position as a royal tax collector – one of the wealthiest men in *ancien régime* France. Despite welcoming the Revolution, and serving as a powerful behind-the-scenes advocate of metric reform, he was executed in 1794 for his participation in the tax-gathering authority of the *ancien régime*.

Adrien-Marie Legendre (1752–1833). One of France's leading mathematicians. He helped found modern statistics using the data gathered by Delambre and Méchain.

Claude-Antoine Prieur-Duvernois, known as Prieur de la Côte-d'Or (1763–1832). A junior military engineer who became a co-dictator of France as a member of the Committee of Public Safety. He was a central force behind the adoption of the metric system.

Etienne Lenoir (1744–1822). France's premier instrument-maker. He fashioned the Borda repeating circle, as well as the definitive platinum metre bar of 1799.

The
MEASURE
OF ALL
THINGS

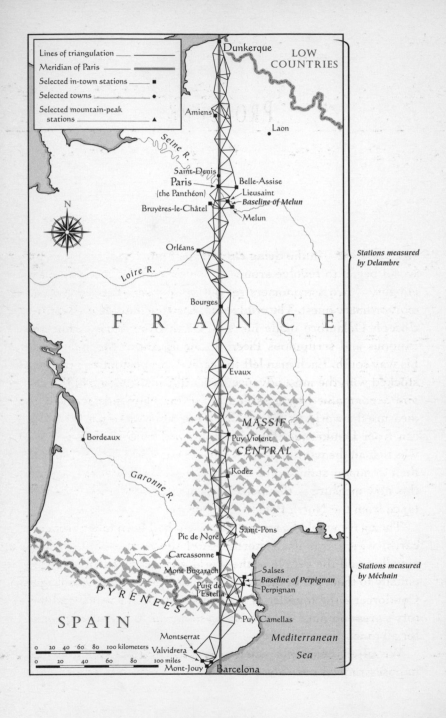

Lines of triangulation ⎯⎯⎯
Meridian of Paris ⎯⎯⎯
Selected in-town stations ⎯⎯⎯ ■
Selected towns ⎯⎯⎯ ●
Selected mountain-peak stations ⎯⎯⎯ ▲

Dunkerque

LOW COUNTRIES

Amiens

Laon

Seine R.

Saint-Denis
Paris
(the Panthéon)
Bruyères-le-Châtel

Belle-Assise
Lieusaint
Baseline of Melun
Melun

N

Orléans

Loire R.

F R A N C E

Bourges

Stations measured
by Delambre

Evaux

Bordeaux

MASSIF

Puy Violent

CENTRAL

Rodez

Garonne R.

Saint-Pons

Pic de Noré

Carcassonne

Mont Bugarach
Puig de
l'Estella

Salses
Baseline of Perpignan
Perpignan

Stations measured
by Méchain

P Y R E N E E S

Puy Camellas

S P A I N

Montserrat

Valvidrera

*Mediterranean
Sea*

0 20 40 60 80 100 kilometers
0   20   40   60   80   100 miles

Mont-Jouy

Barcelona

PROLOGUE

In June 1792 – in the dying days of the French monarchy, as the world began to revolve around a new promise of Revolutionary equality – two astronomers set out in opposite directions on an extraordinary quest. The erudite and cosmopolitan Jean-Baptiste-Joseph Delambre made his way north from Paris, while the cautious and scrupulous Pierre-François-André Méchain made his way south. Each man left the capital in a customized carriage stocked with the most advanced scientific instruments of the day and accompanied by a skilled assistant. Their mission was to measure the world, or at least that piece of the meridian arc which ran from Dunkerque, through Paris, to Barcelona. Their hope was that all the world's peoples would henceforth use the globe as their common standard of measure. Their task was to establish this new measure – 'the metre' – as one ten-millionth of the distance from the North Pole to the equator.

The metre would be eternal because it had been taken from the earth, which was itself eternal. And the metre would belong equally to all the people of the world, just as the earth belonged equally to them all. In the words of their Revolutionary colleague Condorcet – the founder of mathematical social science and history's great optimist – the metric system was to be 'for all people, for all time'.

We often hear that science is a revolutionary force that imposes radical new ideas on human history. But science also emerges from within human history, reshaping ordinary actions, some so habitual we hardly notice them. Measurement is one of our most ordinary actions. We speak its language whenever we exchange precise information or trade objects with exactitude. This very ubiquity, however, makes measurement invisible. To do their job, standards must operate as a set of shared assumptions, the unexamined background against which we strike agreements and make distinctions. So it is not surprising that we take measurement for granted and consider it banal. Yet the use a society makes of its measures expresses its sense of fair dealing. That is why the balance is a widespread symbol of justice. The admonition is found in the Old Testament: 'Ye shall do no unrighteousness in judgment, in meteyard, in weight, or in measure. Just balances, just weights, a just *ephah*, and a just *hin*, shall ye have.'[1] Our methods of measurement define who we are and what we value.

The men who created the metric system understood this. They were the pre-eminent scientific thinkers of the Enlightenment, an age which had elevated reason to the rank of 'sole despot of the universe'. These 'savants' – as the investigators who studied nature were known in those days – had a modern face looking towards us, and an older face glancing back towards the past. In their own minds, of course, they were not two-faced; it was their world which was two-faced, with its burdensome past obstructing progress and a Utopian future waiting to be born.

The savants were appalled by the diversity of weights and measures they saw all around them. Measures in the eighteenth century not only differed from nation to nation, but within nations as well. This diversity obstructed communication and commerce, and hindered the rational administration of the state. It also made it difficult for the savants to compare their results with those of their colleagues. One Englishman, travelling through France on the eve of the Revolution, found the diversity there a torment. '[I]n France,' he complained, 'the infinite perplexity of the measures exceeds all comprehension. They differ not only in every province, but in every district and almost every town . . .'

Contemporaries estimated that under the cover of some eight hundred names, *ancien régime* France contained a staggering 250,000 different units of weights and measures.[2]

In place of this Babel of measurement, the savants imagined a universal language of measurement which would bring order and reason to the exchange of both goods and information. It would be a rational and coherent system that would induce its users to think about the world in a rational and coherent way. But all the savants' grand plans would have remained fantasy had not the French Revolution – history's great Utopian rupture – provided them with an unexpected chance to throw off the shackles of custom, and build a new world upon principled foundations. Just as the French Revolution had proclaimed universal rights for all people, the savants argued, so too should it proclaim universal measures. And to ensure that their creation would not be seen as the handiwork of any single person or nation, they decided to derive its fundamental unit from the measure of the world itself.

For seven years Delambre and Méchain travelled the meridian to extract this single number from the curved surface of our planet. They began their journey in opposite directions, and then, when they had reached the extremities of their arc, measured their way back towards one another through a country quickened with revolution. Their mission took them to the tops of filigree cathedral spires, to the summits of domed volcanoes, and very nearly to the guillotine. It was an operation of exquisite precision for such violent times. At every turn they encountered suspicion and obstruction. How do you measure the earth while the world is turning beneath your feet? How do you establish a new order when the countryside is in chaos? How do you set standards when everything is up for grabs? Or is there, in fact, no better time to do so?

At last, their seven years of travel done, the two astronomers met in the southern fortress town of Carcassonne, and from there returned to Paris to present their data to an International Commission, the world's first international scientific conference.

The results of their labours were then enshrined in a metre bar of pure platinum. It was a moment of triumph: proof that in the midst of social and political upheaval, science could produce something of permanence. Accepting the fruit of their labour, France's new ruler made a prophecy. 'Conquests will come and go,' declared Napoleon Bonaparte, 'but this work will endure.'[3]

In the last two hundred years, conquests have indeed come and gone, but the metre has become the measure of all things. The metric system serves today as the common language of high-tech communications, cutting-edge science, machine production and international commerce. Older forms of measurement have receded as the metric system has made possible trade and economic coordination on a fully global scale. Paradoxically, the leading nation in the global economy remains the sole exception to this rule. Thomas Jefferson failed to convince Congress to make the United States the second nation to adopt the metric system, and every reformer since has met the same fate. John Quincy Adams, asked to consider whether the United States should join the metric system, called it the greatest invention since the printing press and predicted it would save more human labour than the steam engine. Yet he recommended against its adoption. Only in recent years have American manufacturers begun retooling for metric units. Few Americans realize that a silent revolution is finally underway in their nation, transforming their measures under the pressures of the new global economy.[4]

As things stand, of course, this conversion is embarrassingly incomplete. Americans became painfully aware of this fact in 1999 with the loss of the Mars Climate Orbiter. A NASA investigation into the satellite's failure revealed that one team of engineers had used traditional American units, while another had used metric units. The result was a trajectory error of sixty miles, and a $125-million disappearing act.

The revolutionary scientists created the metric system two hundred years ago to avoid just this sort of fiasco. One of their aims was to facilitate communication among scientists, engineers and administrators. Their grander ambition was to transform France –

and ultimately, the whole world – into a free market for the open exchange of goods and information. Today, their goal seems within reach. Over 95 per cent of the world's population now officially uses the metric system, and its success is touted as one of the benign triumphs of globalization.

But behind the public triumph of the metric system lies a long and bitter history. The fundamental fallacy of utopianism is to assume that everyone wants to live in the same Utopia. France, it turns out, was not only the first nation to invent the metric system; she was also the first to reject it. For decades after its introduction, ordinary people spurned the new system, and clung to their local measures and the local economies they sustained. In the face of this revolt from below, Napoleon, on the eve of his disastrous invasion of Russia, returned France to the Paris measures of the *ancien régime*. Now he mocked the global aspirations of the men he had once admired. 'It was not enough for them to make forty million people happy,' he sneered, 'they wanted to sign up the whole universe.'[5] Not until the middle of the nineteenth century did France revert to the metric system, and use of the old measures persisted into the twentieth. It would take enormous scientific effort and years of bitter conflict to make metric measurement banal; just as it had taken a revolution to bring the metric system into being. Things might easily have turned out differently.

What neither advocates nor opponents of the metric system could have known is that a secret error lies at the heart of the metric system – an error perpetuated in every subsequent definition of the metre. Indeed, as I discovered in the course of my research, the only people who could have known the full extent of this error were Delambre and Méchain themselves.

For those who wish to know the origins of the metric system, there is one place to turn: the official account composed by one of the leaders of the meridian expedition, the north-going astronomer, Jean-Baptiste-Joseph Delambre. Delambre wrote the *Base du système métrique* – which we might translate as *The*

Foundation of the Metric System – in order to present all the expedition's findings 'without omission or reticence'.[6] At over two thousand pages, this magisterial work certainly appears thorough enough. But bulky and authoritative as it is, the *Base* is a strange book, with puzzling contradictions. Reading it, I began to get the sense that this was not the complete history of the metre, and that Delambre had himself scattered clues to this effect throughout the text. For instance, in Volume 3 he explained that he had deposited all the records of the metric calculations in the archives of the Observatory of Paris lest future generations doubt the soundness of their procedures.

The records are still there. The Observatory of Paris is an imposing stone structure just south of the Luxembourg Gardens in the heart of modern Paris. In the 1660s, when Louis XIV founded the Royal Observatory and Royal Academy of Sciences, his goal was to couple the glory of his rule with the new heavenly science, and also to supply his savants with the tools they would need to assemble an accurate map of his kingdom here on earth. The building is perfectly aligned along the nation's north–south meridian. Like France, it presents two faces. From the north, it might almost be mistaken for a royal fortress, with austere stone walls guarding a grey plain of mist and gravel that stretches towards the North Sea. From the south, it resembles an elegant residential palace, with octagonal pavilions looking out over a terraced park that seems to step, via an alley of plane trees, down to a remote Mediterranean. During the *ancien régime*, most of France's finest astronomers lodged within its green precincts. Today, the site remains the privileged workplace of its leading astrophysicists.

The Observatory archives are located in the south-eastern octagon, where the papers of the meridian expedition fill twenty cartons. They include thousands of pages of computation in logbooks and on scraps of paper, along with maps, protocols, diagrams and formulae, the seven years of calculation which went into the making of a single number: the length of the metre. Leafing through one of Méchain's logbooks, I found an extended commentary written and signed by Delambre.

I deposit these notes here to justify my choice of which version of Méchain's data to publish. Because I have not told the public what it does not need to know. I have suppressed all those details which might diminish its confidence in such an important mission, one which we will not have a chance to verify. I have carefully silenced anything which might alter in the least the good reputation which Monsieur Méchain rightly enjoyed for the care he put into all his observations and calculations.[7]

I can still remember the shock I felt upon reading those words. Why was there more than one version of Méchain's data? What exactly had been hidden from the public? Part of the answer lay in the one carton which had not been deposited with the rest, but stored separately by Delambre and placed by him under seal as a special precaution. That seal is now broken. Inside, there are no logbooks or calculations. Instead there are letters, dozens of letters between Delambre and Méchain, as well as letters between Delambre and Madame Méchain. Had I stumbled, amid all these dusty calculations, on a scandal of intrigue and deception? Reading through these letters, I began to realize that I had discovered something much more interesting: a tale of scientific error and the agonizing choices it forced upon men and women of integrity. In the margin of Méchain's last letter to Delambre, posted from the abandoned monastery of Saint-Pons in the remote Montagnes Noires (the Black Mountains) of southern France, Delambre had scribbled a final explanatory note.

Though Méchain more than once begged me to burn his letters, his mental state, and my fear that he would one day turn against me, led me to keep them in case I ever needed to use them to defend myself . . . [B]ut I thought it prudent to place them under seal so that they could not be opened unless someone needed to verify the extracts I published in the *Base du système métrique*.[8]

The remaining clues to the mystery lay elsewhere, scattered not only across France and the sources Delambre preserved, but also in the records of the savants' many correspondents in Spain, Holland, Italy, Germany, Denmark, England and the United States, including a cache of Delambre's papers which had mysteriously vanished from a French archive – along with the rubbish, said the archivists – to find its way, via a London auction house, to the library of Brigham Young University in Provo, Utah. And finally, I tracked down something long presumed lost: Delambre's own copy of his magisterial work, the *Base du système métrique*.

Those volumes are located today in the private home of David Karpeles, a collector of rare books and manuscripts in Santa Barbara, California. There, on the title page, in his angular hand, Delambre had inscribed Napoleon's grand prophecy: '"Conquests will come and go, but this work will endure", words of Nap. Bonaparte to the author of the *Base*.'[9] Yet the title page was not the only page on which he had recorded his marginal comments.

Together, these documents reveal a remarkable story. They reveal that Méchain – despite his extreme caution and exactitude – committed an error in the early years of the expedition, and worse, upon discovering his mistake, covered it up. Méchain was so tormented by the secret knowledge of his error that he was driven to the brink of madness. In the end, he died in an attempt to correct himself. The metre, it turns out, is in error, an error which has been perpetuated in every subsequent redefinition of its length, including our current definition of the metre in terms of the distance travelled by light in a fraction of a second.

According to today's satellite surveys, the length of the meridian from the pole to the equator equals 10,002,290 metres. In other words, the metre calculated by Delambre and Méchain falls roughly 0.2 millimetres short, or about the thickness of two pages of this book. It may not seem like much, but it is enough to feel with your fingers, enough to matter in high-precision science, and in that slender difference lies a tale of two men sent out in opposite directions on a Herculean task – a mission to measure the world – who discovered that integrity could carry them in directions as contrary as their carriages. Both were men in their mid-forties, men of humble origins from the French provinces

BASE

DU SYSTÈME MÈTRIQUE DÉCIMAL,

ou

MESURE DE L'ARC DU MÉRIDIEN

COMPRIS ENTRE LES PARALLÈLES

DE DUNKERQUE ET BARCELONE,

EXÉCUTÉE EN 1792 ET ANNÉES SUIVANTES,

PAR MM. MÉCHAIN ET DELAMBRE.

Rédigée par M. Delambre, secrétaire perpétuel de l'Institut pour les sciences mathématiques, membre du bureau des longitudes, des sociétés royales de Londres, d'Upsal et de Copenhague, des académies de Berlin et de Suède, de la société Italienne et de celle de Gottingue, et membre de la Légion d'honneur.

SUITE DES MÉMOIRES DE L'INSTITUT.

Les conquêtes passent et ces opérations restent

TOME PREMIER. *Paroles de Nap. Bonaparte à l'auteur de la Base.*

PARIS.

BAUDOUIN, IMPRIMEUR DE L'INSTITUT NATIONAL.

JANVIER 1806.

Delambre's *Base du système métrique décimal*, Karpeles edition

On the title page of his own copy of the Base, *Delambre wrote: "'Les conquêtes passent et ces opérations restent.' Paroles de Nap. Bonaparte à l'auteur de la* Base.' *This may be translated as: '"Conquests will come and go, but this work will endure." Words of Nap. Bonaparte to the author of the* Base.'*

(Karpeles Museum, Santa Barbara, California. Photo: David Karpeles.)

who had risen to prominence on the basis of talent and a mind-numbing capacity for work. Both had been trained by the same astronomer, Jérôme Lalande, and elected to the Academy of Sciences in time for the Revolution to hand them the career opportunity of a lifetime: the chance to sign their names to the world's measure. But during their seven years of travels, the two men came to have a different understanding of their metric mission and the allegiance it commanded. That difference would decide their fates.

This then is a tale of error and its meaning: how people strive for Utopian perfection – in their works and in their lives – and how they come to terms with the inevitable shortcomings. What does it feel like to make a mistake, and in a matter of such supreme importance? Yet even in failure, Delambre and Méchain succeeded, for by their labour they not only rewrote our knowledge of the shape of the earth, but our knowledge of error as well. In the process, scientific error was transformed from a moral failing into a social problem, forever altering what it meant to be a practising scientist. And the consequences of their labour resonated far outside the realm of science. We can trace its impact in the globalization of economic exchange, and in the way ordinary people have come to understand their own best interest. In the end, even the French countryside they traversed has been transformed.

To come to terms with this history, I set out to retrace their journey. In the year 2000, at a time when France was celebrating the millennium along the Meridienne Verte – a six-hundred-mile row of evergreen trees which was meant to mark out the national meridian, but which was somehow never planted – I set out on the zigzag trail of Delambre and Méchain. I climbed the cathedral towers and mountain peaks from which they conducted their survey, and combed the provincial archives for traces of their passage. It was my own Tour de France. Delambre and Méchain had demonstrated that the judicious application of scientific knowledge might, as Archimedes once boasted, move the world. Where they travelled by carriage and on foot, I substituted a bicycle. After all, what is a bicycle but a lever on wheels? – a lever which allows the cyclist to move along the surface of the world, or, which is much the same thing, move the world.

Chapter One

THE NORTH-GOING
ASTRONOMER

> Fabrice showed them his passport indicating he was a
> barometer salesmen *travelling with his wares*. 'Are they fools!'
> cried the border guard. 'This goes too far!'[1]
>
> STENDHAL, *The Charterhouse of Parma*

The countryside was strangely silent, the roads deserted. The
local militia had been ordered to stop 'any unknown person travelling by foot, horse or carriage, and with the amiability which
Equality and Liberty prescribe, check their identity against their
passports, and, if the passports prove false, conduct them to the
town hall to be judged according to the law.'[2] That afternoon, a
gendarme told a man travelling by carriage with his wife and
daughter to hurry home. The fortress of Verdun had fallen and
eighty thousand Prussian soldiers were crossing the plains of
Champagne, marching towards Paris to restore the French king
to his throne. Everything was in readiness for the onslaught. A
proclamation had gone out from the capital that the people of the
surrounding hamlets should prepare themselves to 'share with
their co-citizens the honour of saving their fatherland – or of
dying in its defence'.[3] Inside the city walls, the gendarme told the
traveller, patriots had begun to massacre all the city's prisoners
lest they rise in aid of the aristocrats.

That same day – 4 September 1792 – in the highest reaches of a
château set on top of the region's most elevated prominence, a man

was bent over a strange apparatus, sighting across the horizon. The man – a savant by all appearances – had set up an observatory inside a lofty twenty-two-foot pyramid that normally served as a belvedere where diners might admire the delightful prospect. At intervals, he lifted his head from the instrument to manipulate the two telescopes on their interlaced brass rings, pivoting them first one way, then the other, as if solving a mechanical puzzle. Then he bent his eye to the eyepiece to take another sighting, while one assistant verified the gauge and another recorded the value. It was a delicate operation, sensitive to the least vibration. The men dared not shift their weight lest the floorboards transmit their motion to the instrument and perturb values destined to serve as the unique and permanent measure of all things.

The château de Belle-Assise was aptly named. It was indeed 'beautifully situated', famous for its view over the fertile valley of Brie. A château had stood on the hill since the thirteenth century. The current owner, the comte de Vissec, had permitted the expedition to labour in his pleasure pavilion. On the western horizon, the savant could pick out twin domes rising from the grey jumble of Paris: the leaden dome of the new Panthéon, and the golden dome of the old Invalides. On the southern horizon he could make out the Gothic church at Brie-Comte-Robert. And on the northern horizon he could identify the church belfry of Dammartin, due to be demolished. Nearer to his position he could see the medieval dungeon of Montjai, from which he had originally hoped to conduct his measurements. His task was to measure the horizontal angle separating these sites with a precision never before achieved.

That evening, just as the savant completed his fourth and final day of observations at Belle-Assise – night had fallen and his assistants were packing the instruments into their carriages in preparation for the post-horses they had ordered from the town of Lagny – a party of militiamen arrived instead. They were well armed with muskets and well fortified with wine. They had secured permission from the local municipal council to search all the surrounding châteaux. Rumours of treason were circulating through the countryside. It was widely suspected that the four visitors to Belle-Assise were spying for the Prussians. Was it not true they had paid the local carpenter Petit-Jean to build a platform on the ruined tower at Montjai, which, as everyone knew, was haunted by the demons of a murderous priest? They would have to show their papers.[4]

The savant presented his passport. It identified him as Jean-Baptiste-Joseph Delambre, 'jointly charged by the National Assembly to carry out, in conjunction with Monsieur Méchain, the mathematical measurement of the meridian from Dunkerque to Barcelona'. Delambre was a solid, well-set man of forty-two, of average height for that time – he was five foot four – with a round face, a strong nose, blue eyes and brown hair swept back from his forehead. It was a frank and open countenance, yet curiously observant, with a mouth inclined towards irony. His blue eyes were disarmingly naked, and on closer inspection it was clear why: Delambre had no eyelashes. He was an observer rather than a man readily observed.[5]

His assistants presented their papers as well. The first was Michel Lefrançais, a twenty-six-year-old apprentice astronomer, nephew of the illustrious astronomer Jérôme Lalande. The second was Benjamin Bellet, a thirty-two-year-old instrument-maker, an apprentice to Etienne Lenoir, whose workshop had built the expedition's newfangled 'Borda repeating circle', the instrument that was to bring unrivalled precision to their survey. And the third was a manservant named Michel.[6]

The leader of the militia seemed satisfied with these documents. But his followers did not agree. They complained that the passports had expired – or more precisely, had been issued by a political authority which had itself expired. In the four months since they had been signed, an uprising had deposed Louis XVI and installed a republic.

Delambre tried to explain that he had been sent on a mission to measure the size of the world. He was a practitioner of geodesy, the science of measuring the size and shape of the earth. Improbable as it sounded at a time of national emergency, the government had assigned his mission its highest priority. His mission was to travel up and down the meridian of France. The Academy of Sciences –

'There is no more *Cademy*,' interrupted one of the militiamen, 'the *Cademy* is no more. We're all equal now. You'll come with us.'[7]

It was not true, not yet; the Academy still existed, as far as Delambre knew. Earlier that week, Antoine-Laurent Lavoisier, the great chemist and treasurer of the Academy, had admonished him not to quit his mission until he had 'exhausted every reserve of strength within him'.[8] Any halt or failure would have to be justified to the National Assembly itself. But at the moment, further resistance seemed pointless. As Delambre wrote to a friend, 'They were armed and we had only reason; the parties were not equal.'[9]

So Delambre and his team accepted the militia's 'invitation' to accompany them across the night-time fields. The mud was thick, the sky black. A heavy rain had begun to fall. 'Luckily I had time to place a frock coat over my clothes,' Delambre wrote. 'And as we marched, we could talk to the men and make them see reason, so that they began to show us some courtesy, warning us of treacherous footing ahead, and giving us a hand when we needed to be pulled out of the muck.' For the next four hours they accompanied the militia on their rounds, searching houses for arms and requisitioning horses. After struggling for six miles through the dark, they finally arrived in Lagny shortly before midnight, just as a squall soaked them to the bone.[10]

The municipal council was in candlelight session. The town was on a wartime footing. Mayor Aublan, a former financial agent for the local abbey (now abolished), had recently congratulated his constituents on overthrowing the 'odious king' and unmasking the 'perfidious proclamations of the corrupt ministers and other vampires of the realm'.[11] Delambre presented his papers to the assembled officialdom. One alderman recognized the signature of the district official on the papers, and argued that Delambre should be released. But Mayor Aublan was more suspicious. He ordered all four members of the expedition escorted by armed guard to the Hôtellerie de l'Ours (the Bear Inn), where they were 'not to consider themselves arrested, *but merely detained*'. In the

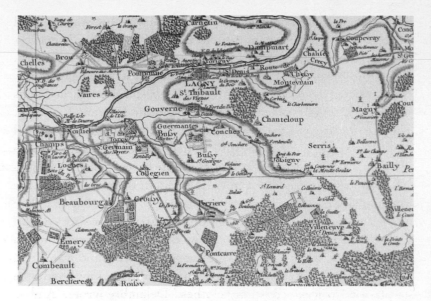

**Cassini map, showing the region around Belle-Assise
to the east of Paris**

*This portion of the great Cassini map of France (1740–95) shows the
area around the château of Belle-Assise. The château is here labelled
Belleassise, south-east of Lagny on the way to Villeneuve. In the
nineteenth century the château came into the possession of the Baron de
Rothschild and was demolished at the end of the century. Its formal
gardens (pictured on the map) are today a tangle of muddy forest. Only
the windmill (likewise indicated) still stands. The town of Lagny is now a
suburb of Paris, and the land to the east of the town is today the valley of
Disneyland, Paris.*

(Earth Sciences Library, University of California, Berkeley. Photo: Custom
Process.)

meantime, Delambre might send a message to the district office so
that they might vouch for the legitimacy of his mission.[12]

'That night we had nothing to change into, no night-clothes,
nothing; and to dry ourselves, only a few sticks of firewood and a
couple of glasses of bad wine.'[13] Their two guards had a worse night
of it, however; they had to spend all night in a draughty corridor,
detaining men who had no intention of escaping. As Delambre

noted in his expedition logbook: 'Consigned to the Hôtellerie de l'Ours, two sentinels on guard at the exits; 4 September 1792, the second year of liberty and the first of equality.'[14]

In the morning, when confirmation came from the district office that the mission was indeed sanctioned by the highest authority in the land, Delambre thought it advisable before leaving town to thank the municipal council in person for their overnight hospitality. As he entered the town hall, the mayor rushed over from his office to apologize for the 'little trouble' of the previous evening – while the impatient militiaman who despised academies stood by with a sullen expression, having apparently slept off his wine.[15] According to the municipal records, Delambre then 'thanked the municipality for so promptly allowing him to continue on his way'.[16]

'And so ends the true and tragicomic history of the memorable arrest of the former *Cademician*,' Delambre wrote that evening to a friend – as if his troubles had not just begun.[17]

Delambre's wry equanimity seems to have been due in part to his late start in science. He did not take up astronomy until his mid-thirties, an age when many scientists are either at the heights of their powers, or already on the downhill slope. He was born in the cathedral town of Amiens on 16 September 1749, the eldest child of modest drapers. The family name Delambre probably derives from *lambeau*, meaning 'rag'. When he was still an infant, fifteen months old, he was stricken with smallpox, which nearly cost him his eyesight and permanently denuded him of eyelashes. If the latter loss ultimately made it easier for him to take up the telescope (lashes tend to get in the way of beginners), his weak eyesight hardly presaged a promising career in observational astronomy. Until the age of twenty he was acutely sensitive to sunlight, and could hardly read his own handwriting. He grew up assuming he would one day go blind. For just that reason, he devoured every book he found. He learned English and German, and studied with the Jesuits until the order was expelled from France, at which point the town brought in three replacement teachers from Paris.[18]

Delambre might have aspired at most to a position as a local *curé* had not one of these teachers put him up for a scholarship at du Plessis, a famous Paris school where adolescent boys absorbed Roman virtues through an endless diet of Latin classics. Graduates included devout theologians, atheist physicians, military republicans and illustrious savants. The high hopes for young Delambre were not fulfilled at exam time, however. He failed his finals because he could not read his exam papers. Without a scholarship for university, his parents urged him to return to Amiens and take up holy orders.

Instead, Delambre stayed in the capital, living on bread and water, studying ancient Greek by day and carousing with demi-monde literati by night. It was the high tide of the Enlightenment. While an elderly Voltaire issued epigrams from Ferney, and moody Rousseau wrote diatribes from the country, their would-be usurpers plotted Utopias in cafés and wrote subversive pamphlets in garrets. Delambre and his friends formed their own literary club. To support himself, he took a temporary position tutoring a nobleman's son in nearby Compiègne. To instruct his pupil, he was obliged to learn mathematics himself. He read Milton's *Paradise Lost* in the original, and composed his own English primer, which included such homilies as: 'To love riches is the property of a base and grovelling soul, as to live [poorly] in comparison of virtue is the property of a noble and generous mind.'[19]

He was certainly poor enough. At the age of twenty-two he returned to Paris to tutor the son of Jean-Claude Geoffroy d'Assy, a member of the prosperous elite who managed the kingdom's finances. For the next thirty years, Delambre remained a part of the d'Assy household. The grateful parents even offered him a sinecure in their financial offices, but Delambre accepted a more modest annuity that would enable him to devote the rest of his life to study. Thus did many a promising young man from the provinces set himself up as a lay cleric in the *ancien régime*, a bachelor scholar on a small pension. In those days, Delambre styled himself the 'abbé de Lambre'. It was his dream fulfilled. He was a cosmopolitan humanist, rigorous in his learning, tolerant in his poverty, a connoisseur of human absurdity. He had narrow eyes, quizzical eyebrows and a mouth framed by sceptical curves. Already in his mid-thirties, he still had no career.[20]

For the past several years, he had been reading ancient Greek science. To supplement his studies, he looked into modern astronomy, and this led him to the standard textbook in the field, *Astronomy* by Jérôme Lalande. While he was at it, Delambre decided to audit Lalande's lectures at the Collège Royal. One day he heard the teacher comment that the Milky Way had the width of the celestial sphere. After class, Delambre informed the professor that this observation had also been made by the Greeks. From then on, whenever Lalande wanted to check to see whether his students had understood his lecture, he called on Delambre, who always supplied the right answer – not surprising really, since Delambre had obtained all his information from Lalande's own textbook. Even two hundred years ago, this was a well-worn student ruse. 'You're wasting your time,' Lalande finally told him one day. 'What are you doing here?'[21] For no other reason, Delambre confessed, than to get to know Lalande.

Everyone knew Lalande. He was France's foremost scientific publicist, enemy of every human prejudice. He was an outspoken atheist. He ate spiders to prove that arachnophobia was irrational. And he had recently calculated the likelihood that a comet would devastate the earth, causing all Paris to panic. He was a small ugly man, and impossibly vain. He liked to boast he was as ugly as Socrates. If he was not the world's greatest astronomer – as he sometimes seemed to think – he was certainly its most famous, having trained many of the world's leading practitioners, the most recent of whom was Pierre-François-André Méchain. In 1783, looking for a new recruit among the dozens of auditors on his astronomy course, Lalande judged 'the abbé de Lambre, already very able'.[22]

Lalande lent Delambre a three-and-a-half-foot sextant and began to incorporate his student's observations into the third edition of his *Astronomy*. Delambre's eyesight had steadily improved with the years. Despite his late start in science, he had become a superb calculator. When he returned for yet another assignment from his *maître*, Lalande refused. 'Don't be a fool,' he told him. 'Work for yourself and to get into the Academy.'[23] In short order, Delambre made himself into one of the nation's leading astronomers. When in 1787 the d'Assy family moved to a new home in the Marais – at 1, rue de Paradis – they built him his own private observatory on the roof.

For the next twenty years, Delambre lived at the d'Assy residence. The elegant neoclassical structure still stands, renumbered more prosaically as 58 bis, rue des Francs Bourgeois, now the administrative offices of the French National Archives. For Delambre, paradise was on the roof. After climbing the ninety-three steps to his bedroom, he had only one short flight more to enter an observatory built to the most exacting specifications and outfitted with state-of-the-art equipment. In 1789, when the observatory was complete, he had every reason to think he had arrived at astronomer's heaven.[24]

The French Revolution, which convulsed Paris that year, over-turned the comfortable hierarchies of the *ancien régime*, dragging its unexamined standards of conduct and deference into the harsh light of reason. Among these were the standards which governed the nation's economic life. The wealth of the d'Assy family, for instance, derived from its marketplace monopoly over the up and coming Temple district of Paris. Any butcher or baker who wished to set up shop in the neighbourhood near the Marais had had to apply to Geoffroy d'Assy for a licence.[25] That monopoly had now vanished, along with the rest of the nobility's legal privileges. French citizens were henceforth free to trade, independent of the personal control of others. Among the unexamined *ancien régime* standards subject to this searching review were measurement standards – and here it was the savants who led the revolution. In 1790, the newly elected National Assembly authorized the Academy of Sciences to design a system of uniform measures. These savants had the courage to look beyond their immediate historical circumstances and place their standards on a permanent foundation. They vowed to choose a set of measures which would 'encompass nothing that was arbitrary, nor to the particular advantage of any people on the planet'.[26] They decided to base their new measures on the size of the earth itself.

In April 1791 the Academy of Sciences confided this meridian mission to three of its members: Pierre-François-André Méchain, Adrien-Marie Legendre and Jean-Dominique Cassini. These three eminent savants were the logical choice for a body which prided itself on its logic. Méchain was an astronomical workhorse, the editor of the *Connaissance des temps*, the annual tables of celestial events which guided French navigators at sea. Legendre was a gifted mathematician who had perfected the calculations for the measurement of the globe. And Cassini – or Cassini IV, as he was known – had every reason to consider the meridian mission his birthright. He had been born in the Royal Observatory, which his father, grandfather and great-grandfather had each directed in turn. The Cassinis were one of the most illustrious examples of sustained family achievement in the history of science. Each Cassini had surveyed the meridian of France with the most advanced equipment of his day. Since his youth, Cassini IV had worked with his father to create the great Cassini map of France, surveying the stations which would serve as the template for the new meridian mission. No one was more capable of following in the footsteps of Cassini III than Cassini IV.[27]

If pedigree, seniority and disciplinary turf mattered in the Academy, it was for such logical reasons as these. Yet Cassini was reluctant to begin the mission. For one thing, his wife's recent death had left him with five young children to care for. Then there was the problem of his royalist sympathies. On 19 June 1791, Cassini had secured a royal audience for the members of the metric commission. At six in the evening, they presented themselves at the Tuileries Palace: Cassini, Méchain, Legendre and a fourth savant, Jean-Charles de Borda, the inventor of the repeating circle, the new instrument which would push the expedition to a new level of precision. In the eyes of history, Louis XVI has earned a reputation as a political simpleton. But he had his talents. He was a skilled watchmaker and, like his grandfather Louis XV and his great-great-great-grandfather Louis XIV, he was a connoisseur of cartography. After all, if the Cassinis owned the map of France, the Bourbons owned the real thing.

The king also took a surprisingly close interest in the spending of the royal purse. 'How's that, Monsieur Cassini?' he asked the savant. 'Will you again measure the meridian your father and grandfather measured before you? Do you think you can do better than they?'

Cassini managed to preserve filial respect while holding out the promise of progress. 'Sire,' he answered, 'I would not flatter myself to think I could surpass them had I not a distinct advantage. My father and grandfather's instruments could but measure to within fifteen seconds; the instrument of Monsieur Borda here can measure to within one second. That is the sum of my merit.'[28]

The king's sangfroid was all the more remarkable because that evening the royal family had been secretly preparing to flee the country. The next morning, the king and his immediate circle set out on the infamous 'flight to Varennes' which ended when a provincial innkeeper spotted the king, disguised as an English merchant, walking with the notorious Bourbon waddle. Louis was hauled back to an enraged capital and confined to his palace under the city's watchful eyes. Cassini henceforth considered himself released from all obligations to an 'illicit, usurping, seditious government of assassins'. If Louis XVI refused to serve France, how could Cassini IV serve her?[29]

But while Cassini dithered, the rest of the government grew impatient. Jean-Marie Roland, chief minister of the nation, was an expert on the new economy transforming Britain. He wanted France to enjoy the advantages of a uniform system of measures. Such a reform would aid the free circulation of grain, and so help resolve the food crisis at the heart of the nation's troubles. A modern nation needed a standard, any standard, and the surest course of action would be to declare the units used in Paris the national units. On 3 April 1792, Roland threatened to do just that.[30]

Roland's demand threw the Academy into consternation. Their dream of a universal measure seemed about to evaporate. At their next meeting, they divided the meridian expedition into manageable sectors – and urged Cassini to set out. One commissioner would take charge of the northern portion, from Dunkerque to Rodez; while the other would take the southern portion, from Rodez to Barcelona. If the northern sector was twice as long as the one in the south, that was because the north had been previously surveyed, most recently by Cassini's father in 1740; whereas the south was more mountainous and included the uncharted Spanish section. This division of labour was only provisional, of course; the two teams were to work their way towards each other as rapidly as possible and meet up where they would.

Yet even as he refused to set out, Cassini asserted his right to command any meridian expedition. The Revolution may have upended the kingdom, but the Academy still stood upon certain formalities. He offered to remain in Paris while an adjutant carried out the actual surveying. In the end, the Academy rejected this proposal. A savant needed to have direct contact with nature, to travel and measure for himself, so that he might personally vouch for the reliability of his findings.

This was where Delambre came in. On 15 February 1792 he had been unanimously elected to the Academy, in part, as Lalande informed him, because the members thought they might need him for the meridian mission. When Cassini refused a final plea to set out, Delambre was elected on 5 May to lead the northern portion of the meridian survey, with Méchain to lead the southern sector. Decades later, Delambre would recall pleading with Cassini to change his mind. The two men had graduated from the same school a year apart. In revolutionary times, Delambre warned his colleague, a citizen must demonstrate his devotion to the national good, if only to shield himself from reproach. But Cassini would not serve a régime he considered illegitimate. For Delambre, it was the sort of career opportunity the Revolution made possible.[31]

As soon as the king's authorization arrived on 24 June, Delambre began to scout out stations in the vicinity of Paris. His plan was to revisit the stations of the 1740 Cassini survey of the meridian, conduct improved measurements with his new instruments, and wrap up his mission by the end of the year, applying to his geodetic measurements of the earth the same thoroughgoing precision he had so recently brought to his astronomical measurements of the heavens.[32]

Delambre was a fast learner – a humanist in his mid-thirties who had, in the course of a decade, become one of the nation's leading astronomers – and the central method of geodesy was in principle quite simple, little more than Euclidean geometry on a curve. The method was known as triangulation and for two hundred years cartographers had been using it to map terrain, and would continue to do so right up to the advent of the satellite. Triangulation relied on an elementary theorem in geometry: if you know all three angles of a triangle, plus the length of any one side, you can calculate the length of the other two sides. Hence, if you know all the angles in a *set* of triangles connected side by equal side in a chain, plus the length any single side, you can calculate the lengths of all their sides (since every two connected triangles share at least one side). The geodeser simply took advantage of this. First he identified a series of observation stations which might serve as nodes (the vertices or 'corners') of his triangles – church steeples, fortress towers, open hilltops, purpose-built platforms – each node visible to at least three other stations such that they formed a chain of triangles that straddled the meridian. He then travelled from station to station measuring the horizontal angles separating adjacent stations. Next he measured along the ground the actual length of one side of one of the triangles – a 'baseline' – typically by placing rulers end to end over the course of several miles, and used this value to calculate the lengths of all the sides of all the interconnected triangles. From this, he could derive the distance along the meridian arc from his northernmost to his southernmost station. Finally, he determined the respective latitudes of the northernmost and southernmost stations using astronomical observations, so that he might extrapolate from the length of that arc to the full quarter meridian. And that gave him the size of the earth.

This was the principle, anyway. But as in any science where extreme precision is sought, the practical challenges were considerable. First, because the geodeser necessarily measured the angular distances from stations that were somewhat elevated, he had to adjust all his values to a common surface-level triangle. Second, because he could not always place his measuring instrument at the exact vertex of that triangle, another correction had to be included. Third, because atmospheric refraction distorted apparent sightings, all the angles had to be adjusted for the bending of light. And fourth, because the angles of a triangle on a curved surface do not quite add up to 180°, this had to be corrected for as well. All these adjustments complicated the calculations. They did not change the basic principles involved.

By revisiting the stations used in Cassini's 1740 survey of the French meridian, Delambre hoped to skip one of the most laborious steps in any triangulation: the identification of workable stations. But first he had to verify that these sites could still serve his purpose. For their station in the capital city of Paris, the surveyors of 1740 had chosen the belfry of the church of Saint-Pierre near the summit of Montmartre, a Benedictine abbey which still stands today, a stone's throw from the current site of the church of Sacré-Coeur. On 24 June Delambre set out with his two assistants to climb the hill of grapevines, quarries and windmills. Even then, Montmartre offered a famous panorama over Paris. From the hilltop, they could look back on the jumble of low grey buildings which swarmed like angry insects around the city's massive royal and religious edifices.

But when they climbed still higher to set up the instrument on the platform of the bell tower of Saint-Pierre, Delambre met with a disappointment. The view was miserable. None of the surrounding sites used in 1740 was visible. Indeed, the church tower barely cleared the roof. In every direction he looked, the view was blocked by surrounding buildings.

Back in the capital an old etching cleared up the mystery. Fifty years before, the church had been capped by a tall wooden belfry, since destroyed. A half-century of urban construction and demolition had transformed the Paris cityscape, levelling steeples, raising palaces, filling empty lots. The abbey of Saint-Pierre would no longer serve, and Delambre would have to locate an alternative station elsewhere in the capital. This, he decided, could best be done from the outside looking in. He decided he would travel anticlockwise around the capital, visiting the peripheral stations which ringed the city, scanning for an appropriate landmark in the centre.[33]

The next few weeks of travel showed him just how much the past fifty years had transformed the French countryside as well. To the south of the city, the observers of 1740 had adopted the tower of Montlhéry, an abandoned medieval fortress which guarded the main route into Paris. The tower still stood, as it does today, pigeons fluttering in its broken interior. Delambre discovered that the first ten steps of the staircase had crumbled away, and though he sent a workman up to verify the view he had no appetite for hoisting himself and his instruments up the ninety-six-foot turret. He settled for an observation point on the overgrown fortification wall below.

Next, Delambre visited the Malvoisine farmhouse, perched on a low ridge twenty miles to the south-east, and which had been used by Cassini III in 1740. The site is still a working farm, its muddy courtyard piled with machinery and patrolled by dogs. But even from the farmhouse roof Delambre could barely make out the adjacent station of Montlhéry. Stands of tall trees had grown up around the property in the intervening fifty years. He secured permission from the owners to add six feet of height to the farmhouse chimney so as to create a workable observation signal there, and continued his travels anticlockwise around the capital.

The Gothic church tower at Brie-Comte-Robert still suited his requirements. But at Montjai, Delambre encountered new obstacles. Even in 1740, Cassini III had hesitated to climb the medieval tower – the eastern twin of the tower at Montlhéry – not for fear of the demons which supposedly haunted its ruins, but because he had been warned that the tower might fall down. Delambre decided the better part of valour was to hire a local carpenter named Petit-Jean to rig a freestanding observation platform alongside the structure. While that work got underway, he continued to Dammartin, a town on top of a steep ridge just outside today's Charles de Gaulle airport. There he learned that the Collégiale chapel, which had served Cassini III in 1740, was about to be sold off as part of the Revolutionary sale of church lands, and the buyer intended to demolish the church for building materials. On the spot, Delambre decided to make this site his highest priority. But first he had to verify that he could see Dammartin from Saint-Martin-du-Tertre, the next station to the north; there too his preliminary observations did not match those of 1740, suggesting that the position of the church belfry had been shifted by several hundred feet at some point during the past fifty years.[34]

Geodesy is a natural science. It is the science which measures the size and shape of the earth, a planet formed by the same gravitational forces that had spun the solar system out of a disc of luminous nebular dust (according to Pierre-Simon Laplace's reigning theory). What is the earth's shape? Even more: what is meant by shape? Our planet's surface is not smooth. It is scarred with mountains and valleys, roiled by geological processes. The imaginary shape that our planet would possess if its surface were everywhere at sea level, which scientists today call the 'geoid', was known in the eighteenth century as the 'figure of the earth'. A meridian, for these savants, was the surface of an imaginary canal that ran unswervingly from north to south; in this case, from the North Sea to the Mediterranean. Yet to measure the length of this canal, and hence the figure of this imaginary earth, geodesers relied on the very geological processes that have distended the surface of the planet, creating the mountains and hilltops from which they surveyed the terrain.

Geodesy is also a science which depends on human history and human labour. Where the surface of the earth is too level, with neither mountains nor hilltops available for triangulation, the geodeser must commandeer man-made structures for his view: church steeples, fortress towers, observation platforms, any lofty site. As Delambre's whirlwind circuit of Paris had taught him, however, the same human purposes that had raised churches, towers and platforms could tear them down. The world would not hold still while he measured it. In such times, with commercial and political revolutions converging at violent speed, the past was a treacherous guide.

On 10 August 1792, Delambre was finally ready to take his first definitive measurements. He set up his delicate instrument in the belfry of the Collégiale at Dammartin and sent young Lefrançais back to Montmartre with instructions to light a parabolic reflector from a roof-top observatory there, so that he might pick out the station amid the hodgepodge of buildings on the hill. At ten o'clock that night Delambre had still not detected a signal from Montmartre. But he did see flames from an unexpected direction: the Tuileries Palace was on fire. That day, unknown to Delambre, some ten thousand Parisians had stormed the royal palace, where the king was a virtual prisoner, set it on fire, and, with the help of the turncoat Paris militia, massacred six hundred of the king's Swiss Guard, some by defenestration, others with cold steel. As the dominant hill to the north of the city, fortified with its own battery of cannon, Montmartre possessed strategic value to those jostling for control of the capital. Later that night, three militiamen were killed at a Montmartre barricade while arresting members of the Swiss Guard. To light signal flares from the hilltop that night, as Lefrançais had planned, would have been suicidal. The next night, with the help of his uncle Lalande, he did manage to set a flare, though it did not burn long enough for Delambre to get a good reading. The uprising of 10 August brought the monarchy to an end. Delambre would never again risk lighting signal flares at night.[35]

This meant giving up on Montmartre, however, and selecting another site as his central Paris station. Delambre had just settled on the dome of the Invalides for that purpose, and had begun retaking his angle measurements using the golden dome as his sighting target, when he received news that the residents of Montjai, armed with muskets, had forced Petit-Jean, the carpenter, to tear down the observation platform he had been building alongside the crumbling tower. Delambre rushed to Montjai to insist that the local town council order their constituents to stop harassing the carpenter. In a republic, the council responded, one might exhort citizens, but could not command them. If Delambre wanted the carpenter to build the platform, he would himself have to persuade the citizenry to allow it. Delambre did make the effort. But his explanations only succeeded in rousing the surrounding villages against his mission. Montjai would have to be abandoned. Searching for a substitute station, he noticed the château de Belle-Assise on a nearby prominence and secured permission from its owner, the comte de Vissec, to use his charming belvedere as an observation station. Four days later, the local militia came and escorted him out of the château to the town hall of Lagny and thence to the Hôtellerie de l'Ours, where he was 'not arrested, *but merely detained*'.[36]

On the morning of 6 September 1792, when they were finally able to leave Lagny, Delambre and Bellet drove their carriages towards the hilltop town of Saint-Martin-du-Tertre, skirting Paris anticlockwise as they continued their circuit of the capital. On the way, as a precaution against further trouble with local militiamen, they stopped in the district office of Saint-Denis to obtain a certificate of safe passage. The district office had recently been relocated to the ancient abbey, the most sacred pilgrimage site in France.

For fifteen hundred years, the basilica of Saint-Denis had served as the ancestral tomb of royal France. Monarch to monarch, and dynasty to dynasty, the dead kings of France had been borne to the crypt at Saint-Denis so that a new king might live. King Dagobert, first of his line, lay beside Queen Nanthilde. Henry IV, assassinated in his prime, lay between his two wives, Marguerite de Valois and Marie de Médicis. The Sun King Louis XIV lay beside Marie-Thérèse. Above their tombs, sculpted effigies rested on marble beds, some robed in stony gowns, others naked in their final agony – while bronze angels and bishops knelt at their feet. Fifteen hundred years of royal succession had ended the previous month.

The town council of Saint-Denis had ordered the fleurs-de-lys effaced from the abbey as a sign of despicable feudalism. They had erected furnaces in the chapels of the basilica so that the bronze statues of Charles VIII, Henry II and Catherine de Médicis might be melted into cannon. And just that week they had debated whether to dig deeper and extract lead from the royal coffins so that the casings of kings might be made into cannonballs to hurl against the enemies of the new republic.[37]

The chief administrator of the district, Denis-Nicolas Noël, signed Delambre's certificate of safe passage, but warned the savant that this piece of paper would not provide much protection. Saint-Denis straddled the main road north from Paris, and every village from here to the frontier had raised barricades to stop aristocrats fleeing the capital. Peasants were digging fortifications. The Prussians were expected at any moment.

The savant, however, refused to delay. He directed his two carriages on to the Route de Poissy that ran north-west along the loop of the Seine. Fifteen minutes out of the city, at a barricade at the approach to the village of Epinay-sur-Seine, the local militia halted their carriages and demanded passports.

Regrettably, their passports made no mention of the strange instruments they were transporting. What was the purpose of this equipment, and why were they bearing it towards the frontier? The instruments seemed designed to spy out actions at a distance. Might they not also have a military purpose?

Delambre explained that these were astronomical instruments that would enable him to measure the size of the earth.

And why would he want to do that?

So by the side of the road, within sight of the reedy banks of the
Seine, Delambre unpacked his instruments and explained his mission. It was a warm end-of-summer afternoon – perfect weather
for an outdoor seminar – and the size of the crowd swelled as
word spread that a scientific sideshow was underway at the edge
of the town. The local militia had stopped some highfalutin carriages bearing mysterious instruments towards the frontier. As
newcomers joined the throng, they insisted on being told what
was going on. Delambre was obliged to restart his seminar several
times. Mayor Louis Beaudoin, a local wine grower, joined the
crowd, as did two land surveyors. Delambre appealed to these
men for help. He showed the mayor his official papers, including
the guarantee of safe passage signed by the district official that
morning. He begged the surveyors to vouch for him. Surveying
and geodesy were brotherly trades. Both measured the earth;
while surveyors measured fields, geodesers measured planets.

The surveyors, however, refused to confirm Delambre's words.
And Delambre could easily understand why. 'They sensed the
mood of the crowd, realized it would be useless to speak in our
behalf, and dared not back me up.' As for the mayor, he too preferred to exercise caution. He ordered the guards to escort
Delambre and his carriages back to Saint-Denis for questioning.[38]

Where the main square of Saint-Denis had been empty that
morning, a thousand exuberant men and women were now
assembled under the mismatched spires of the basilica, the taller
tower on the left draped with the tricolour flag, the shorter tower
on the right capped with a gigantic liberty bonnet. Among the
crowd, several hundred young men wore the insignia of the First
Division of the Paris National Guards, volunteers marching
north to help their comrades repel the Prussian invaders. They
had paused in Saint-Denis only long enough to encourage the
local lads to join them. The fatherland was in danger! The
Prussian army was on its way to restore the king. To save the
republic, the newly formed government had pleaded for three
hundred thousand volunteers. Volunteers! The very idea was
revolutionary. For centuries soldiers had died for pay, glory,

Basilica of Saint-Denis

This early nineteenth-century view by Giuseppe Canella down the main street of Saint-Denis shows the façade of the basilica as it appeared at the time of the Revolution. Street lighting had been introduced in the 1770s. The basilica has since been shorn of the tower on the left.

(Saint-Denis, Musée d'art et d'histoire. Photo: Irène Andréani.)

booty and loyalty to comrades and commanders. That men were now willing to die for an abstraction called a nation meant that for the first time they conceived of themselves as citizens of a nation, rather than retainers of a seigneur or subjects of a king. Some eight hundred young men from Saint-Denis and the surrounding hamlets had answered the call. They had left their bakeries, workshops, farms and families to defend a revolution which promised them liberty and equality. They had assembled in the main square of Saint-Denis to demand that, in return for their sacrifice, the municipal council supply them with firearms, a thousand pounds of bread, and carriages to transport their provisions to the front.[39]

Then, like a miracle, two carriages appeared, escorted by the militia of neighbouring Epinay. As the militia cleared a path through the crowd, they boasted to their comrades of their prize: two suspects caught on their way to the frontier with spying instruments. The square rang with their joyous cries: 'Long live the nation! Behold the aristocrats!' Shouts of abuse were showered on Delambre as he was hustled into the town hall beside the cathedral, and then into the district offices in the half-moon courtyard of the abbey. Inside their offices, the huddled administrators chastised the mayor of Epinay for creating this volatile situation.[40]

Outside, in the square, the people of Saint-Denis thought they had good grounds for suspicion. France was full of traitors. Two weeks earlier, General Lafayette, the hero of the American Revolution, had tried to subdue the capital by force and, when his troops refused obedience, fled to Belgium. Generals and aristocrats had gone over to France's enemies. Now that Verdun had fallen, only the common people could halt the Prussian advance on the capital.

The crowd was growing impatient. While Delambre was conferring inside the abbey with the officials, a fiery band of volunteers stormed the carriages and pulled down the leather cases that contained his instruments. Inside, they also found a cache of fourteen letters, all sealed with the royal signet. This was a stupendous find! Might not these letters bear a message from the imprisoned king to the northern front? Only with great difficulty did the militia persuade the men to reload the cases on to the carriage. In return, the crowd was promised a full explanation of the royal letters. The call went out for Delambre. The crowd was summoning him.

To Noël and the other public officials, this summons sounded like Delambre's death knell. In the early days of the Revolution, the assistant to Saint-Denis' mayor had been stabbed fourteen times in the church tower for refusing to lower the price of bread to two *sous*. All that week in Paris, in the Revolution's worst outbreak of popular violence, ordinary prisoners had been hauled out their cells, accused of participating in aristocratic conspiracies, and murdered by mobs of men, including members of the Paris National Guard, some of whom might have been in the crowd that day outside the basilica. Chief administrator Noël told Delambre to hide in a cupboard before he stepped outside to see what the crowd wanted. Only when he was satisfied that they would not tear Delambre to pieces did he bring the astronomer out to explain his mission, his instruments and, above all, the royal letters. Delambre took his stand on the steps of the town hall. He broke the royal seal.

PROCLAMATION OF THE KING
concerning the observations and experiments to
be performed by the commissioners of the Academy of
Sciences in execution of the law of 22 August 1790,
which ordered that weights and measures be rendered
uniform, dated 10 June 1792 . . .

And so on for three dense pages of royal legalese, in which the king commanded local administrators along the length of the meridian to aid the appointed commissioners with horses, food and lodging, and to permit them to erect signals, scaffolding and reflectors 'on the roofs and exterior of steeples, towers, and fortresses'.[41]

Even at a rapid clip, the letter would have taken fifteen minutes to read. Yet the crowd insisted on hearing every word. Who knew but that some malevolent plot might be concealed among the verbiage? Apparently satisfied that this letter – albeit a royal proclamation – was innocent, the crowd turned its attention to the thirteen other letters, likewise sealed with the royal signet. Delambre was obliged to unseal a second letter so that it too might be read to the crowd. The second letter proved identical to the first. But what of the others? Perhaps a single traitorous letter had been concealed among the innocuous ones? So Delambre agreed to have a third letter read, and then a fourth, and then a fifth. And so an hour passed and more. The reader's voice was tiring. To read all fourteen letters would have taken all night. The September evening was drawing down. Moreover, each broken seal rendered that letter null and void. So Delambre offered his audience a deal. He would agree to read one more letter, selected at random, and if that letter did not match the previous letters word for word, he agreed to answer for it with his life. The deal was struck. A letter was chosen, unsealed and read. It matched.

Still the crowd was not satisfied. (The volunteers apparently still had their eye on those two carriages, ideal for transporting their provisions to the front.) They demanded to know the purpose served by his instruments. It was up to Delambre to explain.

He did his best. As free men and women they had the right to know why this work was being done in their name, and how it was being carried out. His mission may have sounded arcane, remote from their immediate concerns. But its successful completion would one day transform their lives more than any battlefield triumph.

Ought not a single nation have a uniform set of measures, just as the soldier fought for a single *patrie*? Had not the Revolution promised equality and fraternity, not just for France, but for all the people of the world? By the same token, should not all the world's people use a single set of weights and measures to encourage peaceable commerce, mutual understanding and the exchange of knowledge? That was the purpose of measuring the world.

As everyone in the crowd there that day knew, measures in France varied from province to province, town to town and parish to parish. Despite the similarity of their names, they also varied from trade to trade, and for different goods. When a volunteer from Saint-Denis visited Paris to hoist a *pinte* of beer to salute his Paris comrades, he discovered that the *pinte* of Paris held two-thirds the beer of his hometown *pinte*. The bakers in the crowd used a *livre* (pound) that was lighter than the *livre* of the ironmongers. In many parts of France, a pound of bread really did weigh less than a pound of lead. For instance, Saint-Denis' measures were enshrined in masonry just inside the basilica doors immediately behind him: two receptacles for two different types of grains, two for salt, plus an *aune* (an ell, about three feet long) mortised into the wall. The *aune* was used exclusively for measuring cloth, and Paris had three different *aunes* for three different kinds of fabric, while in Delambre's hometown of Amiens his father's shop used one *aune* to buy wholesale and a shorter one to sell retail, while thirteen different *aunes* were used in the surrounding villages. All across France such discrepancies caused endless confusion, disrupting trade, baffling administrators, inviting fraud.[42]

For centuries royal administrators had been trying to wrest authority over weights and measures from the hands of local nobles, guildmasters and town aldermen. Uncertainty about the true value of local measures hindered the state's efforts to extract revenue on sales, implement a fair property tax, assess imports and regulate the supply of grain and the price of bread. The army likewise aspired to uniform measures in order better to coordinate the production of war matériel, fortifications and maps. In the last few decades the mood among the educated population had swung decisively against the diversity of measures. Alexis Paucton, the age's foremost compiler of weights and measures, both ancient and modern, urged reform: 'They are the rule of justice which must not vary, and the guarantee of property which must be sacred.'[43] The enlightened authors of Diderot and d'Alembert's famous *Encyclopédie* bemoaned the 'encumbrance' of France's diverse measures. But like so many others in those days, they had found the fault 'beyond remedy'. Even the king's most capable minister, Jacques Necker, had decided that metrical uniformity was beyond the monarchy's power.[44]

Where faint-hearted kings had failed, however, the Revolution was determined to succeed. The nation, to be a nation, had to be coaxed into a uniformity of weights and measures. In the *Cahiers de doléances*, the famous litany of grievances solicited by the king in 1788, it was the people themselves who had called for the reform of weights and measures. Some 128 of the regional *Cahiers* demanded uniformity, as did 32 of the nobility's, and 18 of the clergy's. And at the local level, thousands of village *Cahiers* echoed the call for 'one law, one king, one weight and one measure'. The townsfolk of Saint-Denis had themselves made this demand in their own *Cahier*, and the mayor of Epinay had himself signed his town's *Cahier* asking that France be governed by a single set of weights and measures.[45] It was up to the Revolution to make good on that demand.

We do not know what the crowd made of Delambre's explanations. We only know that they were primed for battle, not for an impromptu lecture on measurement and geodesy. Delambre himself detected a certain impatience.

> The instruments were spread out on the square, and I was
> obliged to recommence my lecture on geodesy, the first
> lessons of which I had given earlier that day in Epinay. I was
> not heard any more favourably this time. The day was coming
> to a close; it was increasingly difficult to see. My audience
> was quite large. The front rows heard without understanding;
> the others, further back, heard less and saw nothing.
> Impatient murmurs began to be heard; a few voices proposed
> one of those expeditious methods, so in use in those days,
> which cut through all difficulties and put an end to all
> doubts.[46]

Before the hecklers made good on their threats, administrator Noël intervened. Feigning severity, he ordered the suspect carriages placed under seal and towed into the abbey's courtyard for safekeeping. Then he forced Delambre back into the town hall on the pretence that he wanted a more convincing account of his mission. Once inside, he obliged Delambre to spend the rest of the night in the company of the aldermen – for his own safety – while they requested instructions from Paris. Delambre and Bellet slept that night in armchairs in the town hall of Saint-Denis. Only at dawn were they allowed to take lodgings in the nearby inn known as the Trois-Maillets.[47]

The National Convention voted later that evening, 7 September 1792, to make Delambre and Méchain official emissaries of the republic and ordered local authorities to assist them on their route. An expedition licensed by the king had become the people's mission. Lefrançais brought Delambre the decree as soon as it was released, and together they took it to the Sunday morning meeting of the municipal council so that they might get the seals removed from their carriages and continue on their mission. That night the Benedictine monks performed their last holy mass after a thousand years of continual prayer in the kingdom's greatest abbey.[48]

In the interim, the volunteers of Saint-Denis had been marched to a barracks outside Paris for accelerated training, and thence to meet the Prussian invaders at Châlons-sur-Marne. The volunteers of Saint-Denis helped save the Revolution. But the Revolution was not finished with Saint-Denis. In December, a crowd invaded the basilica, not to slaughter the living, but to disinter the dead. The militia protected the royal graves, but popular calls for a mass exhumation multiplied after Louis XVI was guillotined in January 1793. The king's body was dumped in an anonymous grave. His ancestors did not deserve a better resting place.[49]

Again the National Convention scrambled to lead the populist charge. In honour of the first anniversary of the uprising of 10 August, they ordered the tombs of Saint-Denis destroyed so that the royal dead might be reburied in a mass grave at Valois and the lead from their coffins – nine tons in all – refashioned into cannonballs and musket shot. Only the statues of François I and other high Renaissance sculptures were to be saved because of the excellence of their art. Chief administrator Noël turned the first spade of earth. The first royal corpse brought to the surface was that of Henry IV, the nation's most popular king, perfectly preserved, his face black as pitch. A young soldier cut a lock from Henry's beard before laughing onlookers, held it to his chin and declared: 'Well, I'm a soldier too! Now I'm sure to vanquish those English bastards.' When the Sun King was exhumed, a worker sliced open the corpse's belly to the applause of the crowd. To mask the stink, officials burned a mixture of vinegar and gunpowder and closed the basilica.[50]

**Henry IV exhumed and displayed in the
Basilica of Saint-Denis**

*The corpse of the most popular French king was briefly exhibited in the
Basilica of Saint-Denis shortly after its exhumation in 1793.*

(Photo: Roman Stansberry.)

Not long after, the municipality was debating whether to permit
local patriots to use cannonballs to shoot down the basilica's belfry
when the Commission of Weights and Measures intervened. The
tower, they said, was crucial to the survey of the meridian that ran
from Dunkerque to Barcelona. In consideration of its 'great utility'
for determining the new republican measures and triangulating

the territory of the republic, as well as for other cartographic and scientific purposes, the council ought merely to efface those remaining crucifixes and fleurs-de-lys that offended the good patriots of Saint-Denis, and leave the tower standing. So science saved the basilica, even as science itself came under attack.[51]

Chapter Two

THE SOUTH-GOING
ASTRONOMER

'Welcome to [Barcelona], thou mirror, lanthorn, planet, and
polar star of all chivalry in its utmost extent! welcome
valorous Don Quixote de la Mancha, not the false, fictitious
and apocryphal adventurer, lately in spurious history
described; but, the real, legal, and loyal knight . . .'[1]

MIGUEL DE CERVANTES,
The History and Adventures of the Renowned Don Quixote

According to the official account of the meridian expedition –
Delambre's magisterial *Base du système métrique* – Méchain left
Paris for Barcelona on 25 June 1792, accompanied by three aides,
riding in his custom-built carriage, and bearing the scientific
instruments he had waited so long to receive. After years of delay,
the expedition was finally underway. The Academy had originally
hoped to launch the survey when it named its commissioners in
the spring of 1791. Their departure had been postponed while
Lenoir finished building the measuring instruments and Cassini
dithered. By January 1792, Lenoir had delivered the instruments
and Méchain was expecting to leave the next spring. In April the
French Foreign Ministry secured the full cooperation of the
Spanish government, and in May Méchain informed the
Spaniards he would be leaving Paris on 10 June. Yet Méchain
appeared in no rush to begin. On 9 June he announced that he
would actually be leaving the capital on 21 June. On 23 June he

said he expected to leave the next morning. Every day counted; the nation's chief minister was threatening to cancel the expedition. The government was having second thoughts about the cost of measuring a meridian measured several times before.[2]

Yet three days after 25 June – the date on which, according to Delambre, Méchain left Paris – the southbound astronomer was still in the capital. On 28 June a public notary visited the Méchain residence in the grounds of the Paris Observatory so that Citizen Méchain – 'being on the point of leaving for Barcelona in Spain as one of the Commissioners of the Academy of Sciences to determine the length of the arc of the meridian' – could sign a power of attorney over to his wife. The document empowered his wife to collect his salary during his absence, carry out financial transactions in their name and dispose of their property as she saw fit.[3]

Barbe-Thérèse Méchain, née Marjou, was an educated, competent woman who assisted her husband in his astronomical work. They had been married for fifteen years and lived with their three children – two sons and a daughter – in a small house in the grounds of the Paris Observatory. It was a comfortable life. The family had the right to plant vegetables in the plot behind the house, and their front windows looked out on to the rue du Faubourg St Jacques. Lodgings in the neat little house was one of the privileges with which a lucky savant on the royal payroll might supplement his meagre income.[4]

The Marjou family had also made their careers in royal service. Thérèse's father had been a valet to the king's brother at Versailles. Her elder brother earned a generous pension as head cook for the duchesse d'Aiguillon. Thérèse had brought a substantial dowry to her marriage, as well as a cool head for business. The Revolution, however, had wiped out her family's fortune. Her parents had died of old age in 1789, just as her brother's blue-blooded employers had fled the country. Versailles was deserted. The Méchain family now depended on her husband's scanty earnings. The metric expedition would not pay him a salary – as a member of the Academy, Méchain was expected to serve on the meridian mission for honour's sake – and he could not afford to abandon his Paris duties. So Madame Méchain agreed to continue her husband's astronomical measurements during his absence, including a study of lunar eclipses. As she would fulfil her husband's official duties at the Observatory, it is hardly surprising that he granted her a power of attorney as well – whatever the exact date.[5]

Yet exact dates matter. For astronomers, time is sacrosanct. The precise moment of a celestial event is the foundation upon which all heavenly knowledge rests. Why then did Delambre lie about so trivial – and verifiable – a matter? After all, Méchain's presence at the Observatory could hardly have gone unnoticed. Most of the nation's leading astronomers also lived in the grounds. Oddly enough, the solution to this mystery may be found in another falsehood Delambre published in the same official account. There he implied that his own first day on mission was 26 June, one day after Méchain's supposed departure. Yet his private notebook – which he consulted while preparing his magisterial *Base* – indicates that he had been scouting stations as early as 24 June.[6]

Two lies do not make a truth, but they may solve a mystery. With these two fabrications, Delambre established Méchain's seniority on the project. This one-day priority, slim as it was, dignified Méchain as the senior partner on the expedition. Méchain was forty-seven and Delambre was forty-two, but Méchain had ten years of seniority within the Academy of Sciences and had been nominated to the meridian survey two years before Delambre. Scientific life in *ancien régime* France operated upon such slender courtesies. Even today, scientific careers still depend on the roll call of authorship, and colleagues read those rosters with a subtlety worthy of biblical exegesis.[7]

But we lie about time at our peril. An expedition sent forth to measure nature with unsurpassed precision ought not to begin under a cloud of dissimulation, especially when its central purpose is to define the attributes of time and distance for all people, for all time. Méchain left Paris expecting to return in seven months. Seven years would pass before he set foot again in the capital.[8]

Of course, none of this explains *why* Méchain delayed his

departure. Can it be that he had doubts about its wisdom, or about his ability to carry it out during a time of such unrest? Already in 1789 Méchain was having trouble remaining calm amid the 'brusque alarms, continual worries and serious risks' that convulsed his Paris neighbourhood.[9] Two days after the fall of the Bastille a mob of three hundred armed citizens had invaded the Observatory to search for gunpowder, weapons and food. They had forcibly entered his home, terrorized his wife and obliged Cassini to conduct them through the labyrinthine cellars, where they found nothing more deadly than a kitchen rôtisserie. Yet these 'Don Quixotes', as Cassini called them, also stripped the roof of lead to make musket shot.[10] Soon after, Méchain had been drafted into the bourgeois militia to maintain order in his section of the city. 'You can imagine,' he wrote to a colleague, 'that it is not easy to keep one's mind free and clear for scientific work under such circumstances.' Yet for the next several pages of his densely written letter he did just that, analysing the exact dimensions of Saturn's rings.[11]

This single-minded concentration made Méchain an ideal choice for the meridian expedition. Accuracy was his religion. He was born on 16 August 1744, the son of a small-town plasterer from Laon, a medieval city perched on a narrow crescent ridge above the dense soil of Picardy, a lush green landscape punctuated by fortified towns. He had been educated by the Jesuits, and had demonstrated enough mathematical talent to win entry to the Ecole des Ponts et Chaussées, France's pre-eminent school for civil engineers. His father, however, was unable to maintain his son in school, and Méchain was obliged to quit his studies and accept a tutoring position. There, he saved enough money to indulge his youthful passion for astronomy, buying some telescopic equipment. Then calamity struck. His father lost a crippling lawsuit, and the son (or so the story goes) loyally agreed to sell his instruments to pay off the family debt. This setback proved his first stroke of fortune. His instruments were purchased by Jérôme Lalande, France's most illustrious astronomer, with connections at every level of French society.[12]

Lalande secured the young man a part-time position at the navy's cartography department at Versailles. In that capacity, Méchain participated in mapping expeditions along the Normandy coast and prepared detailed military maps of the Mediterranean, using observations gathered by others to chart a coastline he had never seen. For twenty years he spent his days inside the dark bureau working his way through reams of calculations, and his nights scanning the bright northern skies. Lalande was also a generous taskmaster and set the young man to work on his celestial tables.

Méchain was harder on himself than any master could be, the sort of astronomer who favoured the long mathematical route to a sure answer over a quick superior technique which struck him as unproven. Eventually, these qualities made him one of France's leading astronomers, the discoverer of eleven comets, a member of the Academy of Sciences, and the editor of the *Connaissance des temps*. This was France's premier journal of astronomy, a work whose principal merit was its exactitude. Méchain transformed it into a worthy rival of the Greenwich Observatory's exemplary *Nautical Almanac*.[13]

To reward these achievements, he was appointed *capitaine-concierge* of the Observatory. His sponsors assured Cassini that he would show the director 'due deference'. They also pointed out that Méchain was 'very capable, very honest, as well as young, poor and married'. In 1783 the family moved into the cosy residence in the Observatory grounds, with its kitchen, dining room, bedrooms, offices and small courtyard and back garden where the previous tenant, a botanist, had planted exotic trees. His youngest son was born there in 1786.[14]

Méchain was a short dark-haired man, with pale delicate features which would have been regular if they had not been tugged in opposite directions by emotions working just below the surface: thick eyebrows hitched high in entreaty, liquid eyes that searched out commiseration and a gentle mouth that drooped towards self-doubt. He had never hoped for public fame, and luckily (his second stroke of fortune) he had had little need to fret about his career once he married Thérèse Marjou. Still, he had steadily climbed the rungs of his profession, until, in the very last year of the *ancien régime*, he had been named to the Franco-British expedition to survey the difference in longitude between the Paris and Greenwich observatories.[15]

It was this expedition which had first demonstrated both the promise of the Borda repeating circle and Méchain's capacities as a geodeser. The natural philosophers of the world's two greatest powers thought a cross-Channel survey would allow navigators to translate readily between British and French sea charts. They also hoped that a dash of scientific rivalry would spur their two governments to support science more generously. Pitted against Jesse Ramsden's great theodolite, a massive new surveying instrument financed by George III, the French team would deploy their repeating circle, funded by Louis XVI and designed by Borda to be the ideal instrument for geodesy. The Borda circle was easier to build and transport than the Ramsden monstrosity (it weighed twenty pounds, rather than two hundred pounds); it could be adapted to measure both terrestrial and celestial angles; and it promised to reduce errors nearly to zero. The quest for precision demands cooperation; it is also a form of competition.[16]

The British were unwilling to concede the contest. As their leader put it: 'I perceive, Sir, that your small circle measures angles very justly when a mean of many observations can be taken. Our instrument I consider with regard to its construction and divisions as perfectly free from error.'[17]

Throughout 1787 and 1788 the British and French teams triangulated towards their respective coasts and double-checked one another's results across the Channel. In the end, they both achieved such a high degree of precision that they could not agree on their findings. While the French acknowledged that the British instrument produced errors of less than two seconds of arc, they boasted that they had closed their triangles to within 1.5 seconds of arc, a tenfold improvement over the results of the previous decade. Cassini considered this proof that the Borda circle had pushed science towards a perfection which bordered on the sacrilegious.

Usually, in the arts and sciences, the closer one approaches perfection, the more the number of difficulties multiply and accumulate; so that one is sometimes tempted to think that there is a limit beyond which even the genius and hand of man cannot cross, were not that unhoped-for success did not come to reanimate our trust, and prove to us that nothing is impossible for men of inquiry and perseverance.[18]

Yet where the results of the two nations overlapped – such as at Blancnez – their angle values differed by six times as much as their vaunted precision, or a disheartening 12.7 seconds. Who was to blame? Not the French, said the French, and underscored their supreme confidence in their own measurements by pointing out that they had been verified by Méchain.[19]

This confidence in Méchain was not misplaced. Throughout the operation, he had served as Cassini's workhorse, measuring at Dunkerque, Watten and the other stations along the French coast. Thanks to his seniority, Cassini had initially monopolized the use of the Borda circle, relegating Méchain to the secondary role of checking the circle's accuracy against an older instrument called the sector. However, Méchain did get a chance to practise on the instrument, and by the project's end he was proficient in its manipulation. This experience and his renowned exactitude made him an obvious choice for the meridian survey, his third (and final) stroke of fortune – just as the Revolution wiped out his wife's modest income.[20]

It was typical of Méchain's pessimism, however, to read even this opportunity in the worst possible light. He saw little likelihood that the expedition would improve his prospects or enhance his reputation. 'So you see', he wrote to an old mentor upon hearing of his selection, 'that as insignificant as I have been until now, I must still not expect to become anything greater in the future.'[21]

Méchain's exactitude was not cold. He was a man of strong sentiments: anxious, melancholic and acutely aware of other people's feelings – especially when their troubles mirrored his own. Though born to the lower orders, Méchain identified with the institutions of the *ancien régime*, which had, after all, treated him rather well. When some of his colleagues, inspired by the new democratic ethos, suggested updating the Academy along more egalitarian lines, he sided with the traditionalists. He was a cautious man, a safe man; anxious to do the right thing. He had been thrust into the senior role in the metric expedition, and he had accepted that responsibility. That was what his honour required.

Sometime after 28 June 1792, Méchain finally left the capital, accompanied by his aides and bearing the two repeating circles he had waited so long to receive. His primary adjutant was a military cartographic engineer named Jean-Joseph Tranchot. The mountainous terrain of Catalonia had never been surveyed, and Méchain needed a skilled and hardy assistant. Tranchot was thirty-seven and a native of north-eastern France, but he had spent half his life triangulating the Mediterranean island of Corsica, France's most recent territorial acquisition – a terrain as rough and barren as any in Europe. The two men had already worked together to determine the position of Corsica relative to the map of the Mediterranean coast, and Méchain had personally instructed Tranchot in the finer points of astronomical observation and calculation. Méchain's other assistant was the instrument-maker Esteveny, trained in Lenoir's workshop. He was also accompanied by a manservant named Lebrun.[22]

The team met only one obstacle on their voyage south. On the first day's ride out of Paris, they were halted at a barricade near the town of Essonne, the sort of roadblock Delambre encountered at every turn. The local militia mistook their astronomical instruments for high-tech weaponry, and detained them while they conferred with local officials. In those days, however, before the fall of the monarchy, their royal proclamation saw them through. Past this barrier, they made good speed through a quiet countryside.

All was still calm when Méchain arrived a week later in Perpignan, the southernmost large town in France. The purple-walled Moorish city lay on a coastal plain of scorched vineyards and salt lagoons, locked in the long crooked arm of the Mediterranean as it reached out from Spain to embrace Italy. At the town's back, a range of hulking blue mountains, dominated by the Massif de Canigou, rose out of the parched lowlands like a dark muscular shoulder. These were the Pyrénées, where Méchain would begin his operations. The border with Spain lay along the crests of the mountains.

After presenting himself to the municipal assembly of Perpignan, Méchain and his team took the Grande Route towards Barcelona, then the major highway between the two kingdoms and still the route of a modern six-lane motorway today. Slanting away from the coast, the road traversed rich farmland before it climbed up hills broken by intermittent seasons of sun, rain and frost towards a low mountain saddle, where it passed under the guns of the massive French fortress of Bellegarde and entered Spain. Thereafter, the French king's magnificent highway became a 'natural and miserable road', descending through a desolate terrain of loose sandy soil, with cork trees growing on the upper slopes and olive trees cultivated sparsely below. Only as it once again approached the coast did the signs of human industry multiply. The air became fragrant with flowering shrubs and aromatic herbs. The road was bordered by hedges of aloe, Christ-thorn and wild pomegranate. Chain pumps irrigated fields of corn and orange groves. The number of towns increased. Soon they had passed through the gates of Barcelona, a metropolis seized by the expectation of change.[23]

Eighteenth-century Barcelona had prospered under the watchful gaze of its Castilian overlords. The Catalan town boasted silk manufactures, an Italian opera and a half-mile-long quay that docked a hundred ships simultaneously. Gold streamed in from the Americas, and textiles and manufactured goods went out to the colonies. With the commercial boom, Barcelona also became an intellectual capital, in part because of its relative openness to its neighbour to the north.

This influx of French ideas was not always appreciated. With prosperity, the town's population had tripled to 120,000. Many of the new arrivals were French, comprising nearly one-eighth of the residents by the end of the eighteenth century. These immigrants irritated the town's residents and its Castilian rulers. Artisans viewed the newcomers as competitors, and the Castilians worried about radical ideas. The Revolution only confirmed their suspicions. The French were blamed for the rising price of bread and the downward spiral of wages. For several decades the writings of Enlightenment authors such as Voltaire and Rousseau had been smuggled into Spain, along with political pamphlets, anticlerical tracts and pornography, sometimes all folded together in one scintillating read. Madrid tried to staunch this flow of subversive works, even banning the scientific *Journal de physique* in 1791 for its purported atheism. Now, on top of these disturbing tracts, came aristocrats and priests fleeing the godless Revolution. Not that these émigrés found a warm welcome themselves. The Governor-General feared that revolutionaries disguised as priests were fomenting trouble. In July he ordered the army to stop fugitives at the border. Refugees had to swear that they would remain in Spain and observe the Catholic religion.[24]

Yet the Spanish crown also wished to profit from the latest innovations in geodesy. For the past few years, the two nations had begun to cooperate on a venture to define their common border. The Spaniards were especially eager to have a look at the Borda repeating circle, so ideally suited to this task.[25]

Immediately upon his arrival in Barcelona on 10 July, Méchain met with Spanish officials and their team of scientific collaborators. Their leader was Lieutenant José Gonzales, commander of the frigate *Corzo* (*Roebuck*) and an expert in celestial navigation. Méchain was familiar with his work. He was seconded by Ensign Alvarez, as well as ship's lieutenant Francisco Planez. Like most scientific men of the day, the Spaniards spoke French. They agreed to spend the rest of the month equipping their expedition. All told, they needed supplies for sixty men for several months.[26]

During his stay, Méchain met the elite of the Catalan Enlightenment, savants in close touch with French ideas and thinkers. He was a man with a remarkable gift for friendship. At times melancholy, even petulant, he also inspired admiration and affection: a man of honour in a calling of integrity. His self-deprecating manner had its own charm. He had astronomical correspondents around the world – from Pisa to London and from Copenhagen to Madrid – men with whom he traded celestial data and discoveries. So it was perfectly natural that he befriended such Catalan intellectuals as the polymath general, Antoni Martí i Franquès, an astronomer, mathematician and chemist who was the first to calculate the correct mix of gases in air (revising Lavoisier's estimates), and that he struck up a friendship with the medical innovator, Doctor Francesc Salvà i Campillo.

Méchain had Barcelona's artisans construct conical tents that would shade the repeating circle while marking the exact position of the station so that it could be located from afar. The tents could also double as shelters for the expedition at night. Méchain had designed the tents in the shape of teepees. A vertical spine composed of a heavy wooden rod having the dimensions of a carriage axle would be driven several feet into the earth, supported by three or four strong pieces of wood, then draped in canvas. Where the rod rose above the tent, it was capped with a double-backed cone like a giant child's top, painted white to serve as a target for sighting. The bizarre design inspired rumours in talkative Barcelona, already abuzz with news of the tension between Bourbon Spain and Revolutionary France. As usual, the gossips got the story half right. A local grandee heard rumours that the tent signals were to be planted on mountain tops and fortresses from Barcelona to the frontier to relay nightly news of war preparations against the French.[27]

Once the tents were ready in early August, the team could set out north towards the mountains. In this first pass up the meridian, Méchain's goal was to reconnoitre a workable chain of stations through the uncharted region between Barcelona and the high-mountain border, so that he might then double back south and measure the stations accurately with his repeating circle. The distance was not far as the crow flies: not much more than eighty

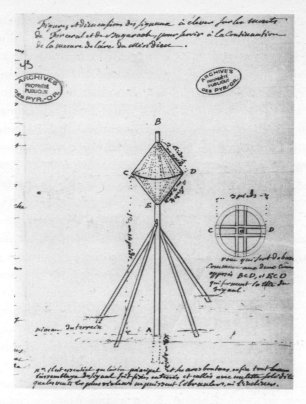

Méchain's signals in Catalonia

This drawing, in Méchain's hand, shows the signals he designed for his triangulation through Catalonia. The lower conical section could be draped in canvas to form a tepee. The upper double cone was painted white to serve as a target for sighting from afar. The overall height was about twenty feet.

(Méchain to Llucía, 6 October 1793. Photo: Archives Départementales des Pyrénées-Orientales.)

miles. The terrain, however, was tortuous; the roads medieval. No carriage could negotiate the tracks; and Méchain had abandoned his custom-built carriage in Perpignan. Not even horses could negotiate the high-mountain trails. To complicate matters, a two-week search in the Spanish archives had failed to turn up a single accurate map of Catalonia. The team hired mules to carry their supplies, and local guides to lead them through the up-valley pastures and the staggered crests of the pine forests.

In those days, the mountainous region straddling France and Spain was a zone of ambiguity and danger. The high country of the Pyrénées was uncharted, the border porous and ill-defined. The Pyrénées march from the Atlantic to the Mediterranean in a series of interlaced chains that encompass fertile valleys and a population accustomed to moving freely between temperate farms, up-country pastures and neighbouring villages. Spain had ruled both sides of the range until the French conquered Roussillon (the part of Catalonia which lies north of the Pyrénées), and a 1659 treaty formally designated the border between the two kingdoms as lying along 'the crests of the mountains'. But the French only considered the mountains a natural border as long as it suited their interests. In the 1710s, the Sun King's armies had marched south towards Barcelona in an attempt to bring all Catalonia under Versailles' rule. The conclusion of this war had seen a Bourbon placed on the Spanish throne and reaffirmed the border as the crests of the Pyrénées, without specifying which crests. At the end of the eighteenth century, the inhabitants on both slopes still spoke Catalan, and they asserted a fierce independence from both Paris and Madrid, now closely allied under the Bourbons. Smugglers and bandits – *miquelets*, in the local dialect – plied a dangerous trade in tobacco, firearms and illegal books, harassing travellers, traders and border patrols. This was another reason the French and Spanish governments were eager to cooperate on a meridian survey to Barcelona: the triangulations would help surveyors define the border between the two nations in scientific terms, so that trade between them might be policed, regulated and taxed.[28]

To expedite their search for workable stations, Méchain divided the men into two parties: Méchain and Gonzales took half the men and Tranchot and Planez the rest. Each party advanced in parallel, planting signals the other party might sight. They began at the top of the Valvidrera ridge which defines the western rim of the modern metropolis of Barcelona. From there, they worked their way north through the dry pine forests to the isolated monastery of Montserrat.[29]

This medieval pilgrimage site was slotted into a narrow eyrie halfway up a tremendous rack of cylindrical rock formations which resembled organ pipes and gave the monastery its name: 'the serrated mountain'. Or as a Catalan poet put it: 'With a saw of gold, the angels hewed twisting hills to make a place for you.' It took Méchain and his party three hours by mule to mount the thousand-year-old switchback trail, and even then the view was not grand enough. Méchain eventually planted his signal on the portico of the solitary Notre Dame chapel on the top of the greatest of these stone organ-pipes, which rose another twelve hundred feet above the monastery and four thousand feet above the valley floor. From this vertiginous peak, he saw a full 360° panorama: from the cool northern curtain wall of the Pyrénées to the shimmering southern island of Mallorca. Directly below, in the terraced valley, he could trace out a tumult of broken forms: walnut trees and olive groves segregated by sinuous stone, red-roofed villages clustered along the river bank, dark ridges receding into barren mountains, and, of course, the monastery itself directly below. 'The peaks so closely surround the monastery that they seem about to crash down and destroy it,' he wrote.[30]

At Montserrat, Méchain and his party were housed in a clean room and served good food and wine. Elsewhere in the region, the inns were wretched – three boards laid on trestles sufficed for a bed, and the windows were without glass. The team spent the next month traversing the desolate northern ranges of Catalonia, zigzagging their way further inland towards the high Pyrénées crests. The fields here were fallow or given over to hemp. The mountains were a savage country. Bears from the high country frequently attacked cattle and sheep, killing them by mounting their backs and smashing their heads. In winter, wolves attacked the bears. The shepherds carried firearms, and everyone smuggled.[31]

By the time they approached the border in September the season was far advanced. Méchain had originally hoped to establish stations on the frontier mountains of Costa Bona (elevation

7500 feet) and Massanet (elevation 6000 feet). But snow had already put their summits out of reach. In the valleys it was raining heavily. More damaging still, political tension was rising along the border. News of the overthrow of the French monarchy had unleashed a violent reaction across southern France, and its repercussions had crossed over into Spain. The French feared a Spanish invasion; the Spaniards feared that the French Revolution would contaminate their virtuous kingdom. On the frontier near Costa Bona, Revolutionary enthusiasts had planted a liberty tree. Bands of *miquelets* operated with impunity, and neither side could count on their loyalty. Under the circumstances, were Méchain's joint party of French and Spanish officers to have begun taking telescopic sightings along the frontier it might well have been considered a provocation to war. It might also get them killed.[32]

Indeed, on the same day that Delambre was delivering his impromptu geodesy lecture in Saint-Denis, Méchain was hunkered down in the Spanish hillside town of Camperdon, on the back slope of the Pyrénées. The Governor-General of Catalonia had just ordered the expedition's Spanish officers to move away from the border. And because Méchain's passport obliged him to travel with his Spanish hosts, he would have to retreat too. This precaution was sensible. Just that week, he and his men had narrowly avoided an ambush. Twelve French partisans from the cross-border town of Prats de Mollò had been lying in wait for them on a little slip of French territory which happened to reach across the road between two stations. Luckily the expedition had taken another route. The price had been a three-day detour through 'the roads of hell'. It had probably saved their lives.[33]

Méchain was thoroughly frustrated. Did not these people understand that he was engaged in a peaceful scientific expedition? He wrote to the administrators of Perpignan to ask that a copy of his commission be posted in all the mountain villages. His mission was a scientific study, sponsored by both nations and dedicated to mankind's highest aspirations for universal knowledge and peaceable commerce. Méchain acknowledged that his official commission – the same one Delambre was at that very moment reading out loud to the volunteers of Saint-Denis – had been signed by a king who no longer ruled. But it was all he had.[34]

In the meantime, Méchain had no choice but to turn his back on the frontier and work his way south, station by station, back towards Barcelona. This time, he would conduct the definitive angle measurements using the repeating circle. He had brought two of them for the purpose: one ruled in the traditional 360° scale and the other ruled in the new decimal 400° scale, an expression of the new spirit of rationalization. The repeating circle was the brainchild of chevalier Jean-Charles de Borda, one of Méchain's senior colleagues in the Academy. Borda was France's leading experimental physicist, as well as a seasoned naval commander who had helped coordinate the French fleet's campaign to liberate the American colonies – France's first and last victory against the British at sea – with Borda himself commanding the sixty-four-gun *Solitaire*, captured in an action against overwhelming odds. In the mid-1780s, back in France, he had transformed one of his navigation instruments into a new device for the measurement of the earth. Elderly now, but as rigorous as ever, the stern aristocratic commander had worked with the dwarfish Etienne Lenoir, France's finest maker of scientific apparatus, to create an instrument precise 'beyond any ever conceived'.[35]

The ingenious principle behind the repeating circle allowed the geodeser to take multiple readings of the same angle without resetting the instrument. This repetition promised virtually to annihilate any errors due to the uncertain sense perceptions of the observer or deficiencies in the manufacture of the angular scale. The Borda repeating circle was composed of two scopes set one above the other on brass rings which rotated independently against a precision-ruled circular scale. To measure the angular distance between two points on the earth's surface, the geodeser set the plane of the circle in the plane defined by the two points. He then zeroed the top scope by sighting the right-hand station, tightening the screws that held that ring in place. He then switched to the lower scope and used it to sight the left-hand station, clamping down that ring as well. At this point, the geodeser could have simply read off the angle between the two stations on the circular scale and called it quits. Instead, he did something counterintuitive. He returned directly to the *lower* scope and rotated it in the opposite direction, clockwise this time, moving both rings and both scopes together until he had sighted the right-hand station. In doing so, he had necessarily rotated the top scope that much further clockwise as well. Now he loosened the ring for the top scope and rotated it alone anticlockwise until it sighted the left-hand station. This meant that in total the top scope had traversed the double of the angle he wished to measure. Indeed, by repeating this procedure again, he could add another double angle, and so on. Ten such additional doublings could take as little as fifteen minutes if the stations were easy to observe, or all day if they were difficult to make out. Finally, he noted the final position against the graduated scale and divided it by the number of doublings. The repeating circle's great advantage was that by multiplying readings it sliced error ever more finely. Uncertainties of ten seconds in the observer's sightings or in the instrument's manufacture, if spread over enough readings, would be diminished, in the words of its inventor, to the point where 'an observer of sufficient patience should be able to eliminate nearly all error'.[36]

Méchain began his measurements at the hermitage of Notre-Dame-du-Mont. He measured on the summits of Puig-se-Calm, Matagall and Roca-Corra, where Alvarez slept in a tent on the peak until the station was surveyed from all the surrounding sites. He measured at Puig Rodos, where he and Gonzales lodged in a cowshed, making over a dozen four-hour trips up the mountain in hope of finding clear weather for their observations. He measured at Mount Matas, at the chapel above Montserrat, and, in late October, over the Valvidrera ridge down into the basin of metropolitan Barcelona. On the night of 28 November 1792, residents in the city reported lights burning on Mont-Jouy, above the General's Palace in the hilltop fortress on the southern edge of the city, and in other high spots around the perimeter of the town. Méchain and his colleagues were lighting parabolic mirrors to take accurate night-time readings across the basin.[37]

With these angles in hand, Méchain had every reason to consider the 1793 season a success. He had closed seven stations in
less than three months, traversing with meteoric speed nearly half
the total distance to his final rendezvous at Rodez – and the most
difficult sector of the meridian to boot. If anything, he felt he had
been somewhat rushed in his labours. The Spaniards, he complained, did not like to linger at the more 'arduous' stations. Had
he been working on his own, he would surely have conducted
many more observations, despite the adverse conditions. Méchain
was striving for a degree of precision that had eluded all other
investigators, a precision that would be the best guarantor of the
metre's universality. He was not about to let icy winds or steep
mountain climbs diminish the accuracy of a measure meant for all
people for all time. But compromise was the price one paid for col-
laborative work.[38]

It was true that he had been as yet unable to measure the high
mountain stations along the crest of the Pyrénées. But these frontier stations could always be approached later from the French
side of the border. And once he returned to France he would have
Cassini's maps and precedent to follow, and would be able to
observe to his own satisfaction. This left him only one remaining
task in Spain: the latitude measurements that would fix the southern terminus of his arc in Barcelona. For this, Méchain decided to
conduct his measurements at Mont-Jouy.

The outcrop of Mont-Jouy, with its castle fortress perched at the
sea-cliff's edge, is visible from everywhere in central Barcelona.
The six-hundred-foot ascent from the Mediterranean is unassail-
able. Even on the city side, the slope is a stiff climb. Today, if you
prefer, you can ride up in one of the bright red gondolas that
shuttled spectators to the 1992 Olympic Games. Or you can drive
up the switchback road through the terraced parkland, a legacy of
the World's Fair of 1929, and still the route of one of Spain's most
famous bicycle races and Grands Prix competitions. On quieter
days, however, when the amusement rides are abandoned and
stray cats sleep amid the dishevelled palms and pines, you can

The Borda repeating circle (above)

The Borda circle is here shown in its horizontal configuration for geodetic triangulation.

(Delambre, *Base*, 2, pl. VII. Photo: Roman Stansberry.)

The Borda circle in operation (opposite)

This diagram illustrates the method of reiteration which made the repeating circle a precision instrument. In the example that begins with the top-down view in fig. 1, the sighting targets G (for gauche, left) and D (for droit, right) are separated by 10°, as indicated by the placement of scopes F and L. Both scopes are initially coupled to the graduated circle. In fig. 2 the lower scope F has been rotated clockwise to fix on the target D, thereby moving the upper scope L clockwise the same distance. In figure 3 the upper scope L has been decoupled from the circle and independently rotated to focus on G, such that it must pass through twice 10° relative to its point of origin. Then in fig. 4 the scope L has been recoupled to the circle, so that when it is refocused on D it moves the lower scope F as well. Which means that in fig. 5, when the scope F is independently refocused on G, it adds yet another 10° to the gauge, while bringing the observer back to the original situation — except that instead of having only measured the 10° separation once, twice the original angle has been added to the original reading. This process can then be reiterated as many times as the observer desires (see figs. 6 and 7) without any need to reset the gauge. The final cumulative angle is then divided by the number of iterations to give a precise reading of the angular separation. This method has the advantage of reducing the uncertainties inherent in any single angle observation and of minimizing the impact of irregularities in the manufacture of the calibrated gauge. The side view shown in figs. 9 and 10 demonstrates how the same process of iteration could be used to measure the vertical height of a star S with scope AB relative to the horizontal defined by the level-scope MN. In this situation the process begins by rotating the entire circle 180° horizontally on its axis, and then sighting the star again with the scope AB (not pictured).

(Jean-Dominique Cassini IV, Pierre-François-André Méchain and Adrien-Marie Legendre, *Exposé des opérations faites en France en 1787, pour la jonction des observatoires de Paris et de Greenwich* (Paris: Institution des Sourds-Muets, 1790), pl. 3. Photo: Houghton Library, Harvard University.)

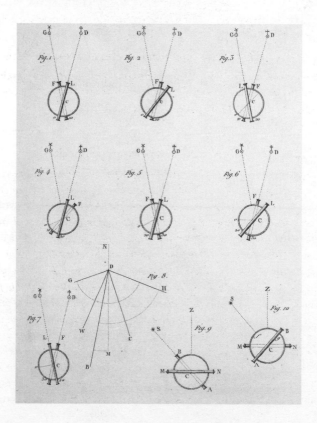

easily imagine the hillside two hundred years ago, when eighteenth-century tourists hunted for fossils in the limestone, or
unearthed heavy fragments of Hebrew gravestones, remnants of
the pre-exile necropolis that gives the site its Catalan name:
Montjuïch, or 'mountain of the Jews'. In French it is known as
'Mont-Jouy', evoking a hillside of play and pleasure. In Latin, it
is 'Mons Jovis', the miniature Olympus where a temple to Jove
once stood. In sum, it is a place of layered histories.[39]

For centuries the Mont-Jouy site had been occupied by a lighthouse. Then in the seventeenth century, as France and Spain
fought for control of Catalonia, and the city periodically rebelled
against both kingdoms, the hill had been fortified. In 1714 the
Spanish monarch began building a fortress along modern lines,
one that English visitors considered so 'perfect in its kind' as to
make Barcelona 'untenable by an enemy'. Massive walls were
arrayed in a pentagon so defenders could train maximum fire-
power on assailants. The fort could accommodate three thousand
men and 120 cannon at one time. It enclosed a large parade
ground and a series of well-appointed barracks for officers and
men. Between the Mont-Jouy castle to the south, and the massive
new Ciutadella fortress to the north, the town was protected from
attack and dissuaded from rebellion. From the heights of impregnable Mont-Jouy, where Méchain and his team were lodged for
the winter, the town's rising clamour against the French was a distant din. It was a fortress poised between the city and the sky.[40]

The view from the Mont-Jouy battlements takes in a vast
sweep of the Costa Dorada. Far to the south, where the oceanic
blue fades to a pale horizon, you can see, if the day is clear and
you have the right kind of equipment, a yellowish break in the
hazy line between sea and sky. That is Mallorca, the largest of the
Balearic Islands, and if you can see it, then you are looking around
the curvature of the earth, relying on the refracted light that bends
through the low damp air of the Mediterranean.

That long view was itself a temptation. In the flush of ambition,
Méchain dreamed of surpassing the Academy's mission and
extending the meridian as far south as Mallorca. The means,

Barcelona harbour, looking towards Mont-Jouy

(Alexandre Laborde, *Voyage pittoresque et historique de l'Espagne* (Paris: Didot, 1806–22), pl. I. Photo: Special Collections, Golda Meir Library, University of Wisconsin-Milwaukee.)

From the heights of Mont-Jouy, overlooking Barcelona

(Alexandre Laborde, *Voyage pittoresque et historique de l'Espagne* (Paris: Didot, 1806–22), pl. IV. Photo: Special Collections, Golda Meir Library, University of Wisconsin-Milwaukee.)

moreover, were to hand. Commander Gonzales had offered to sail his *Corzo* across the one-hundred-mile straits to light flares on the island peaks. Méchain wrote to France and secured permission from his colleagues to try the extension. In December he put that plan into action. While Gonzales set off with his crew to plant the special reflecting mirrors on top of Mallorca's five-thousand-foot Sillas Torellas peak, and Tranchot and Planez scouted locations down the mainland coast, Méchain took up his position on the battlements of Mont-Jouy. On the evening of 16 December he spotted through his telescope a faint light trembling on the horizon. Unfortunately, as he informed his colleagues back in Paris, the resolution was insufficient, and this geodetic triangle, the largest ever attempted, would only be feasible if his repeating circle were refitted with more powerful scopes and he were given larger parabolic mirrors. In the meantime, he assured them, his first priority was to complete the mission he had been sent on.[41]

The rest of December was spent in erecting an astronomical observatory alongside the fortress tower of Mont-Jouy overlooking the sea. The observatory was a wooden cabin, fifteen by twelve feet, with a retractable roof that could be raised by pulleys, and windows that opened out on to an unimpeded view south to the sea and north to the mountains. In this cabin, Méchain positioned his repeating circle in such a way as to capture the transit of stars that crossed the celestial meridian. This was night-time labour, the work of a few experts. The mountain team was disbanded. While Gonzales counted off seconds from a pendulum clock, and Tranchot held a lantern to verify the level of the instrument, Méchain measured the stars.[42]

The latitude measurements to anchor the end-points of the survey were the most delicate operation of the entire mission; the slightest stumble here would distort the final result. Ordinarily, the determination of latitude posed no great challenge to eighteenth-century astronomers. Unlike the longitude problem which had stumped the world's navigators for centuries, mariners had long known how to calculate their distance north or south of the equator by measuring the height of the sun, Polaris (the pole star), or some other celestial object. Latitude was routine. But science thrives on transforming old routines into new problems. Méchain hoped to determine the latitude of Mont-Jouy with a degree of precision hitherto unmatched in the history of astronomy. His goal was to pinpoint its position on the globe to within one second of a degree, a distance of about one hundred feet, an accuracy comparable to today's commercial global positioning systems. This degree of precision transformed a routine measurement into an awesome challenge.[43]

In this quest for celestial precision, Méchain's sole armament was the same Borda repeating circle. But Méchain felt he was up to the challenge. 'As Monsieur de Borda says, it depends only upon the patience of the observer to ensure that errors are in the end totally eliminated.'[44] So far, Méchain's forbearance and patient industry had surpassed all expectations. He wrote to the inventor himself to explain that he had just completed some preliminary calculations of his geodetic results and the sum of angles in his survey triangles never deviated more than 3.5 seconds from 180°, a stunningly small 0.0005 per cent discrepancy. Now he was ready to put the instrument to its astronomical test.

For Méchain, the pursuit of precision was a moral quest as much as a scientific one. Its consummation proved that the investigator possessed the patience, skill and rectitude to reveal nature's predictability and lawfulness. But the pursuit of precision, like all moral quests, is a hazardous affair. Zooming in on nature's fine structure can produce unexpected explosions. If the results do not converge, who will take the blame? Nature or the investigator?

For this celestial operation, the plane of the repeating circle was flipped into a vertical position so that it stood erect like a stop sign aligned in the direction of the meridian. One of the scopes, the one that came equipped with a calibrated air-bubble level, was then directed towards the horizon and its position carefully monitored, while the other scope was angled towards the expected height of the transiting star. As the star approached the line of the meridian, the savant tracked its movement across the wire grid in

the lens, listening out for the beat of the pendulum clock to mark
the exact time. Prowess at this eye-and-ear method was one of the
most basic and demanding astronomical skills, but the repeating
circle demanded something more. Although the angle between
the two scopes, when read off the graduated ring, gave a preliminary assessment of the stellar height, accurate results from the
repeating circle once again depended on multiplication. To accomplish this, the savant rotated the entire instrument 180° on its
vertical axis, so that the stop sign faced the opposite direction,
aligned once again along the meridian. Then he loosened the
screws and swung the working scope back around the graduated
ring until he could sight the star in the wire grid once more. In
doing so, he had doubled the angle traced out along the graduated
ring. This action, repeated, gave a value of four times the stellar
height, and so on. The whirl of dexterous activity, with the
astronomer alternately spinning the plane of the instrument one
way and the scope the other, can be likened to a master mathematician solving a Rubik's cube in the hopes that the right
sequence of high-speed moves will unlock the secret combination
of the heavens.

When Méchain refused to let his Spanish hosts take a crack at
solving this puzzle themselves, they were understandably irritated. Planez complained that Méchain had relegated him to a
'mere spectator', insinuating that the Spanish king would not
appreciate having his officers treated so high-handedly. Yet
Méchain refused to relinquish the eyepiece. The task of measurement, he insisted, was his alone. Only he had the requisite skill
and experience. He alone was answerable to the Academy for the
results. This was the savant's responsibility and his prerogative. It
was the way Cassini had treated him on their 1788 expedition
from Paris to Greenwich. In the end, Méchain had Planez dismissed and replaced with a more conciliatory assistant, the
military engineer Captain Agustín Bueno.[45]

Over the course of the next three months, Méchain took 1050
sightings of six different stars, each comprised of ten repeated
observations. It was a Herculean effort. At night the deep celestial

cold caused buckets of water to ice over. All through the Christmas holidays and into the new year, Méchain, Tranchot and Gonzales worked under the black Mediterranean nights high above the city, pinpointing the latitude of the fortress.[46]

Occasionally he made other sightings. At 6.15 on the evening of 10 January 1793, he reported the discovery of a new comet. Comet-hunting was Méchain's favourite activity and the basis of his scientific reputation, as all his colleagues knew. So lest they think he had been diverted from his official duties, he also reported that the comet was a bit to the west of Mizar, one of the stars he was using for his latitude measurements. 'It's not my fault,' he wrote. 'I wasn't looking for it.'[47] This innocent discovery was written up in the *Diario*, Barcelona's daily newspaper, where the editors felt compelled to add that a comet was a natural celestial body and its appearance did not, as the common people believed, foretell the coming of 'war, pestilence, or the death of kings'. That very week the National Convention began its deliberations in Paris on the fate of Louis XVI. Ten days later, on 21 January 1793, the French king would be executed. War soon followed.[48]

On the evening of 20 February, in plain sight of Mont-Jouy – but beyond the reach of its guns – a French privateer sacked a Spanish convoy on its way into Barcelona harbour with a treasure of American gold. Huge crowds gathered along the sea wall to watch helplessly as the French pirates absconded with 80,000 *duros* due the town's merchants. Mobs clamoured for vengeance, and stoned to death a citizen of Genoa whom they mistook for a Frenchman. The French consul feared that the populace would seize French goods in retribution. 'The Catalans,' he wrote, 'are reckless, bold and vindictive; money is their god.'[49] Méchain was distressed to learn that the frigate *Corzo*, under the command of his friend and scientific collaborator José Gonzales, had been ordered to give chase to the pirates. Three days later Gonzales returned empty-handed. The citizens of Barcelona were enraged. Indeed, some historians have called this incident one of the provocations that led to war between the former allies.

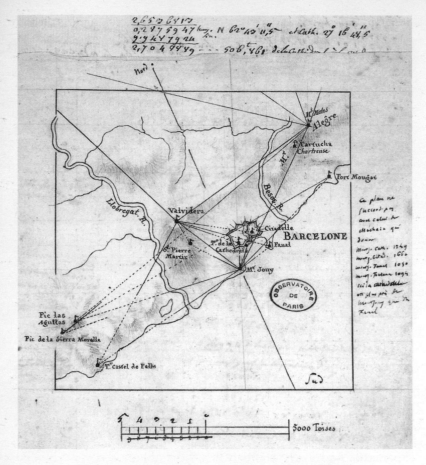

Méchain's triangles around Barcelona, 1792–4

This map of Méchain's stations in the Barcelona basin is not drawn to scale, as Delambre explains in a marginal note on the right. In the city of Barcelona proper, Méchain placed his main station on Mont-Jouy. But he also conducted auxiliary sightings from the northern tower of the cathedral, the Fanal (lighthouse) and the Citadelle (a massive fortification destroyed in the nineteenth century). The unlabelled point located just to the east of the cathedral is the inn known as the Fontana de Oro.

(Photo: Observatoire de Paris.)

The time had come to wrap up the Catalan mission. On the day of Gonzales' return, Méchain wrote for the first time to his northern collaborator, offering Delambre words of encouragement and sympathy. 'Monsieur and dear colleague,' he wrote, 'I heard of your misadventures, and felt for you; I learned of your successes, and they gave me much joy.' He suggested that they swap notes and advice. For instance, how did Delambre set up his signals? How had he positioned his repeating circle? For his part, Méchain described his procedures, and, with elaborate modesty, assured his collaborator that his own southern triangles would never match the precision of Delambre's northern ones. His own ineptitude, he hastened to add, was only partially to blame. Measurements in the mountains presented all sorts of difficulties such as one did not encounter in the lowlands. Up there, heavy clouds obscured stations; trails were treacherous; the cold hampered the instrument. Still, he had been ably seconded, and once his latitude data were complete he intended to return to France 'sometime next month'.[50]

Delambre answered from Paris with best wishes for Méchain's 'health, courage and patience'. On a more practical note he also enclosed updated passports – without the seal of the former king.[51]

But before Méchain could leave Barcelona, the Spanish authorities insisted that he report to them on his geodetic triangles and the latitude of Mont-Jouy. As equal partners in the expedition, the Spaniards had every right to this information. The data would enable them to draw the first accurate map of Catalonia, as well as locate their fortresses with pinpoint accuracy. Cajoling these data out of Méchain tried Gonzales' patience, but at his insistence Méchain spent the month of March preparing a summary report for his Spanish collaborators. He also mailed a précis of his results to Borda in Paris.[52]

What Méchain did not know – because no resident of Barcelona yet knew it – was that a new era was on its way. France and Spain were already at war, though the city had so far seen only signs of rising tension. In early March, the Spanish authorities ordered French citizens to leave the country, and Méchain had to petition to stay long enough to complete his report. Then, in late March, Méchain's Spanish collaborators, military officers all, were ordered to report to their regiments. Commander Gonzales set out to sea in the *Corzo* to escort a supply convoy to the fortress port of Roses. This ended any hope of extending the meridian measurement to Mallorca. Méchain successfully pleaded to keep Captain Bueno with him. Without an escort he would never be allowed to visit the two frontier stations linking Spain and France.[53]

The two kingdoms had been close allies throughout the eighteenth century, each ruled by a branch of the Bourbon family, with common enemies in England and the German states. The stirrings of the Revolution had not altered this basic equation, at least not initially. Carlos IV, newly ascended to the Spanish throne, promised not to join the coalition against the republic so long as his cousin Louis remained unharmed. The French republicans decided to treat this promise as an ultimatum. Not long after they executed Louis XVI, the French pre-emptively declared war on Spain. Spain reciprocated two weeks later, then waited another two weeks before summoning Barcelona's citizens with church bells to hear the crown's vow to extirpate the Revolution. That same day, 4 April, the Spanish military command ordered Méchain to leave Mont-Jouy castle and dismantle his observatory. The army could not tolerate the presence of an enemy citizen in the most important fortress on Spain's Mediterranean coast. Obliged to take up residence in the city proper, Méchain rented a room in the inn known as the Fontana de Oro (the Golden Fountain). The hotel was located just off the renovated La Rambla promenade on the Carrer dels Escudellers, a street known in the eighteenth century for its fine hotels and potters, and famous today for its prostitutes and pottery shops.[54]

Initially, the war on France's southern frontier attracted scant attention from Paris, which was understandably preoccupied by the Prussian and Austrian armies two days' march from the capital. Still, the fighting on the southern front was fierce. Tens of thousands of soldiers fought, and thousands died: regular troops for the Spaniards, and a mix of raw recruits and militia for the French. For more than two years the seesaw campaign would range up and down both sides of the Pyrénées. In the summer of 1793 Spain would seem on the verge of reconquering its former province of Roussillon, along with its capital city of Perpignan. In 1794 France would appear to be on the verge of conquering all of Catalonia, even Barcelona. For two years, troops, militia and *miquelets* – some owing allegiance to France, some to Spain and some to neither – would attack through high passes, lay siege to valley towns and fight for coastal fortresses. These battles would remind both sides that the frontier was not a natural border, but a line drawn in blood, defended with arms and as yet unmarked by science.

In April 1793, however, Méchain still had reason to be satisfied, despite the disastrous turn of international events. His work in Spain was complete, and to his own exacting standards. Under the circumstances, he saw no harm in agreeing to take a long-delayed excursion with his friend, Doctor Francesc Salvà i Campillo. Salvà was one of Barcelona's enlightened savants: a physician who had campaigned for inoculation against smallpox and for healthy fruit consumption, an inventor who had designed proto-railroads and submarines, a natural philosopher who corresponded with the leading physicists of Europe. He had long urged Méchain to visit an innovative pumping station on the outskirts of Barcelona, admired by English visitors for its mechanical ingenuity.

When they arrived at the pumping station, however, they found it closed for the May Day festivities, with no horses in harness to drive the beam. Méchain pronounced himself content to examine the machinery inert, but Salvà insisted on putting the pump in motion. With the help of his manservant he set to work turning the eight-foot beam, while Méchain gauged the water level in the reservoir. Sudden sharp cries roused Méchain from his post. A rise in the water pressure had forced the beam to reverse course, and it was now dragging the doctor and his servant backwards across the floor. Méchain leapt to their aid just as they released the bar. Struck in the chest, he was hurled against the wall, where he slumped to the floor, apparently dead.[55]

A horrified Salvà and his manservant carried Méchain's unconscious body to a nearby house, where the doctor – the leading physician in Barcelona – succeeded in reviving his pulse, though he remained unconscious. Fearing the worst, they lifted him into the carriage and drove back to town. They arrived at midnight and immediately summoned Dr Sanpons, the town's best surgeon. The profusion of blood pouring from Méchain's right ear made it seem doubtful that he would survive the night. None the less, they bled him still more to prevent any build-up of fluid in his brain, wrapped him in sheep's fleece, and put off any treatment of his wounds until morning. The next morning, Méchain was still alive – though unconscious – so they proceeded with their examination. The entire right side of his chest had been caved in, his ribs crushed, his collarbone broken in several places. They wrapped him in bandages and placed him under close surveillance. Three days later his fever broke and he regained consciousness. Later he would recall with rueful gratitude that 'without Doctor Salvà my misfortune would never have occurred, and without his presence, I would not have survived'.[56]

Méchain was immobilized. And even if he had been able to rise from his bed, he had nowhere to go. Whereas a month before he had been obliged to petition to stay in Spain, he was now told he would not be allowed to leave until the war was concluded. The new Governor-General, appointed to direct the war effort against France, feared that Méchain's geodetic data might aid his Revolutionary adversaries. On top of this, the Spanish Treasury had impounded French assets for the duration of the war, and Barcelona's bankers refused to supply Méchain with additional funds.[57]

No wonder a note of self-pity crept into the letter he dictated to his French colleagues – even as he downplayed the extent of his injuries. 'I am unable to write this letter myself and it may lack coherence because I was badly injured two weeks ago . . . though I certainly expect to be back on the job two weeks from now.'[58]

He neglected to mention that he had been injured on a trip unconnected to his mission. And he refused to admit the full extent of his injuries, for fear his Paris colleagues would replace him. Not only did he fear to lose his chance at glory, he could have no guarantee that his replacement would continue the mission with the same exquisite care.

News of Méchain's accident took a month to reach Paris, where it caused great concern. Thérèse Méchain was beside herself with worry. She had been anticipating her husband's return by summer's end. Now she wanted to travel immediately to join him in Barcelona. Lavoisier and the other academicians dissuaded her. It would be difficult, to say the least, for her to cross the border where battles were raging. She spoke no Spanish. Besides, the news from Barcelona was already a month out of date; perhaps her husband had already recovered and was on his way back to France. In any case, her place was with her three young children.[59]

For his part, Lavoisier wrote to Méchain to assure him that fresh funds were on their way. Already he had deposited 34,000 *francs* with the Barcelona banks in 1793. As Treasurer of the Academy, he would insist that the Spanish bankers honour drafts for a project 'which concerns the commerce of all nations'.[60] Moreover, he would personally guarantee the sum – and everyone knew Lavoisier was one of the richest men in France. In the meantime, Méchain should spare no expense in taking care of himself. Indeed, Lavoisier gently scolded his colleague for skimping on comforts, as he was wont to do. Such frugality was not commensurate with the supreme importance of his mission.[61]

In truth, we should reproach you for having practised too
great an economy – one might even say parsimony – with
regard to your expenses. You must not forget that you are
carrying out the most important mission that any man has
ever been charged with, that you are working for all nations
of the world, and that you are the representative of the
Academy of Sciences and all the savants of the universe.[62]

Above all, Méchain should preserve his health and treat himself as 'a person precious to science, to France, to his colleagues and to his friends'.[63]

Lavoisier wrote these noble words of consolation, exhortation and financial relief in mid-June. The difficulty was to convey them to Barcelona. No letter of credit could cross the border because seals were being broken to prevent the transmission of military information. The circuitous mail route through neutral Geneva and Genoa was uncertain. The best Lavoisier could do was ask that the letter be transmitted directly from the French general in charge of the army of the Pyrénées, to his opposing general on the Spanish side. Adversaries at war always maintained a channel of communication, and surely both parties would wish to aid an injured savant engaged on a mission approved by both governments and dedicated to the pursuit of knowledge for the common good. After all, as Lavoisier noted while pleading Méchain's cause, 'The sciences are not at war.'[64]

Chapter Three

THE MEASURE OF REVOLUTION

'A Global Scheme! Ah knew it!' Dixon beginning to scream.
'What'd Ah tell thee?'

'Get a grip on yerrself, man,' mutters Mason. 'What
happen'd to "We're men of Science"?'

'And Men of Science,' cries Dixon, 'may be but the simple
Tools of others, with no more idea of what they are about,
than a Hammer knows of a house.'[1]

THOMAS PYNCHON, *Mason & Dixon*

I fear that the mathematicians, who have not yet troubled the
world, will trouble it at last; their turn has come.[2]

LOUIS-SÉBASTIEN MERCIER, *The New Paris*, 1800

The French Revolution was not one thing but many things,
although it was principally a contest to assert just what one thing
the Revolution was. Paris was the arena where this contest was
the fiercest. But across the nation, villagers, townspeople and
peasants also asserted just what the Revolution meant to them.
For some, the Revolution was a noble struggle between a virtuous
people and a parasitic aristocracy. For others, the Revolution was
a civil war between godless usurpers and a loyal flock. Some proclaimed the Revolution a war of national defence; others, a
movement of international liberation; still others, a war of foreign
conquest. The Revolution, some said, was an attempt by Paris to

bring the countryside to heel; others said it was an attempt by the provinces to establish their autonomy. Some saw the Revolution as a movement to liberate commerce; others, as a social struggle to guarantee a just price for bread. And always, everywhere, the Revolution was a chance to launch a career, make a fortune, join a parade, set out on an adventure, or bemoan the passing of the good old order. Electric promises of equality and panicked warnings of betrayal travelled in waves from Paris to the nation's boundaries – and were returned distorted, amplified, transformed. Like a body of water pelted with stones, the nation was roiled by waves that seemed occasionally to cancel each other out, while under the calm, pressure built for another surge.

The moment the authorities unsealed his carriages and supplied him with new passports, Delambre rushed to Saint-Martin-du-Tertre to sight the Collégiale at Dammartin before the church was demolished for scrap. Thanks to the National Assembly's declaration that the meridian expedition was a mission of national importance, he could move freely through the republic. It helped too that two weeks later the French armies stemmed the Prussian tide at Valmy. The countryside calmed down for the harvest.

Yet no sooner had human events broken his way than nature conspired against him. A northern grey autumn settled on the land. Every day that dismal September Delambre climbed the steeple of the church at Saint-Martin-du-Tertre to peer across the shrouded farmlands. But even though his angle calculations told him exactly where to look, he could pick out neither the Collégiale of Dammartin nor the Paris dome of the Invalides through the mist.

Conditions in the church steeple hardly approximated to the laboratory ideal. Wind and icy rains chilled the bones and clogged the repeating circle. Every toll of the bell seemed likely to shake the narrow belfry to pieces. Even a slight shift in Delambre's weight unsettled the floorboards and rattled the instrument. Eventually, he turned the observations over to his young aides, Lefrançais and Bellet, who took fixed spots in the high cramped tower so they could alternate scopes without having to shift positions. Lefrançais was smaller than Delambre, and fitted into the confined space more easily. And Delambre's weak eyesight obliged him to adjust the focus every time he swapped scopes, enough to disturb the instrument.[3]

Delambre and his team would wait three weeks for a clear view of Paris from the narrow church tower at Saint-Martin-du-Tertre, only to discover, after a rainstorm finally cleansed the atmosphere, that a low hill had been blocking their view of the Invalides dome all along. This left the Panthéon as the only plausible Paris station, meaning they would have to redo all the measurements they had conducted since the beginning of their expedition in June. After four months of labour and danger, they had yet to travel more than forty miles from Paris, and were no further advanced than when they had begun. It was progress worthy of Don Quixote.

With the days growing shorter and the northern skies darkening Delambre pressed on, reversing his route of the previous summer to work his way clockwise around the capital. He completed the Collégiale at Dammartin just before ownership transferred to the demolition men. He returned to the château de Belle-Assise, the scene of his earlier arrest, to retake his measurements there, this time without incident. (In the nineteenth century, years after these measurements were done, the château would come into the hands of Baron de Rothschild, before being destroyed at the end of the century – all but an ancient windmill which still stands watch over the valley of Disneyland Paris today.) Delambre observed from the church at Brie-Comte-Robert in early November. And by the end of the month he was back in Montlhéry, where he had begun his journey half a year earlier, in what was now another historical era. This time he decorated his signal pole as a liberty tree to placate local suspicions. The winter was advancing. The team lit an open fire at the foot of the ancient tower to warm themselves as they worked.[4]

Delambre had hoped to call a halt at that point and return to the capital for the winter, but the owners of the Malvoisine farm, who had kindly allowed him to elevate their chimney as a signal, feared the structure would collapse during the next winter storm and asked him to remove it. This meant he would first have to complete his observations from the Malvoisine roof and from all its adjacent stations. Among these Delambre had selected a site dear to his heart, one he knew with astronomical intimacy: the château de Morionville, the country residence of his patroness, Madame d'Assy, located just outside the town of Bruyères-le-Châtel. On summer days before the Revolution he had often set his astronomical equipment on the terrace outside his rooms there, or conducted observations from his open window. Viewed from the rooftop of the farm at Malvoisine, Madame d'Assy's garden door appeared in his scope like a black stain against a white background.[5]

These were the last observations he conducted that season. As he surveyed his way south towards the outskirts of the royal forests of Fontainebleau, snow began to fall on the woods. In these conditions, geodetic measurements became impracticable. No sooner had Delambre finally worked his way out of sight of Paris than it was time to return to the capital. He would spend February and March at the top of the dome of the Panthéon – the central geodetic station for the city of Paris – triangulating all the sites surrounding the city.

The Panthéon, commissioned by Louis XV as the Eglise de Sainte-Geneviève, had been on the verge of completion when the Revolution erupted. It had since been converted into a temporary warehouse for the thousands of old weights and measures sent in by provincial towns to await comparison with the new republican measures. Now the nation's leaders wanted to transform the warehouse into a mausoleum for the nation's greatest heroes. The building's monumental grey dome dominated the capital's highest hilltop, and the revolutionaries felt that its pure neoclassical form made it a suitable architectural rebuke to the Gothic obscurantism of the royal tombs at Saint-Denis.[6]

But first the Panthéon had to be stripped of its churchly attributes. First to go was the cross on top of the cupola, replaced with an illuminated globe. It was this globe that Delambre had used as his sighting target while he circled the city. The architects had still grander plans, however, and in November – just as Delambre slipped out of sight of the capital – they had replaced the small illuminated globe with a much larger half-globe pedestal, slightly flattened, upon which they planned to set a colossal female statue dedicated to *Renommée* (Fame). Already a sculptor had prepared a full-scale model of the bare-breasted thirty-foot figure, with a fifty-foot wingspan and a trumpet at her lips that would sound (if only metaphorically) 'to the frontiers of France, and be noted from the middle of the ocean'.[7]

It was an ideal spot from which to measure the world. Delambre obtained permission to build a temporary four-windowed observatory high up in the cupola where he could work sheltered from the icy winds. The architect even talked of building a permanent astronomical observatory in the hollowed out half-globe under Fame's feet, thereby 'uniting the beauty of art with usefulness to science'. This goal was consonant with the building's greater dedication to the triumph of Truth over Falsehood. In the pediment over the mausoleum's motto – 'AUX GRANDS HOMMES, LA PATRIE RECONNAISSANTE' ('TO GREAT MEN, A GRATEFUL NATION') – a frieze depicted the sort of heroes entitled to admittance. As the architect explained, it was by 'the conquest of Error' that Genius, in the guise of a robust young man, might manfully seize the laurels of immortal remembrance.[8]

Thus far, the men deemed worthy of being considered for this honour were Mirabeau, the nation's greatest politician; Voltaire, its greatest literary lion; Rousseau, its greatest political thinker; and Descartes, its greatest savant. Mirabeau was the first actually to be buried there. Soon after, Voltaire's remains were borne to the Panthéon in a stately procession witnessed by one hundred thousand spectators. And preparations were now underway to accord Rousseau the same honour. But one man's hero is another man's scoundrel, and just as Delambre returned to the capital, a radical young politician named Maximilien Robespierre demanded that Mirabeau be disinterred for not living up to republican ideals.

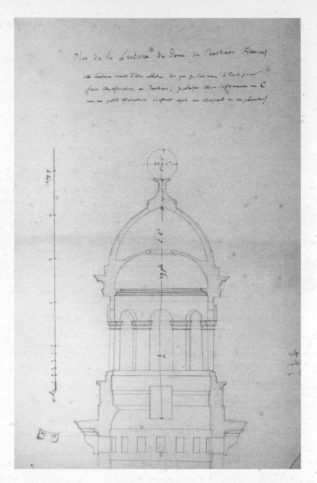

Delambre's crow's-nest observatory on the Paris Panthéon

This drawing of the cupola on top of the dome of the Panthéon was annotated by Delambre and shows the position of the illuminated globe that he used as his sighting target during his triangulations of 1792–3, as well as the special observatory the architects rigged for him at 'C'. The illuminated globe was a temporary replacement for the cross that had once capped the dome, and in February 1793, just before Delambre returned to the capital, it was in turn replaced by a half-dome pedestal for a statue of Renommée *(Fame). Note that Delambre has pencilled in the various dimensions in the old Paris measures.*

(Photo: Observatoire de Paris.)

The Paris Panthéon crowned by *Renommée*

*In this drawing of 1794 the artist imagines how the Panthéon would
appear were it to be capped by the statue of* Renommée. *Although a full-
scale model of the statue was made, it was never installed on its pedestal.*
(Bibliothèque Nationale de France, Coll. Destailleur, Ve Rés. 1047. Artist: De
Machy.)

The decision on Descartes was postponed. In the meantime, the
interior of the Panthéon remained stacked with provincial bushels,
pints and pound weights sent in for comparison with the new
metric measures.

In January and February 1793, during the month it took the
architects to build Delambre his neat little crow's-nest observatory
227 feet above the pavement, the king of France was tried, judged
and executed, war was declared on Great Britain, and food riots
ravaged Paris. Then, from February to March, while Delambre
climbed the cupola daily to triangulate all the stations surrounding
the capital, war was declared against Spain, the west of France
rose up in a counter-revolution, and preparations began in the

National Assembly to create a Committee of Public Safety to organize the nation's defence. His was an admirable perch above a city in turmoil. Delambre sighted the new semaphore telegraph station on Montmartre; it could relay battle news from the northern front in thirty minutes. He sighted his own personal observatory at 1, rue de Paradis. He sighted the stations at Saint-Martin-du-Tertre, Dammartin, Belle-Assise and Montlhéry. When he came down from the Panthéon for the last time on 9 March 1793, he had finally completed the stretch of the meridian that traversed the area around Paris. This represented less than one-tenth of his assigned sector, whereas Méchain in that time had completed nearly half his sector, and was wrapping up his latitude measurements at Mont-Jouy.

Delambre wrote immediately to Méchain to tell him where things stood in the north. Try as they might, his experience of the past year gave him little reason to hope that they would be able to link their triangles this year. 'Too many obstacles will get in our way,' he predicted. Already the new administrators of radical Paris were ignoring his request for a passport, despite the proclamation of the National Assembly approving their mission. (He apologized for the fact that the proclamation listed his name first, but 'I was not there to tell them that your entry into the Academy predates mine . . .') 'Farewell, my dear colleague. I wish you health, fortitude and patience. Be assured of my sincere friendship.'[9] By the time the letter arrived, Méchain's progress had also come to a halt; he was confined to bed, immobilized, out of funds and detained by Spanish law.

The French government was running out of patience. Several years had passed since the Academy had first promised that the survey of the meridian would be done within a year. The actual expedition had been underway for a year now, yet both Delambre and Méchain had stalled far short of their rendezvous point. Various proposals for the reform of measures had been in circulation since the earliest days of the Revolution, and the radical legislature was impatient to bring France into the metric age.

Officials had begun to ask if the meridian expedition was really worth the wait. It was a question that some people had been asking from the start.

Paradoxically, one of these people was Jérôme Lalande, the savant who had given both Delambre and Méchain their start in astronomy. Lalande had been the first person to take advantage of the French Revolution to place before the nation's representatives a proposal to reform its weights and measures. In April 1789 – before the taking of the Bastille, before even the representatives summoned by the king had constituted themselves as the National Assembly – Lalande had denounced the 'unconscionable and multiple abuses of the diversity of measures', and urged the representatives to create a uniform system of measures by simply declaring the Paris measures to be the national standard. In this speech he also demanded a moratorium on the slave trade, the substitution of free public education for the *ancien régime*'s religious schools and the 'liberation' of all monks and nuns.[10]

Lalande had always been a man ahead of his time, yet curiously *ancien régime* in his obsessions. He was not only France's foremost astronomer, he was its foremost popularizer of science – in an age when the popularization of science was the principal weapon against intolerance, superstition and injustice. Lalande had also trained with the Jesuits and had nearly taken up orders, only to become France's most notorious atheist. His father, a Burgundian postmaster and tobacco merchant, sent him to Paris to study law, but every night he slipped out of his student chambers in the hôtel de Cluny to join the astronomers on their rooftop observatory. In 1751, when Lalande was nineteen, his mentor sent him on an astronomical mission to Berlin to help determine, by parallax, the distance from the earth to the moon. In Berlin, Lalande dined with Frederick the Great (Europe's greatest benign despot), lodged with Leonard Euler (its greatest mathematician) and conversed with Voltaire (its greatest wit). On his return to Paris in 1753, he was unanimously elected to the French Academy of Sciences. He was twenty-one years old.

Immediately he was plunged into controversy. Lalande did not agree with his mentor about how to correct his results for the shape of the earth. Lalande was vindicated by the Academy, and his mentor refused to speak to him again. Polemics and controversy pursued him for the next fifty years.[11]

In 1773 a rumour circulated through Paris: Lalande had calculated that a comet might swing close enough to the earth in 1789 to drive the seas from their beds and devastate the earth. The Archbishop of Paris recommended forty hours of prayer and the kingdom's police chief asked the Academy of Science to repudiate Lalande's findings. The Academy replied that it could not repudiate the laws of astronomy. Then, when Lalande finally published the paper, he gave such a low estimate of the odds of disaster (about one chance in 64,000) that many readers assumed that the government had suppressed the truth. In the countryside, word of the impending apocalypse caused public repentances and (supposedly) stillbirths. (If only Lalande had predicted the end of the kingdom instead of the end of the world, one wag later remarked, he would have been the greatest prognosticator of the century.) Some good at least had come of the panic, according to Condorcet. Court ladies and market women had confessed their sins, and the bakeries had experienced a run on unleavened bread, boosting the local economy.[12] Where the laity had once heeded necromancers, they now hearkened to scientific prognosticators. 'The time of prophets has passed; that of dupes will never end.'[13]

Lalande had an insatiable thirst for fame. He cultivated friendships with the great minds of the age and published incessantly. He wrote about paper and platinum, canals and calendars, music and morals. He wrote eulogies for the dead, tributes to the living (including his own estranged mentor), and predictions of astronomical events to come. He was a prodigious travel writer. In Italy he catalogued antiquities, met the Pope, and lobbied him to remove the writings of Copernicus and Galileo from the *Index* of prohibited books. In England he visited the Greenwich Observatory, exchanged pleasantries with King George III, and helped smuggle out the first description of Harrison's famous chronometer designed to determine longitude at sea. He dismissed balloon flight as impossible; then, when the Montgolfiers proved him wrong, claimed he had predicted success and demanded to be taken on the next flight. Later, he set out on a three-hundred-mile balloon voyage to attend a scientific conference in Germany, got as far as the Bois de Boulogne, and declared victory in the form of bad verse. He even wrote the first history of the secretive International Brotherhood of Masons and co-founded its infamous Lodge of the Nine Sisters, which claimed as brothers such luminaries as Condorcet, Danton, Diderot, d'Alembert, Voltaire and Benjamin Franklin.[14]

Such boundless enthusiasm always attracts naysayers – or at least it does in France. One wit diagnosed Lalande as suffering from 'dropsy of celebrity'.[15] Lalande confessed the fault, but excused himself on the basis of his 'innate truthfulness and love of virtue'.[16] Eventually, his mania for fame itself became a theme in the commentaries on his doings, generating still more publicity. Voltaire praised him 'for having found the secret of making the truth as interesting as a novel', and the poet even knocked off a couplet in the astronomer's honour.[17]

> Your glory is known throughout the universe,
> And only when it ends, will your name disperse.[18]

As for his notorious taste for insects and bugs, he reported that spiders had the flavour of fine hazelnuts, whereas caterpillars tasted more like peaches. He dined on the latter regularly at a friend's house, directly after Saturday meetings of the Academy. The friend noted, 'As my home opened directly on to a fine garden, Lalande could easily find enough caterpillars there to appease the first pangs of his hunger; but as my wife liked to do things properly, she would gather a goodly number in the afternoon so she might serve them to him as soon as he arrived. Because I always left him my portion of this ragout, I cannot tell you except by hearsay how they tasted.'[19]

Jérôme Lalande

*This pastel portrait of Lalande was drawn by Joseph Ducreux in 1802,
when the astronomer was seventy years old. It shows him in the uniform of
the new post-revolutionary Academy of Sciences.*

(Musée de Versailles; Réunion des Musées Nationaux. Photo: Art Resource.)

He was a supremely ugly man, and proud of it. His aubergine-
shaped skull and shock of straggly hair trailing behind like a
comet's tail made him a favourite of portraitists and caricaturists.
He claimed to stand five feet tall, but as precise as he was at calculating the heights of stars he seems to have exaggerated his own
altitude on earth. He loved women, especially brilliant women,
and promoted them in both word and deed. His longtime mistress,
Louise-Elizabeth-Félicité du Piery, was the first woman to teach
astronomy in Paris. He sang the praises of women astronomers
such as Caroline Herschel and Madame Lepaute. When he was
appointed director of the Collège de France during the
Revolution, his first official act was to open classes to female students. He dedicated his *Ladies' Astronomy* to his mistress, and this
serious primer, with its examples of active women researchers,
was still being published sixty years later.[20]

Yet he loved women in ways they did not always find agreeable. He never married, though he bragged of refusing advantageous offers. When at the age of forty-four he finally did propose, the fourteen-year-old girl refused him. He had a salacious turn of mind. He noted in his diary that 'Monsieur de V— loved his pretty wife so much that he invited the most agreeable young men to his home and let them have sex with her in front of him; Helvetius was among them'.[21] He made an unwelcome pass at the gifted young mathematician Sophie Germain and wrote an abject apology the next morning – including an apology for slighting her scientific knowledge. As he liked to tell pretty women: 'You have the power to make me happy, but not the power to make me unhappy.'[22] He loved women, he said, but not so much as to distract him from his greatest love: the stars.

In short, he was a shameless self-promoter, a trait which made him especially effective at the calling he prized above all others: teaching. Lalande was a gifted pedagogue, a missionary for his science. His *Astronomy* became the field's standard textbook. His lectures at the Collège de France attracted two hundred students, with auditors from across Europe, who then joined his world-wide network of correspondents. By the 1780s, he had trained France's next generation of astronomers, among them Delambre and Méchain, putting them all to work in his scientific workshop, the Lalande family enterprise.[23]

For Lalande, astronomy was the family business. He had brought his young cousin Lefrançais in from the country, adopted him as his nephew, trained him in astronomy and then married him to his illegitimate fifteen-year-old daughter, Marie-Jeanne-Amélie Harlay, whom he had trained in mathematics to calculate his celestial data. He was very attached to his 'niece' and 'nephew' – as he called them – and easily moved to tears where they were concerned. Amélie performed the bulk of the calculations for the massive navigation tables he published in 1793, 'tiresome calculations, but here acquiring a more noble character by the aid they offer navigators, connecting remote parts of the universe'.[24] He could be a hard taskmaster. He complained to his mistress, 'When I'm not around, my shop goes on holiday.'[25] 'Wash out my nephew's head if he's not working.'[26] At the start of the Revolution, he announced a new goal for his astronomical family: a monumental catalogue of thirty thousand stars that would surpass the old celestial survey. His students were likewise expected to contribute to this enterprise.[27]

Méchain had already done the lion's share of the work for the second edition of Lalande's *Astronomy*, which appeared in 1781. The prize competition of the Academy that year was to chart the trajectory of the comet Lalande had predicted might devastate the earth in 1789. Méchain wrote a paper that showed that his mentor had erred by confusing the appearances of two different comets. When the Academy of Sciences awarded its 1781 prize to Méchain, Lalande had the good grace to be pleased. In 1782 he got Méchain elected to the Academy, then looked around for a new recruit.[28]

In 1783 Delambre began supplying Lalande with data for his third edition of *Astronomy*, and Lalande soon ran out of words of praise for his abilities. 'Monsieur Delambre . . . is currently the most able astronomer of any country in the world . . . We must encourage so valuable a recruit, and bind him to a science in which he performs prodigious feats without hope of any position or advantage.'[29] Delambre's first astronomical coup likewise came at his mentor's expense. In 1786, he was one of only two astronomers in Paris to record the last flicker of Mercury during the planet's transit across the sun. This was not merely an observational coup, it was a theoretical triumph. As was his custom, Lalande had publicized the time of the planet's transit in advance. However, the overnight clouds persisted beyond the appointed hour, and the country's leading astronomers all quit their posts. When the clouds broke at eight in the morning, Delambre, still at his telescope, saw Mercury exit the sun forty minutes after Lalande had predicted it would. The only other observer to capture the transit had been looking for sunspots. Delambre had stayed at his post because he doubted Lalande's calculations.[30]

For all his boundless vanity, Lalande never took offence when his students contradicted him. 'I am waterproof to insults, and a sponge for praise,' he said.[31] Perhaps he remembered the repudiation he had suffered at the hands of his own *maître*. In any case, there was work enough in heaven for all. Science was a collective triumph, even if the race was run by men hungry for personal fame. He presented Delambre's Mercury results at the next meeting of the Academy of Sciences – with Delambre in attendance – then promptly used the data to update his own tables.[32]

The Mercury coup exhibited the scientific virtues which would sustain Delambre on his seven-year pursuit of the meridian: his patience and perseverance, his precision and scepticism, his ability to marry observation and theory, and his confident willingness to show up his elders. Méchain, despite tremendous preparation, had missed the observation of Mercury. As he noted ruefully, he had believed Lalande's prediction.[33]

Delambre also knew how to parlay success into greater opportunities. At that same Academy meeting, Pierre-Simon Laplace presented another instalment of his *System of the World*, his life's work synthesizing Newton's cosmology. Laplace was the same age as Delambre, but he was already his era's leading physicist, a theoretician who claimed that, were the position and motion of all the world's particles known, the entire future of the universe could be predicted by a being of infinite intelligence. This particular paper refined a technique for tracking the perturbations which one planet caused in the orbit of another. Delambre was dazzled by the ability to calculate orbits so precisely. Several years before, Méchain had supplied Laplace with some preliminary data for the newly discovered planet of Uranus. Delambre now proposed to confirm Laplace's theory using Uranus. This involved tremendous labour: for nearly two years Delambre observed at night for eight hours at a stretch, then spent equal time on daytime calculations. His achievement did not go unrewarded. At the prompting of Laplace and Lalande, the Academy of Sciences announced that the topic for its 1789 prize competition would focus on the precise calculation of the orbit of Uranus. As Delambre remarked to a friend, he could be sure his memoir would be the best, since it was the only one submitted. The prize committee consisted of Cassini, Lalande and Méchain. As Delambre privately admitted, it would have been hard to find judges more 'favourably disposed'.[34] In the report which awarded his student the prize, Lalande praised Delambre as 'an astronomer of wisdom and fortitude, able to review 130 years of astronomical observations, assess their inadequacies, and extract their value'.[35]

How did Méchain feel about this rising star? Did he resent Delambre's Mercury coup, his one-upmanship on the Uranus data? Was he jealous of his *maître*'s new-found favourite? Méchain's feelings mattered to Delambre. Méchain not only sat on the prize committees; as a member of the Academy, he could have blocked his rival's election. The two astronomers had occasionally observed the stars together. Lalande asked them to collaborate on yet another edition of his *Astronomy*. And Méchain published Delambre's work in his astronomical journal. Yet relations between the two remained formal and Delambre still thought it wise to ask his friends to put in a good word about him with Méchain.[36] For his part, Méchain always addressed the up-and-coming astronomer with elaborate courtesy as the 'abbé de Lambre'.[37]

Delambre would later acknowledge that 'one can trace certain similarities between the early years of Delambre and Méchain'.[38] Both were men of modest background, sons of provincial Picardy, educated first by the Jesuits, rejected by the *ancien régime*'s institutions of higher learning, and later employed as tutors. And from there, their lives converged ever more closely: the same scientific discipline, the same *maître*, the same expedition. But a career trajectory does not decide one's fate.

Lalande's proposal that the nation adopt the Paris measurement standards received little attention until the juridical revolution of August 1789, when the nobility renounced all its legal privileges, including its authority over weights and measures. From that point onwards, a flood of proposals poured in from the members of provincial learned societies, under-employed state engineers and enthusiastic citizens of all stripes. Each of these pamphleteers had their own pet notions about the form the nation's new system of measures should take.

Yet in the end, the metric system that emerged was essentially the creation of a core of savants from the Paris Academy of Sciences. The French kings had traditionally referred questions of measurement to their Academy. In the run-up to the Revolution, Lavoisier and other academicians had been invited on to royal commissions to consider the advantages of uniform measures. Now savants such as Condorcet, Lavoisier, Laplace, Borda and Legendre quickly formed a Commission of Weights and Measures to hammer out the specifics of the metric reform. Some of these men, like Condorcet, themselves sat in the legislative assembly. Other members of the legislature, such as a young engineer named Claude-Antoine Prieur-Duvernois (known as Prieur de la Côte-d'Or), helped promote the reform from within the ranks of officialdom. Over the next four years, these men transformed the citizenry's simple plea for a uniform system of weights and measures into a hyper-rational system, combining features which had long been sought by savants. Each of these features was proposed independently, and each proved controversial.[39]

The single demand which united all the savants, legislators and pamphleteers was the expectation that the new weights and measures would be applied uniformly throughout France. It was true that in 1788 some of the complainants in the *Cahiers de doléances* had thought it sufficient to demand regional measures based on the standards of provincial market towns. But merely regional measures quickly came to seem inadequate to the leaders of the new national government, and in February 1790 the first proposal for metric reform to reach the National Assembly reiterated Lalande's plea that the legislators adopt the Paris standards throughout the nation. It was a proposal bound to appeal to a nation newly conscious of its unity. It was also the logical culmination of a thousand years of French centralization. In any other nation at any other time this proposal would undoubtedly have become the law of the land – in which case the metric system would almost certainly never have existed in anything like its current form.

But it was not just any nation and it was not just any time. It was a nation conscious of its place at the vanguard of history at a time when history called for actions of universal significance. Even the men who proposed the Paris standard recognized the thrum in the air. They knew that some of their comrades had grander ambitions, and they feared that those ambitions would scuttle any chance for a more limited success. 'Do not,' they pleaded, 'take us beyond our desires and our hopes.'[40] The savants, however, were determined not to let slip this chance to design a truly rational system of weights and measures.

One month later, in March 1790, Charles-Maurice de Talleyrand offered the legislature a far grander proposal. While the adoption of the Paris measures might seem expedient, it failed to rise, he said, to 'the importance of the matter, nor the aspirations of enlightened and exacting men'.[41] In its place, the former bishop, sometime revolutionary and perennial master of French foreign policy advanced the proposal favoured by the savants, especially by Condorcet. In place of a measure derived from history or the fiat of kings, he asked that the legislature derive its fundamental measure from nature, the common heritage of all mankind. Only a measure derived from nature, he declared, could be eternal because only such a standard could be reconstructed should its man-made physical embodiment suffer the ravages of time. For instance, the Paris *toise* – which equalled, by definition, six times the length of the royal *pied* (foot) – was actually a bar of iron mortised into the wall at the foot of the staircase of the great Châtelet courthouse. Yet as everyone knew, the original bar had become badly bent as the building settled and had been replaced in 1666. By 1758 even the new bar, equal to half the width of the entrance to the royal Palais du Louvre, had begun to show its age. Surely so ephemeral a standard would not suffice for a new régime founded on the rights of man. Only a measure taken from nature could be said to transcend the interests of any single nation, thereby commanding global assent and hastening the day when all the world's peoples would engage in peaceable commerce and the exchange of information without encumbrance.[42]

Talleyrand – again, at Condorcet's prompting – also proposed an additional feature for the new system of measures: that its various units (length, area, capacity, weight, et cetera) be rigorously linked by an interconnected *system*. The idea was that once the unit of length had been derived from nature, all the other units might be defined in relation to it. This would ease all kinds of calculations and comparisons, especially for those professionals – engineers, doctors, savants, artisans – who transformed nature into useful things. This provision was reiterated in all subsequent proposals, although the savants would themselves disagree about how to define these relationships, especially for the unit of weight. Lavoisier and the crystallographer René-Just Haüy set to work early in 1793 to define the *grave* (as the gram was then called) as a cubic centimetre of rain-water weighed in a vacuum at the melting point of ice (0°C). But without a definitive determination of the metre, their findings were necessarily provisional. Ultimately, in 1799, the chemist Lefèvre-Gineau would define the gram as one cubic centimetre of rain water weighed in a vacuum at the temperature of maximum density (4°C).[43]

Not longer after Talleyrand's proposal had been voted into law, the legislature authorized the addition of a third feature, one long desired by savants. They declared that all the metric units would be divided by a *decimal scale*. The idea for decimal measures went all the way back to the proposals of Simon Stevin, the Flemish engineer who had invented the decimal point in the Renaissance. In the seventeenth century, the advantages of decimal measures had been echoed by the English philosopher John Locke and the French military engineer Sébastien Le Prestre de Vauban. More recently, in his textbook for the new chemistry, Lavoisier had urged that decimal measurement be adopted by all the world's savants. Given the near universality of the decimal scale in arithmetic, the savants pointed out, a complementary system of measures would ease calculation, not only for learned folk, but for everyone engaged in trade, commerce or construction. The decimal system could even be considered a kind of natural scale because human beings have ten fingers. To cap this reform, the

National Assembly was also considering a proposal to decimalize the new national currency, as the American republic had done a few years before.[44]

Yet even this proposal proved controversial. Several pamphleteers suggested that the metric system be designed around a duodecimal scale. Because of its many divisors, a base-12 system would allow buyers and sellers to divide and subdivide goods easily, enabling butchers to chop sausages into halves, thirds or quarters. The admitted drawback of the duodecimal system, its incompatibility with the usual arithmetic scale, could be solved by switching our arithmetic to a duodecimal system and adding two new single digits for '10' and '11'. The Revolution was a chance to rethink all old assumptions. Then again, other pamphleteers preferred a scale derived from base 8 because it would enable commodities to be divided in half again and again and again, like a pie. Yet another proposed instead division by base 2. And one great mathematician even toyed with the idea of a scale built around a prime number, like base 11, since from a mathematician's point of view a fundamental unit ought not itself to be divisible.[45]

The fourth and final addition made by the savants – and certainly the one that most baffled their countrymen – was their proposal for a nomenclature of prefixes. Only gradually did the Academy come around to the view that the new measures needed new names. In May 1790 the citizen Auguste-Savinien Leblond was the first to propose the neologism 'the metre' for the fundamental unit of length, 'a name so expressive that I would almost say it was French'.[46] And for the next few years, the reformers continued to assume that the multiples and subdivisions of the metre would go by their own simple names, like the *perche* (10 metres) and *stade* (100 metres), or the *palme* (0.1 metres) and *doigt* (0.01 metres). The idea that one might use Greek and Latin prefixes – kilo to mean 1000, and milli to mean 0.001 – first surfaced in a report by the Commission on Weights and Measures in May 1793. In spite of a counterproposal that the prefixes would be more authentically French if taken instead from Low Breton, this system of Classical prefixes was the final element to be added to the metric system as we currently know it today.

Each of these features was controversial, added in succession, and debated in turn during the early years of the Revolution. Yet no single feature caused more consternation, frustration and second-guessing than the proposal to base the fundamental unit of length on the measure of the earth. 'Was it really necessary,' one critic asked, 'to go so far to find what lay so near?'[47]

Indeed, Talleyrand had initially proposed that the fundamental unit of length be derived from the length of a pendulum beating one second. This was an idea with a long pedigree, going back to the early seventeenth century when Galileo had first demonstrated that the period of a pendulum's beat was determined entirely by its length, so long as its swing was not too wide. In the 1620s, the Dutch savant Isaac Beeckman and Father Marin Mersenne of Paris had discussed a natural standard for length calibrated against the length of a pendulum beating at one-second intervals. In 1775, the reform-minded chief minister Turgot had asked Condorcet, the rising star of the Academy of Sciences, to draw up a plan for a scientific system of weights and measures based on the one-second pendulum.[48]

Talleyrand, on Condorcet's advice, now proposed that the French government invite two savants from each of the world's nations to participate in a joint experiment to determine the length of a pendulum beating one second. Talleyrand further announced that he was in contact with Sir John Riggs Miller, a member of the British Parliament who had introduced similar legislation in the House of Commons. Talleyrand considered this a hopeful sign, and wondered if it were 'permissible to see in the concourse of two nations interrogating Nature together the principle of a political union by the mediation of science'.[49] If successful, such a measurement system might extend beyond Europe and around the globe. A savant from France's young sister republic across the Atlantic had sent word that he too was interested in this project. Thomas Jefferson, in his capacity as America's first Secretary of

State, had been asked by President Washington to report on the reform of American weights and measures, and he had likewise agreed to coordinate his proposals with the French. Condorcet privately predicted that the French, the British and the Americans – 'the world's three most enlightened and active nations' – would all be employing the same measures in short order.[50]

There was only one catch. In the two centuries since Galileo's day, the savants had learned that the length of the one-second pendulum was also sensitive to the latitude at which it was measured, because gravity varied slightly with latitude. Hence, Talleyrand reminded the legislators that this pendulum experiment would have to be conducted at some specific location. The equator might have seemed the most natural choice, positioned as it was, equidistant from the poles. But the equator was unfortunately remote from the scientific nations. So the second most natural location, Talleyrand argued (on Condorcet's advice), would be the mid-point between the pole and the equator – at 45° of north latitude – where the pendulum possessed its average length. And since the experiment ought to be carried out at sea level and far from any disturbing mountains, the most plausible site on earth was on the outskirts of Bordeaux in south-west France.

Needless to say, this aspect of Talleyrand's proposal did not meet with international approval. Miller of Britain plumped for a measurement in London. Jefferson of Virginia argued for a measurement on the 38th parallel, both the median latitude of the United States and conveniently downhill from his estate at Monticello. And some Parisians dared to suggest that the experiment might most easily be carried out in Paris. The achievement of universality, it seemed, would require some delicate diplomacy. But Talleyrand was a master diplomat. He saw to it that the law passed by the National Assembly on 8 May 1790 gently suggested the pendulum measurement be taken at 45°, 'or whatever other latitude might be preferred', and invited the Academy of Sciences to form a Commission of Weights and Measures to carry out this plan.[51] In England Miller praised this concession before the House of Commons, and Jefferson redrafted his final report to the US House of Representatives to tout the advantages of cooperating on the measurements at the 45th parallel 'with the hope that it will become a line of union with the rest of the world'.[52]

Yet after all these delicate negotiations were concluded, when the Commission reported back one year later, on 19 March 1791, they urged that the pendulum standard be dropped altogether in place of a metre based on one ten-millionth of the distance from the North Pole to the equator, as established by a survey of the meridian that ran from Dunkerque to Barcelona.

It was Borda, as chairman of the Commission, who justified this change on scientific grounds. The problem with the pendulum, he noted, was that it would make one fundamental unit (the length of the metre) depend upon another unit (a second of time). What would then happen if the units of time were themselves to change? Even as he spoke, the Academy was considering whether the arbitrary division – inherited from the Babylonians – of a day divided into 24 hours of 60 minutes of 60 seconds each should likewise be converted to the decimal scale, so that the day might be more rationally be divided into 10 hours of 100 minutes of 100 seconds. By contrast, there could be nothing simpler or more natural than basing the fundamental unit of length (the metre) upon another unit of length (the size of the earth).

Besides, it was only fitting that a measure for all the world's people be based on a measure of the world. It was consonant with the universal aspirations of the Revolution. As Laplace would later point out, a metre based on the size of the earth would entitle even the most humble landowner to say, 'The field that nourishes my children is a known portion of the globe; and so, in proportion, am I a co-owner of the World.'[53]

Obviously, the circumference of the entire earth would make an awkward unit of length for ordinary purposes. But a measure of length based on the quarter-meridian divided by ten million would come out to very near the length of the *aune* of Paris, a three-foot length comfortably on the human scale and familiar to many

French citizens. To determine that length with the necessary precision, the National Assembly need only to authorize a new
expedition to measure a meridian – or at least a portion of one.

Borda explained how the French academicians had selected
just such a meridian on the basis of rational criteria that would
'exclude all that was arbitrary'. First, the selected arc would have
to traverse at least 10° of latitude to allow for a valid extrapolation
to the full arc of the earth. Second, the selected arc would have to
straddle the 45th parallel, which, as the intermediate distance
between the pole and equator, would minimize any uncertainty
caused by the eccentricity of the earth's shape. Third, its two end-
points would have to be located at sea level, the natural level of the
earth's figure. And fourth, the meridian would have to traverse a
region already well surveyed so that it could be quickly measured. There was only one meridian in the entire world which met
all these requirements: the meridian that ran from Dunkerque
through Paris to Barcelona. He assured the legislators that 'there
was nothing in this proposal that would give any nation the least
pretext for reproach'. He also assured them that the task could be
completed in a year.[54]

As Borda noted, the idea of basing a natural unit of measurement on the circumference of the earth had long been a cherished
dream of savants. Centuries before Columbus set sail west from
Spain, learned folk had known that the earth was round.
Eratosthenes, director of the fabulous Library of Alexandria in the
third century BCE, was also the father of geodesy and had measured the earth's circumference to within 10 per cent. Eratosthenes
knew that in the Egyptian town of Syène (Aswan), located 5000
stades due south of Alexandria, the sun stood directly overhead at
noon on the spring equinox because its light then reached the
bottom of a deep well. So one spring equinox around 240 BCE, he
simultaneously measured the solar height at noon in Alexandria
by means of the shadow cast by an upright stick, and found it to
be 7.2° from vertical, or approximately one-fiftieth of the total
circle of 360°. From this, he deduced that the earth's circumference was fifty times greater than the 5000 *stades*' distance between

the two towns, or 250,000 *stades*. Not a bad estimate, given what we know about the length of the *stade*.

Two millennia would pass before anyone came up with a better one. Jean Fresnel, doctor to Henry II of France, took a comparatively crude approach in the sixteenth century. He measured the distance from Paris to Amiens by counting the number of times his carriage wheel turned along the route (a mechanical ticker inside his carriage helped him keep track of the rotations). Because he knew Amiens was located one degree of latitude due north of the capital, and the road between the two ran straight, he simply multiplied the number of rotations by the circumference of his wheel and again by the full 360° of the globe. Considering his methods, he was not far wrong either. The modern technique for using triangles to measure earthly distances, however, was introduced in 1617 by Willebrord Snell, 'the Dutch Eratosthenes', on the frozen fields outside Leyden, and his method persisted for the next 340 years.

One of the earliest official acts of the French Academy of Sciences had been to remeasure Fresnel's itinerary from Paris to Amiens by triangulation, and this measure of the globe's regularity had inspired Gabriel Mouton, a chaplain-astronomer in Lyon, to propose that the earth serve as the basis for all human measurements. In 1670 he suggested that the fundamental unit of length, which he dubbed the *mille*, be set at the length of one degree of the earth's arc, with all sub-units determined by decimal division, such that the *virgula* (about the length of the king's *pied* or foot) would equal one ten-thousandth of the *mille*. The marvel of nature's regularity suggested that human activities be aligned according to the same metric.

But all these learned folk, from Greek astronomers through European scholars, had assumed that the earth was a perfect sphere until Isaac Newton – without ever setting sail from Cambridge – announced that our round planet was slightly flattened at the poles. His hypothesis of an oblate earth began as a theoretical prediction. Calculating the effect of rotation upon a homogenous liquid sphere (our earth) in which all particles attract one another all the time, Newton estimated that centripetal forces would produce an earthly eccentricity of ½₃₀. In other words he suggested that the earth's radius at the pole was 1/230th shorter than its radius at the equator. Newton then corroborated this prediction by several choice pieces of evidence. He reanalysed the French Academy's meridian survey to show that the earth had flattened slightly as the triangulation proceeded north. He noted that the pendulum clock carried by another French savant to the Caribbean in 1672 had beaten more slowly as it approached the equator, suggesting that gravity weakened slightly as the earth bulged, because the point was further from the earth's centre. Finally, he pointed out, astronomers had noticed that Jupiter was flattened at the poles. On earth, then, as it is in heaven. And to cap it all, Newton made one final startling deduction. The bulge around the earth's middle, he surmised, explained a phenomenon that had baffled astronomers for two thousand years. The pull of the sun and moon on the earth's bulgy middle was responsible for the precession of the equinoxes, the slow but steady 26,000-year swivel of the earth's axis of rotation. Newton had banished the earth's spherical perfection along with the planets' circular orbits. We do not live on a perfectly round orange, but on a flattened tomato. Nature's perfection lay not in childish geometry, but in nature's forces deeply hidden – and by Newton revealed.

The century-long debate that followed proved to be the golden age of geodesy; that is to say, an age of bitter controversy and world-shaking reversals. A survey of the meridian of France, undertaken by Cassini I in 1700, seemed to confirm Newton's hypothesis, until his son, Cassini II, reviewed the data and dared to suggest the opposite, that Newton had in fact erred and the globe, if anything, was elongated at the poles: prolate rather than oblate, a long lemon rather than a flat tomato. The question was not just academic; it affected mapping projects on land and sea. The difficulty was that a 1 per cent error in the determination of the latitudes was enough to flip the earth from oblate to prolate, from tomato to lemon – or the other way round.

To resolve the question, the Academy of Sciences launched an expedition to Peru to determine whether the earth bulged at the equator. It also dispatched rival savants to Lapland to measure the earth's curvature as it approached the pole. These stirring voyages cast science in a heroic light, bringing Newtonian physics to public attention, and entertaining the salons of Paris with the splenetic quarrels of the academicians. In 1740 the French crown also sponsored Cassini III's survey of the meridian from Dunkerque to Perpignan to help settle the controversy – and recast the map of France. On his return, Cassini abjured his father's prolate earth and acknowledged that we live on Newton's flattened globe – although no two savants could quite agree on just how flattened it was.[55]

As one chieftain of eighteenth-century science confessed, this controversy had seemed to dash the 'flattering dream' of a universal measure based on the perfection of nature.[56] But natural philosophers do not give up so easily on nature's perfection. Several savants – Laplace in particular – remained convinced that the earth, flattened though it was, might still serve as the basis for a perfect metre. As Condorcet explained to the National Assembly, these arguments had convinced him to switch his allegiance from a simple pendulum standard to a geodetic mission. And he pleaded with the Assembly likewise to embrace this revised plan. The meridian project was based on the soundest science, he noted, of such universal principles that in future years no one would even be able to say which nation had performed the task. And he went on to urge them, in somewhat contradictory terms, not to wait for 'the concourse of other nations' before settling on a standard. As the representatives of a great and enlightened nation, one whose vision reached out to all people and all times, it was incumbent on the French to reject the easy path, and instead 'approach perfection'. On 26 March 1791, despite some grumbling about the likely cost and delays, the National Assembly adopted this meridian standard.[57]

This decision had lasting repercussions. In the short run, it ended any chance of international cooperation. To savants outside

France the meridian project smacked of self-interest. Those savants who favoured the pendulum standard refused to concede its inferiority. Geodesers, they pointed out, also relied on many other units like time and angles to measure the globe, so no unit could ever be truly fundamental. The leaders of London's Royal Society accused their French colleagues of seeking to 'divert the attention of the European public from the true amount of their proposal, which in fact is that their measurement of 9 or 10 degrees of a meridian in *France* shall be adopted as the *Universal* standard'.[58] Jefferson likewise withdrew his support from the metric system when he learned that the French would survey their own meridian. As he pointed out: 'If other nations adopt this unit, they must take the word of the French mathematicians for it's [sic] length . . . So there is an end to it.'[59]

For the French savants themselves, however, the expedition paid handsome dividends. Thanks to the extravagance of the meridian project, the budget for the creation of the metric system was revised upwards to 300,000 *livres*, roughly three times the annual operating costs of the entire Academy under the *ancien régime*. Government funds also flowed into the coffers of instrument-makers like Lenoir, who had been hit hard by the disruption of the luxury trade at the advent of the Revolution. Almost every one of the Academy's savants who worked in the physical sciences found employment on the metric-system project. All, that is, except for Lalande, who refused to participate in a project he considered pointless – though he wished his former students well.[60]

Even within France, some criticism was heard. The literary critic Louis-Sébastien Mercier thought the meridian expedition smelled of charlatanism. The savants, he said, had 'preserved their pensions and salaries . . . under the pretext of measuring the arc of the meridian'.[61] Other commentators were more scathing. Jean-Paul Marat, bilious enemy of the Academy ('those cowardly lackeys of despotism') deemed the 300,000-*livres* budget 'a little gâteau they will share out among confederates'.[62] There were even some savants who (privately) ascribed the change in plans to ulterior motives. Delambre himself later speculated that Borda had pushed the meridian project to enhance the reputation of his repeating circles. Others wondered whether Laplace and the other physicists had primarily promoted the project as a way to pin down the exact shape of the earth rather than as an attempt to ascertain the length of the metre.[63]

Of course, as many commentators have pointed out then and since, the decision to base the metre on one ten-millionth of a quarter of the earth's meridian was itself arbitrary. To begin with, it was not even a real distance, but a calculated distance along a portion of the surface of an imaginary sea-level geoid that would have to be extrapolated from one small segment of the arc to the whole. And metric *kibitzers* suggested many alternative ways of slicing up the globe. Some preferred a metre based on the circumference of the equator. Not only was the equator unique, it was apparently circular and unchanging. A meridian, by contrast, was arbitrary, elliptical and possibly subject to change over time. Still others agreed with the choice of a meridian, but wondered why the savants hadn't chosen their standard to equal one hundred-millionth of the total meridian (rather than one ten-millionth of the quarter meridian) to make the metre more nearly come out to the length of the foot, a more manageable size for daily use. Occasionally – as the meridian project suffered one delay after another – politicians and ordinary citizens even had the temerity to ask whether a natural standard was necessary at all. Nature was changeable and irregular, some said. 'Everything in nature is unequal,' another complained. Even the shape of the earth might change over time, as Laplace himself admitted – though surely the measure of the meridian would not take *that* long to complete.[64]

All along, Lalande, the old iconoclast, stuck to his preference for a physical standard, like the copper *toise* of Paris which the Academy had in its keeping. Such a standard could be defined with much greater ease and accuracy, whereas any attempt to find a standard in nature would prove ephemeral because too many factors influenced the investigation of natural phenomena. For instance, there were many factors which might influence the length of a pendulum besides the latitude at which it was measured, from the arc of its swing and the ambient temperature, to the air resistance. Moreover, he noted, savants could not even be sure that the pendulum's periodicity was the same at every spot along the *same* latitude, since the tug of nearby mountains or other deformities of the earth might affect its oscillation. The same factors would affect any measure based on geodesy. Given these uncertainties, Lalande predicted that the progress of science (in which he most fervently believed) would produce more accurate results twenty years hence. And what would happen then? Would the natural standards have to be periodically revised to take account of the improved results? Under the circumstances, what was the point of striving for precision now?[65]

Eager as he was to resume his campaign as soon as the spring weather allowed – delay might give the bureaucrats an excuse to cancel the mission, and the cost of transportation was rising – Delambre still needed official permission before he set out. The barricades of the previous year had taught him the value of a valid passport. In March, he petitioned the Paris municipal council for permission to move freely throughout the republic. The council – now in the hands of the radical sans-culottes party and hostile to the Academy as an elitist institution – voted unanimously against his petition. 'Is it possible,' he wrote back, 'that in Paris, at the centre of enlightenment and the arts, the executors of a law applauded across Europe find themselves stopped in their tracks?'[66] He resubmitted his petition, countersigned this time by prominent administrators. The council voted unanimously to issue him a passport. As an extra precaution, he wrote to all the towns along his route to assure them that his mission was benign.[67]

The republican government had proclaimed France a nation of 'one law, one weight and one measure'. It had promised to end the shameful inequalities of *ancien régime* justice and taxation. It had promised to open careers to talent and to liberate commerce. But the Revolution had also shattered the royal authority which governed France from the centre. Placing sovereignty in the hands of the people had made every town its own master. Markets were in disarray, and food prices were rising. The towns were suspicious of the countryside; the peasants mistrustful of the towns. At every step Delambre had to present his papers. When he passed through his hometown of Amiens an old friend had to prepare a dossier of official documents for him, all signed, stamped and dressed with fancy seals.[68]

At least Delambre had a strategy for this campaign season. Rather than circle Paris in futility, he and his two collaborators, Lefrançais and Bellet, would begin at the beginning. They would start at Dunkerque, the northernmost station, and work their way south. The strategy was logical, but circumstances have their own logic. The optimal season for geodetic campaigning coincided with the optimal season for military campaigning. All that spring, while Delambre had been waiting for a passport in Paris, the Prussian–Austrian army had been massing on the frontier for another drive to restore the monarchy. By the time he arrived in the north country, the plains of Flanders had become a battlefield, with the invaders again advancing towards Paris.

In mid-May of 1793, Delambre hurried to Dunkerque before the French defences collapsed. There he was assisted in the bell tower by Monsieur Garcia, whose family had been tower masters for three hundred years. Sometime during that long interval, the 162-foot tower had been separated from the main body of the church by a road that still serves as the main thoroughfare through the city centre. From the top of the red-brick belfry – a climb of 264 steps, 'and we counted' – the team had a view of several nations: France, the Low Countries and, across the Channel, England. Cassini and Méchain had used the belfry in their 1788 survey to link Paris and Greenwich. Nearer to view, Delambre could see the dunes along the beach, the port made idle by war with Britain, and the long low coast that swept into the grey mists. Inland, he could see the French armies manoeuvring along the border.[69]

From Dunkerque, Delambre made steady progress south through Picardy, his home region. It was an ideal landscape for triangulation, and summer was the ideal season for geodesy. The corrugated countryside was laced with low ridges, and each town boasted an elegant church steeple. Excellent stations were abundant, though each presented its particular challenge. In Watten, a small town a dozen miles inland on the Aa river, the church tower was not high enough to be seen from afar, so he capped it with a crown of white planks. At Cassel, to the immediate east, the summer heat in the steeple was suffocating. At Mesnil, he had to wait four days to erect his signal until the local carpenters had finished celebrating their village festival in the *cabarets*. At Fiefs, he had to wait for permission to punch holes in the church steeple so that he might have a clear view in all directions. At Bayonvilles, he had to chop down several trees to open up a line of sight. By mid-July he had closed ten triangles, accomplishing more in one month than in all the previous year. By his own account, this was his happiest portion of the meridian, and the most well favoured. Behind him, the battle was turning against the French. The British had laid siege to Dunkerque and the Hanoverians were closing in on Lille. But by then Delambre was approaching Amiens.[70]

Lefrançais never made it there. In mid-July he had to rush back to Paris. His wife (Lalande's daughter) was approaching her due date. On 27 July, she gave birth to a girl, Uranie, though the baptism was postponed until Delambre could stand in as her godfather. Delambre wrote to congratulate the young mother: 'I admire you for having resumed your astronomical labours so soon; in giving us a Uranie you have accomplished enough and could have rested yourself a bit longer.' He had six more stations to complete before he could return to Paris for the baptism of 'our new muse'.[71] As for Lefrançais, grandfather Lalande wrote that he would return to the mission as soon as he was elected to the Academy in his own right, probably at the meeting set for 7 August.[72]

Lefrançais would never return to the mission. On 8 August the Academy was abolished and Lalande put his nephew back to work on his all-important celestial chart.

Delambre was setting up a signal in the cathedral spire of Amiens when he learned of the Academy's demise. 'I don't know if I still have the right to call you my colleague,' Lavoisier wrote, 'though I send you this letter as a fellow believer in the progress of science.' The good news was that the savants had managed to preserve the metric reform, and the meridian survey along with it. 'The suppression of the Academy ought not in any way disrupt your labour, nor diminish your indefatigable activity.'[73] The bad news was that there was no money to pay Lefrançais, and the continuation of the survey had been ransomed with a dangerous concession: the establishment of a 'provisional metre'.

The suppression of the Academy did not come as a complete shock to Delambre. For years the academicians had been attacked as self-appointed elitists who disparaged popular inventors and thinkers. In the past months, radical politicians had called for the dissolution of all royal institutions. Some legislators had tried to exempt the Academy of Sciences because of the transcendent truthfulness of science and the useful services it provided the nation – especially with regard to the reform of weights and measures – but to no avail. In the end, some academicians even came to agree that the Academy was undemocratic and applauded its fall. When Cassini IV tried a procedural motion to delay the final closing, they echoed the same phrase the drunken militiaman of Lagny had hurled at Delambre: 'There is no more *Academy*!'[74]

This time they were right. Everything was now reversed. Instead of having the Academy of Sciences sponsor a meridian expedition to define the metric system, the creation of a metric system had become the main justification for the state-funding of science. On 1 August 1793, one week before the dissolution of the Academy of Sciences, a new law codified the metric system as we know it today and gave the French people one year to prepare themselves for its obligatory use. Of course, the meridian expedition would not be completed by then, as everyone recognized. Hence, the law established a 'provisional' metre which state administrators and commercial enterprises might use while they waited for the meridian survey's 'definitive' results. The value for this provisional metre had been coaxed out of the Academy under some duress.[75]

Even before Delambre and Méchain had set out, Borda had privately estimated that the metre would come out to about 443.5 *lignes* in the old Paris units. (A *ligne* was one-twelfth of a *pouce* (inch), so that a *pied* (foot) contained 144 *lignes*.) It was a quick back-of-the-envelope calculation based on what everyone already knew about the size and shape of the earth. In public, however, Borda said nothing. To announce this estimate might have under-cut the efforts to measure the meridian properly.[76]

Several state agencies were impatient to know this value, how-ever. The plan for a new national map, which would enable the government to tax accurately every piece of landed property in France, had been stalled because the surveyors were expecting to use the new standard of length. Nor could the Treasury decimal-ize the currency without some sense of the weight of the new silver coins. In January 1793 the Finance Committee begged the Commission of Weights and Measures to make a serious estimate of the likely length of the metre. To oblige them, Borda, Lagrange and Laplace, three of the most illustrious mathematical physicists of all time, did so in three easy steps. They assumed that the length of one degree of arc at 45° of north latitude was average for the entire quarter meridian; they took the value for this distance from Cassini III's survey of 1740; then they multiplied this number by 90 (for the 90° of the quarter meridian) and divided by 10 million. Their guesstimate came to 443.44 *lignes*. Nothing could be simpler.[77]

Yet only when the Academy was threatened with dissolution later that year did the Commission cough up this value. By then, control of the nation's legislature had been seized by the Jacobin party, who had vested executive power in the hands of a Committee of Public Safety. This Committee included not only political radicals like Robespierre and Saint-Just, but also military engineers like Lazare Carnot and Prieur de la Côte-d'Or, whose task it was to direct the war effort and organize the production of war matériel. The law of 1 August 1793 was intended to implement the metric system as soon as possible, using this provisional metre as the standard. Not long thereafter, Lalande wrote to Delambre to tell him there was little point now in pressing on with the mission. 'The new measures are being adopted for commerce independent of the new measure of the earth; so there's little need for you to push yourself too hard to bring your results in now.'[78]

Borda's folding provisional metre

*This iron metre stick was built to match the specifications of the 1793
provisional metre. It reads: 'Metre stick equal to one ten-millionth part of
a quarter of the earth's meridian, Borda, 1793.'*

(Musée des Arts et Métiers-CNAM. Photo CNAM.)

Delambre spent that week in Amiens, conducting his observations from the second storey of the cathedral's spire, the loftiest in France. The interior of the spire was encumbered with heavy carpentry and massive bells. The steeple also inclined slightly to the west, which marginally skewed his observations. Below, the red-brick town appeared calm, and no one was without bread – although food riots had disturbed the city the previous month, and bakers were again running low on provisions. On 9 September, shortly after Delambre left town, officials arrested sixty-four priests who refused to swear allegiance to the state.[79]

Though he rarely went home or commented on politics, Delambre had joined an Amiens political society in 1791, one co-founded by his brother-in-law. The Société des Amis de la Constitution preached moderation, despite its motto: *Vivre libre ou mourir* ('Live free or die'). Delambre shared the Society's principled moderation. Amid the passion, he dared suggest to his hometown newspaper that both democrats and aristocrats repudiate their extremist factions, and discuss their differences at a nightly educative assembly. 'To be reasonable,' he urged, 'one must be without passion.'[80] This modest proposal was blasted by another local citizen, Gracchus Babeuf, the radical politician who would one day be called the world's first Communist. Delambre, he sneered, had failed to understand that 'a man without passion is incapable of noble enterprises; great deeds are beyond his reach; he is without energy and hence contemptible'.[81] Delambre's response was to emphasize the modesty of his proposal, and to express the hope that his opponent, by venting his bilious tirade, had at least improved his health. This was as close to political commentary as Delambre would get in thirty years of service to a half-dozen régimes: the *ancien régime* monarchy, the constitutional monarchy, the republic, the Directory, Napoleon's empire and finally the Restoration monarchy. Throughout his decades of public service he maintained a careful ambiguity about his political views.

His duty lay to his mission and on that basis he was determined to proceed. He could manage without Lefrançais, so long as Bellet remained with him. The young instrument-maker had proved himself an excellent observer and a cheerful companion. He accompanied Delambre at every step of the mission, and would do so until its final triangle. Delambre paid him a 500-*livres* bonus out of the savings from Lefrançais' salary. And now that the Academy was abolished, he himself could collect a daily wage as a Commissioner for Weights and Measures. It came to the princely sum of 10 *francs* a day, roughly the salary of a competent artisan.[82]

In early October Delambre at last connected the new season's chain of triangles with the Paris chain of the previous year. This meant that he had now formed a continual lattice of triangles from Dunkerque through the Ile-de-France, about one-third of the distance to his rendezvous in Rodez. In late October he passed to the south of the capital to pick up where he had left off the previous winter.

Working in the Orléans forest just north of the Loire, Delambre was caught up in the political tensions he had so far evaded that year. The church tower at Cour-Dieu which had served as Cassini III's signal in 1740 was completely hemmed in by trees, and no plausible substitute could be found amid the rolling terrain of tall oaks in the old royal forests, a favourite hunting spot of the Bourbon kings. Delambre's only option was to construct an observation tower on a low hill called Châtillon. Where nature offered no elevated view, and belfries were unavailable, the geodeser had to build from scratch.

The construction of this sixty-four-foot wooden tower took over a month and attracted unwanted attention. The citizens of the surrounding hamlets wondered about the strange doings in the former royal forest. 'They reported seeing three or four hundred brigands building scaffolds and piercing holes in church towers . . . undoubtedly in preparation for a counter-revolutionary uprising.'[83] This would have been amusing but for the fact that the local citizens had called for six hundred soldiers to attack the site. Fortunately, when the time came, they vented their anger elsewhere. On 27 December, just as the tower was nearing completion, the local popular society unanimously voted instead to destroy a nearby stone obelisk erected in honour of the 1740 survey by Cassini as an 'odious sign of extinct despotism in the guise of a stone pyramid called the meridian and built by the one-time seigneurs as a sign of their greatness'. The obelisk was torn down to be used as paving stones at the same time that the eminent jurist Malesherbes, on whose land the obelisk stood, was executed for acting as the king's lead counsel in his final (futile) defence.[84]

On New Year's Eve, Delambre and Bellet climbed their high tower platform at Châtillon for the first time and began to hoist their precious circle into position with ropes and pulleys. The observation deck had been boxed in to provide shelter from storms and snow. Unfortunately, this protection also gave the wind a broader surface area to push against. The circle had just arrived safely when a tremendous gust shook the tower, forcing the observers to scramble back to earth, a fifteen-minute ordeal because the circle had to be lowered with care. The next day, the wind was calmer and they remounted the tower. But the cold was painful, the day short and their observations were of poor quality.[85]

Yet when the devastating blow came two days later, it emanated from neither the local citizenry nor the weather, but from the supreme power in the land. On 4 January 1794 Delambre received a letter from the Commission of Weights and Measures notifying him that by order of the Committee of Public Safety he had been purged from the meridian survey along with several of his colleagues. The letter informed him that he was to hand over all his field notes, calculations and instruments so that a successor might take his place 'should the meridian survey continue'.[86]

Whatever this meant for the future of the mission, to quit there would have negated the months of labour that had gone into building the Châtillon tower. If a winter storm toppled the tower, all the surrounding triangles would have to be redone. At a minimum, the survey ought to terminate at fixed stations – such as the church towers along the Loire at Châteauneuf and Orléans – so that his successor, should one ever be appointed, might start his labours from a secure foundation. Moreover, Delambre estimated that he would need at least three months to put his notebooks in order and complete his calculations. He wrote to the Commission, begging for a chance to implement this plan, while setting furiously to work to carry it out before they refused.[87]

Their sealed response arrived a few days later in the hands of the engineer Gaspard Prony, Delambre's former colleague in the former Academy, and, as it turned out, his replacement on the Commission. Yet Prony always found an excuse not to hand over their answer. Instead, he assisted Delambre with his observations at Châtillon, and even accompanied him to Orléans to finish the triangles on the banks of the Loire.[88] The crucifix which had once stood at the top of the Orléans cathedral spire – and which would have made an ideal signal, like the crosshair on a telescopic sight – had recently been replaced with a misshapen cast-iron liberty bonnet. The cathedral, now known as the Temple of Reason, had just that week witnessed an even greater sacrilege. The belle Rosalie, a young prostitute who worked the rue Soufflet, had been costumed like a goddess, with a pike in one hand and a *bonnet rouge* on her head, so that she might be paraded through town on a tremendous chariot bedecked with tricolour flags and pulled by twelve white horses led by six young men in togas. All the town's citizenry had followed, wearing Roman attire. At one point, the float had to squeeze under a low portal and the goddess was heard to shout, 'Hey, you bastards! Hey, buggers! Stop, you fuckers, I'm falling off!' before she hopped down into the crowd so as to clamber back up on the other side.[89]

In a year and a half of labour, Delambre had covered nearly half his assigned itinerary from Dunkerque to Rodez, surveying a two-hundred-mile arc from the North Sea coast to the banks of the Loire. In doing so he had zigzagged more than twelve times that distance, or some 2400 miles, on the hard roads of France.[90] On 22 January 1794, he made a final note in his expedition logbook: 'It began to rain and there was no time to redo the angles.'[91] Later that day, Prony handed over the Commission's response, now three weeks overdue. The cover letter read:

Citizen,

The Commission of Weights and Measures has sent one of its members to bring you the decree of the Committee of Public Safety regarding your request, and to invite you to conclude your operations in such a way as to ensure that your temporary signals become unnecessary. It further enjoins you to complete the transcription of your calculations and observations as you suggest.

Lagrange, President, Commission of Weights and Measures[92]

In ambiguous language, his friends on the Commission had honoured Delambre's request, allowing him to keep his expedition logbooks for the time being. The enclosed order in the hand of Prieur de la Côte-d'Or was written on the imposing stationery of the Committee of Public Safety. It was dated 23 December 1793, now a full month past:

The Committee of Public Safety, considering how important it is for the improvement of public morale that government officials delegate their powers and functions solely to men known to be trustworthy for their republican virtues and their

abhorrence of kings . . . decrees that from this day forth
Borda, Lavoisier, Laplace, Coulomb, Brisson and Delambre
cease to be members of the Commission of Weights and
Measures, and that they immediately hand over to the
remaining commissioners all their instruments, calculations,
notebooks, with a full inventory of the same. And
furthermore, that the remaining members of the
Commission . . . apply Revolutionary enthusiasm to bring the
new weights and measures into use among all citizens.

C.-A. Prieur, B. Barère, Carnot, R. Lindet, Billaud-Varenne[93]

The next day Delambre packed his equipment to return to
Paris. 'Even though, for the life of me, I cannot understand why I
have been recalled, I will return without complaint to those occu-
pations from which I was regrettably torn away.'[94] On his way he
had one personal matter to attend to. His patron, Geoffroy d'Assy,
was being sought by the Revolutionary police. Delambre needed
to stop at the d'Assy country residence in Bruyères-le-Châtel,
where d'Assy was living in retreat.[95]

The Revolution had entered the phase known as the Terror,
when the Jacobin state declared an emergency military draft,
imposed wage and price controls and enforced its decrees with
imprisonment and execution. The world's first war of mass mobilization was being fought. On the frontiers of France, a coalition
of Prussians, Austrians, English and Spaniards was ranged
against the republic. From within, the republic was being undermined by defiant aristocrats, reactionary peasants, grain-hoarding
merchants and recalcitrant priests. Lavoisier had been arrested
earlier that month along with the rest of the financiers of the 'tax
farm' that had once collected so many odious and unfair levies on
the king's behalf. And just as Delambre arrived at the d'Assy residence, his patron was likewise hauled off to the Luxembourg
prison.

Later that winter, a storm felled the Châtillon tower.

Chapter Four

THE CASTLE OF
MONT-JOUY

There is almost nothing right or wrong which does not alter
with a change in clime. A shift of three degrees of latitude is
enough to overthrow all jurisprudence. One's location on the
meridian decides the truth, that or a change in territorial
possession. Fundamental laws alter. What is right changes
with the times. Strange justice that is bounded by a river or a
mountain! The truth on this side of the Pyrénées, error on the
other.[1]

BLAISE PASCAL, *Pensées sur la religion*

Cut off by war on the far side of the Pyrénées, Méchain knew
little of these developments. For nine months he heard no news
from France. The most recent letter was dated March 1793,
from before his accident at the water-pumping station. For two
months after that injury he convalesced in bed, until the summer
sun lured him out of his dark room on to the terrace of the
Fontana de Oro. The summer solstice was approaching, and
Méchain insisted that he be carried out, not in search of a solar
cure, but in search of solar knowledge. He was borne out into
the dazzling Mediterranean summer and propped up on pillows
under the repeating circle. The salt breeze breathed across the
paving stones. The noon-day heat silenced the town. Just out of
sight, the sea sloshed against the quays. The fashionable crowd
on the Rambla had retreated indoors. Only mad dogs,

Englishmen and solar astronomers go out in Barcelona's noon-day sun in summer.

Tranchot had prepared the instrument for him. For four thousand years star-gazers had sought to define one of astronomy's most fundamental constants: the obliquity of the ecliptic, or in other words, the angle of the earth's tilt relative to the plane of its orbit around the sun. With the Borda circle to hand, and the summer solstice upon them, Méchain had an ideal opportunity to make the definitive measurement of this constant.

This was painful work under the best of circumstances, but Méchain insisted that he alone take the readings. While Tranchot held the smoked lens to the astronomer's eye, Méchain tracked the sun until it reached its maximum altitude. Then Tranchot rotated the Borda circle for him, while Méchain fine-tuned the position of the scope. Working together, they managed to take a few preliminary readings before the heat of the sun began to distort the circle's brass gauge. Méchain was having difficulty fine-tuning the scope with his left hand. For a man recovering from a shattered chest – for a right-handed man with a dangling right arm – the effort was too much. They were forced to break off. A relapse followed. For twenty years he had laboured in the dark bureaux of naval cartography to map a Mediterranean coast he had never seen. Now its light suffused his mind, even when his eyes were shut.

Salvà was worried and proposed a medicinal cure at the thermal springs of Caldas. Chastened, Méchain took his advice. The hot baths and showers were comforting. But six months after his accident, his right arm still hung limp at his side. The doctors told him he might never recover its use. 'Time has done more than art,' he would conclude a few years later when his arm had regained its strength.[2]

By the time he returned from the spa – still impaired, if somewhat more capable – Spain was on the verge of a military victory that would make her mistress of both slopes of the Pyrénées, unifying Catalonia for the first time in 150 years. France may have declared war first, but Spain struck first. In May, General

Ricardos, the supreme Spanish commander, ordered his main body of forty thousand troops through the saddle west of the Bellegarde fortress where Hannibal had attacked two thousand years before, while three columns of 3500 of Ricardos' soldiers spilled through the high inland passes that Méchain and Tranchot had triangulated the previous summer. Overpowering the French garrison at La Garde, they marched down the Tech valley to join the main body of troops occupying the plains of Roussillon. Had they pressed their advantage then, the Spanish might well have conquered Perpignan. Instead they paused to fortify the heights, set siege to Bellegarde and surround the city. All April, May and June, within sight of the panicked citizens of Perpignan, they bombarded Bellegarde from nearby Puig Camellas, reducing the mighty fortress to rubble. When the brave garrison finally surrendered, one thousand prisoners were marched to Barcelona, to be incarcerated in the Mont-Jouy castle where Méchain had conducted his celestial observations the year before. The captives were lodged in a cellar and guarded by cannon charged with shrapnel 'in order to avoid insolence from so evil a people'.[3]

Eighteenth-century warfare was suffused with its own contradictions. Courtesy coexisted with brutality, even between the defenders of the Catholic monarchy and the apostles of Revolutionary liberation. The Spanish generals allowed the captured French officers to spend a night in Perpignan before marching them off to prison. The Revolutionaries tried to persuade the Catalonians to adopt their cause. Like eighteenth-century scientific rivals, enemy officers often had more in common with one another than with the leaders who sent them into battle. With the advent of this new kind of mass war, however, the tug of nationalism began to pull even science apart.

Two months after the fact, Lavoisier at last wrote to Méchain to inform him that the Academy had been abolished, but that as a member of the Commission on Weights and Measures he was entitled to a salary of 10 *francs* a day, which his wife might collect on his behalf. At least his family was now getting paid for his labour. Yet Méchain never received this letter. A few months later, at a time when Delambre was being purged from the meridian project and his former colleagues were being imprisoned and threatened with execution, Méchain was still writing to Paris to say that he would defer to the Academy's wishes. He did hear rumours of the demise of the Academy, however, and believed (with some reason) that the Parisian administrators were seeking to replace him. Indeed, he seems to have been spared from the general purge only because the Committee of Public Safety worried that any such threat would cause him to seek permanent asylum in Spain, along with his precious repeating circles and his detailed geodetic data.[4]

So when he was offered a prestigious scientific appointment by the Spanish crown, Méchain was understandably tempted. Tens of thousands of French men and women had fled their country. Thousands had taken up arms against their own nation. The dead king's younger brothers were leading armies against their own people. Compared to this, what harm was there in measuring a few geodetic triangles for Spain? How could scientific work be traitorous? Lavoisier had said it best: 'The sciences are not at war.' Moreover, Méchain was penniless. The Barcelona bankers had frozen his account; his French paper money was worthless; and a French law prevented his colleagues from sending him hard currency.[5]

Most of all, Méchain despised the radical turn his country had taken since 1792. Even the comparatively mild upheavals of 1789 had distressed him painfully, while recent events had horrified him. But if the Revolution had frayed the knots of his patriotism it had not loosened his sense of duty to his colleagues and to his mission. Méchain's virtue was exactitude – a prosaic virtue perhaps, and one rarely associated with genius – but with it came a fierce determination to complete what he had begun. The frontier stations he had once hoped to approach from the French side of the border were now deep in Spanish-occupied territory. Delambre had offered to hurry down from the north of France to help measure them if that were necessary. If, on the other hand, Méchain wanted to measure them himself, this might be his only chance.[6]

Fortunately, his arm had begun to heal, and he had the capable Tranchot to help him. So early in that autumn of 1793 he secured permission from General Ricardos to complete his triangles along the peaks of the Pyrénées. The general would now allow him – the emissary of an enemy power – to conduct sensitive geodetic measurements in a war zone. Already the Frenchmen had calculated the pin-point locations of Catalonia's major fortresses. In return, Méchain gave his solemn word that no member of his team would leave the country without official blessing, nor provide their data to the French until the war was over.

That September the two Frenchmen, accompanied by Captain Bueno, ventured back into the Pyrénées in a bold attempt to complete the high mountain stations. At the massive fort of Figuères – which Méchain wove into his triangles – they split into two parties, each taking one of the repeating circles. Méchain and Bueno angled toward Puig Camellas, the hilltop from which the Spanish cannon had pounded the Bellegarde fortress into rubble, while Tranchot struck out on his own for the high inland mountains.

Tranchot's goal was Puig de l'Estelle, a 5800-foot peak on top of an old iron mine. The summit lay in the shadow of the Massif de Canigou, the glaciated blue behemoth which dominated the eastern Pyrénées. All this had once been French territory. Indeed, just as Tranchot arrived at Puig de l'Estelle the French army, beefed up with reinforcements, broke free of their encirclement at Perpignan. Striking inland from the fort at Salses, just north of Perpignan, they began to drive up the Tech valley, forcing the Spaniards to regroup at Boulou, the strategic town where the Grande Route crossed the Tech river. Outnumbered, nine thousand to twenty-nine thousand, the French were none the less supported by several thousand *miquelets* who travelled along the flanks of the army, assailing the Spanish and terrorizing peasants. In response, the peasants organized into protective bands. The countryside was in turmoil.

Above the cool pine forests, the high dry air offered superb views of the neighbouring geodetic stations, as well as the two armies jostling for position in the valley below. Each side was trying to manoeuvre their cannon on to the dominant hill. Over the course of twenty-four days the French probed the Spanish positions with eleven skirmishes and three general attacks. On 22 September alone they lost three thousand men. On 1 October, French reinforcements arrived in the face of a Spanish barrage and took a hilltop outpost. On 5 October, the French let loose a cannonade of their own to protect a cavalry charge. On 6 October, they established a new battery on the heights, which opened fire the next day on the main Spanish camp. Standing on the bare mountain peak beside his bizarre conical signal, sighting with his double-scoped instrument, Tranchot made a visible target.

He had been taking intermittent measurements for a week – the weather was changeable and strong winds threatened to topple his circle – when, on the morning of 7 October, just as the French battery at Banyuls opened fire on the Spanish positions below, a band of six villagers from the nearby hamlet of Vallmagne ambushed him in the name of the Revolution. Tranchot protested that he too was a French loyalist, an ardent Revolutionary, and on a mission from the National Assembly. He showed them his papers, his passports and a copy of his commission, along with the newly certified documents sent by Delambre. But the villagers would not allow an unknown engineer to 'conduct surveillance' along the front.[7] Puig de l'Estelle looked straight down on the valley where the French army was advancing. They bound Tranchot, gagged him, fastened a rope around his neck, and led him by garrotte to their town, where the local mayor advised them to conduct him to the district capital. From there he was escorted to Perpignan.

This near disaster turned out to be a stroke of good fortune. Francesc-Xavier Llucía, the chief administrator of Perpignan, was familiar with the meridian mission; he secured Tranchot's immediate release and more. A few weeks earlier, Méchain had begged Llucía to plant signals on the peaks of Mount Bugarach and Mount Forceral, well behind French lines, so that he might sight them from the frontier. With Tranchot suddenly on the French side of the front, Llucía authorized him to carry out this task.

Two weeks after his arrest the signals had been planted, and Tranchot was climbing back up the Puig de l'Estelle to continue where he had left off, just as the battle of the 'Battery of Blood' threatened to engulf his mountain eyrie. For several days, six thousand French troops assaulted Puig Singli, where the Spanish cannon commanded the heights over Boulou. The first seven attacks were repulsed, the next three gained the position temporarily, and the eleventh effort, with the Spanish out of ammunition, won the day – until a Spanish counterattack massacred the lot of them the next morning. Calm above the chaos, Tranchot performed his exacting labour.[8]

This would be the only station Méchain ever allowed Tranchot to observe with the circle on his own, and one might well wonder why. It was not due to any lack of experience. Tranchot had toiled for two decades on the triangulation of Corsica, a country as rough as Catalonia. Born in Koeur-le-Petit, a hamlet of Lorraine, where French mingles with German, he had survived a difficult birth to grow into a vigorous man. He may have lacked a formal education or an academic title, but he was among the nation's most capable cartographers, a man of proven integrity. Towards the end of the Corsican project, accusations of scientific fraud had been bandied about. Tranchot's final measures had resolved the controversy. Méchain had himself signed the Academy report which singled out Tranchot's contribution as 'infinitely precious for the precision of geography'.[9] From Méchain, there could be no higher praise. By this date, moreover, Tranchot had mastered the repeating circle. Méchain himself admitted, 'I could rely on him as I relied on myself.'[10] Yet he always supplied Tranchot with prepared data sheets, and never let him perform his own calculations or look into the expedition notebooks. By way of contrast, Delambre allowed Bellet, a mere instrument-maker, to take observations, record data and double-check calculations.

It is hard to imagine that Méchain's scruples were justified on technical grounds. In 1790 the two men had collaborated on the navy's charts of the Mediterranean coast. He knew Tranchot was a geodetic surveyor of consummate skill. Yet collaboration never came easily to Méchain. Despite his time in Lalande's astronomical workshop, despite the assistance he accepted from his wife, and the aides he employed on his journal, Méchain remained an astronomer who worked best alone. His virtue lay in his ability to know the earth, the planets and the stars. He had less knowledge of his fellow man.

Or perhaps he knew his fellow man all too well. Perhaps Méchain learned what Tranchot had really been up to during his excursion across the border – that at some time during his foray into France Tranchot had met with his military superiors and supplied them with the plans and geodetic locations of all the fortresses the team had triangulated in Catalonia: Figuères, Girona, Roses, Barcelona and Mont-Jouy – all the major military installations of north-eastern Spain. As a captain of military cartography, Tranchot was obliged to supply his commanders with this information. Failure to do so would have been treason, at a time when treason meant immediate execution. Besides, Tranchot was a patriot and committed to the Revolution; and these plans would help the republican cause.[11]

For Méchain, though, this would have been a betrayal. Scientific knowledge gathered for the benefit of the world's people ought never to be used for harmful purposes. Méchain valued allegiance to science over allegiance to nation. At a minimum, Tranchot's foray into France broke the promise Méchain had made to the Spanish general. His honour demanded that he keep his promise. A savant's reputation was the outward sign that he remained true to his science. To betray one's own honour was worse than treason.

And this suggests the real reason Méchain did not trust Tranchot. Having betrayed the mission, what was to prevent Tranchot from betraying Méchain and usurping his command of the southern expedition? Tranchot deserved credit for many of the mission's successes to date. And he was the most likely replacement for Méchain should the astronomer be declared incapacitated. It was this haunting thought which drove Méchain, despite his wounded right arm, back into the Pyrénées.

Méchain and Commander Bueno had their own view of the battle of the Tech from their station at the summit of Puig Camellas. From there, they could see across the battle lines into the besieged town of Perpignan, where the French generals were directing their break-out. They could see south into Spanish Catalonia and the district outside Figuères, where a crumbling sea-blackened turret, the Tour de Mala-Vehina (the tower of the 'Bad Neighbour') stood on a crest of land belonging to Captain Bueno himself. They could see north towards Bugarach and Forceral, deep in French territory, where Tranchot had placed his signals. And on the bright, clear morning of 25 October, as they turned their scopes across the Tech valley, they could see a figure on the summit of Puig de l'Estelle, standing beside the double-cone signal, a dark figure against the blue sky hunched over a brilliant brass circle – Tranchot adjusting and readjusting the scopes of his circle while the battle boomed in the valley between them.[12]

When Méchain completed his measurements ten days later, Tranchot was still triangulating. And when another week passed, and another, and Tranchot still had not returned, Méchain grew anxious. He was less concerned about Tranchot's safety than about the possibility that his aide would remain in France. With these angle measurements done, their mission in Catalonia had ended. Lalande, for one, expected Tranchot to remain in France, and even urged Méchain to slip across the border to join him.[13]

This was to underestimate Méchain's scrupulousness. He sent a message to Tranchot demanding that he return to Spain immediately. 'It is for the sake of my duty and my honour to enjoin you not to leave for France without my permission, neither by any route, nor by any means. It is for this reason, and not for the continuation of our mission, that I insist so strongly. It would disgrace us greatly, and deservedly, for you to conduct yourself in any other fashion.'[14] At stake was something greater than the success of a scientific project – 'the most important any man has ever been charged with'. At stake was Méchain's reputation as a man of his word.

The Spanish army chose just that moment to counterattack. They took advantage of their superior numbers and once again drove the French back down the Tech valley, conquering the coastal towns of Collioure and Bagnols and renewing their siege of Perpignan. In the process, they sealed off Tranchot. He was trapped in France, he said, by 'force of arms'.[15] Perpignan now became the scene of an all-out political struggle between moderates and radicals. Another failed French general was guillotined. Individuals suspected of counter-revolutionary activities – especially those with ties to the aristocracy, the Church or the party out of favour – faced summary execution. Among these was Llucía, the French Catalan revolutionary who had once declared, 'It is time to electrify all souls.'[16] He had saved Tranchot, but he could not save himself. Some fifty heads fell that month in Perpignan alone. Only the weather stopped the Spanish advance this time. The November rains ended both the military and the geodetic campaigns. In the Tech valley, the soldiers slept in mud. Méchain returned to Barcelona. And some time that winter, Tranchot slipped back across the frontier to rejoin him.

This proof of the Frenchmen's integrity did not persuade General Ricardos. Despite petitions and personal pleas, he insisted that Méchain and Tranchot remain in Barcelona until a peace was concluded. Nor would Méchain be allowed to send home any more communiqués with numerical data; these would be confiscated at the border as encoded letters. No military leader could knowingly allow this information to fall into enemy hands, although Ricardos was raising the drawbridge after the moat had been breached.[17]

That spring, the fortunes of war shifted once again. In March, while visiting Madrid, the victorious General Ricardos died. His replacement, General La Unión, was the youngest general in the Spanish army, a devout Catholic of high moral sentiment, repelled by the populist atheism of the French Revolution. Soon after, a new French general, Jacques-Coquille Dugommier, fresh from his victory at Toulon (where he had commanded the young Napoleon Bonaparte), took charge of the Revolutionary army in the region. Dugommier quickly set in motion the republic's plan to liberate – or, rather, subjugate – Catalonia. Catalonia, he proclaimed, was ripe for revolution. The province was rich in mines and industry. The people loved liberty and hated their Castilian overlords. If they embraced equality and became an autonomous republic, the province might serve as France's boulevard to the rest of the Iberian peninsula.[18]

Dugommier attacked as soon as the season allowed. By mid-June, the French had recaptured the high mountain passes and had begun to push their way down the southern slopes of the Pyrénées. The Spanish retreated to their massive fortress at Figuères on the Grande Route, positioning nine thousand soldiers and thirty-two artillery pieces to hold their right flank at the Tour de Mala-Vehina on the lands of Méchain's cartographic collaborator, the good Captain Bueno. Should Figuères fall, as seemed likely, the road was open to Gerona, and beyond that, Barcelona.

For the past nine months Méchain had heard no news from his colleagues in France, only rumours. He wrote them a long letter anyway. He had failed to secure passage on a vessel out of Barcelona, and the Spanish were holding him in 'unjust detention'. With the Academy abolished, as reported in the public papers, the meridian project had no doubt been cancelled as well, and the metre would be determined with a pendulum, as originally planned. If his mission had indeed been cancelled he begged to be informed of this at once. If not, he could see a way to complete the meridian survey by the end of the year. He had the scenario all worked out. As soon as the Spanish general released him, he would return to France and triangulate his way north towards Delambre. Because the French terrain had already been surveyed and mapped by Cassini, he would simply revisit those old angle measurements with the repeating circle. If he began next month, he could triangulate as far as Evaux by July. Evaux was the halfway point of the meridian arc, well north of Rodez where he had been scheduled to meet Delambre. With luck, he might even triangulate as far as Bourges, which would mean that he would have surveyed two-thirds of the total arc, instead of the one-third he had been assigned. Then the two astronomers would together measure the two baselines in August, one in Delambre's northern sector and one in his own southern sector, and have the mission wrapped up by the end of the summer.

There was only one snag. The Spanish general refused to let him leave. 'But alas, where am I? In irons! And yet I speak like a man free to indulge his passionate zeal for the success of this superb mission. No matter; at least I have tried to make my slavery useful, if not to the mission itself, then at least to astronomy.'[19]

Méchain was too obsessive an astronomer to remain idle for long. They had barred him from Mont-Jouy, but they had not forbidden astronomical work at his Barcelona hotel. So in December he reorganized his observatory on the terrace of the Fontana de Oro. This time he would take advantage of the winter solstice to measure the angle of the earth's rotation relative to its orbit around the sun. For this, he would also need an exact determination of the latitude of his hotel terrace; last winter's data, gathered at Mont-Jouy, would not serve his purpose. Mont-Jouy, although readily visible to the south of the city, was over a mile distant. Equipped with the world's most exact astronomical instrument, he intended to make this measurement with greater precision than any investigator in the past four thousand years. As an added bonus his observations would also offer a double-check on his latitude results for Mont-Jouy.

Méchain's motives for undertaking these observations – the results of which were to haunt him for the rest of his life – appear to have been mixed, as motives often are. Certainly he wished to prove to his Paris colleagues and Spanish hosts that he remained the same meticulous astronomer as before, and that the accident of the previous April had not diminished his abilities. This would silence any talk of replacing him on the meridian expedition. It would also demonstrate his diligence at a time when individuals who refused to serve the public good risked execution.[20]

There was also something about his earlier results that nagged at Méchain. To calculate the latitude at Mont-Jouy, he had measured the heights of six different stars: Polaris, Thuban, Kochab, Mizar, Elnath and Pollux. More was always better. Thoroughness was always rewarded. For his final analysis, he had used the results of the first four of these stars, those for which he had gathered the most data. Of these, the results for the first three converged to a remarkable degree, giving average latitudes of: 41°21'44.91" (Polaris), 41°21'45.19" (Thuban), and 41°21'45.19" (Kochab). The total spread in these values came to an infinitesimal 0.3 seconds of a degree of arc. This suggested that Méchain had determined the global location of the castle tower of Mont-Jouy to within thirty feet. It was a stunning display of astronomical virtuosity, the kind of precise result that had won him the leadership of the southern expedition.

Results based on the fourth star, Mizar, however, diverged from this pattern, and indicated a latitude of 41°21'41.00", which differed from the others by four seconds of arc, or some four hundred feet. This anomaly irritated Méchain. Why did the readings for this one star differ tenfold from the rest? It was a natural question for a natural philosopher to ask. Yet even then, he might have let sleeping data lie. Only a decade before, a discrepancy of only four seconds would have been a remarkable achievement. It came to little more than 0.01 per cent of the six-hundred-mile arc from Dunkerque to Mont-Jouy. Besides, he had already summarized these astronomical results for his Spanish hosts and sent a précis to Borda in Paris.

On the other hand, Méchain had a hypothesis that might explain this discrepancy. Ah, that other hand! Why is it always with that 'other hand' that scientists open Pandora's box? They do not open it to make their lives more difficult. More often they are simply seeking to reassure themselves, to confirm what they think they already know to a finer degree of certainty. But for better or worse, they do not always know what they think they know. Sometimes they even have the good fortune to be mistaken. And then, as Enrico Fermi once said, they may make a discovery.

The problem with the Mizar data, Méchain hypothesized, was refraction. The corrections for the bending of light had been worked out by astronomers in London and Paris. Perhaps their corrections did not apply to southerly towns like Barcelona where the circumpolar stars crossed the meridian closer to the horizon and the higher temperatures distorted their sighting through the atmosphere. Of all the stars he had measured, Mizar crossed the meridian closest to the horizon. The corrections were small in the first place, of course, and any adjustment would necessarily be smaller still. But the meridian expedition was operating with an unheard-of degree of precision. The repeating circle promised precision limited only by the patience of the observer. And Méchain refused to believe the fault lay with the stars.

So he spent the winter of 1793–4 on his back on the terrace of his Barcelona hotel, taking night-time observations. While Tranchot held the lantern and verified the spirit level, Méchain hunkered down as before, whirling the circle, then the scope; listening for the clock to beat the moment of the star's meridian transit; then whirling the circle, then the scope, and repeating his eye-and-ear measurement. He conducted observations on Christmas Eve, on the night of Christmas, on New Year's Eve, and on the first night of the new year, plus every clear night in December, January, February and March. He took 910 stellar readings, each with ten or more repetitions, for a Herculean total of some ten thousand observations. Then, inside his hotel, during the daylight hours, he calculated his way through this mass of data, his refraction tables and logarithmic tables continually by his side. By early March he had determined the north latitude of his hotel to be 41°22′47.43″ (based on Polaris), 41°22′48.38″ (based on Kochab) and 41°22′44.10″ (based on Mizar). Once again, the results for the first two stars (those in which he had the most confidence) agreed to within an impressive one second of arc (or one hundred feet), making the Fontana de Oro the most accurately located hotel on the face of the planet. Yet once again, the Mizar data gave discordant results, differing some four hundred feet from the others.

One final step would clarify the mystery. Méchain would now need to compare his new latitude results at the Fontana de Oro with his old results for Mont-Jouy by subtracting the distance between them. Calculating this distance, of course, was just the sort of task his expedition had been equipped to perform. He laid out a triangulation which included his hotel, Mont-Jouy, and the cathedral of Barcelona, and to make doubly sure, a second triangulation which included his hotel, Mont-Jouy and the Lanterna that served the port as a lighthouse. There was only one snag: to carry out the triangulation accurately, he would need to take angle measurements at each station, and the Mont-Jouy castle was closed to him as a Frenchman. It was so near, and yet just out of reach.

By mid-March, with Tranchot's help, he had taken measurements from his hotel, the cathedral and the Lanterna. In the meantime, he apparently persuaded the Mont-Jouy commander to grant him a single day at his old observatory tower at the castle. On Sunday 16 March 1794, a slightly overcast spring day, Méchain climbed the hill of Mont-Jouy to perform a final triangulation – while several hundred of his fellow citizens languished in the prison below. Then he returned to his hotel to calculate.[21]

The numbers were quickly tallied. According to the triangulations, Mont-Jouy was located 59.6 seconds of arc from his hotel, or a distance of 1.1 miles south. Comparing this distance with the two latitude measurements was a matter of simple subtraction. After subtracting 59.60 seconds from the average latitude of the Fontana de Oro (41°22'47.91"), the result should equal the average of his most reliable latitude data from Mont-Jouy (41°21'45.10"). It was a moment's work.

Imagine then his horror when the results fell short by 3.2 seconds of arc. Not 3.2 seconds to be folded into the six-hundred-mile arc from Dunkerque to Mont-Jouy, which would have been an insignificant difference of 0.01 per cent, but 3.2 seconds over the course of a 1.1-mile arc, for a stunning discrepancy of 5.4 per cent. Instead of explaining away the anomaly *within* his Mont-Jouy results, Méchain now confronted an anomaly of horrific proportions. Having pinned down his latitude within forty to one hundred feet on two different occasions, he had now discovered

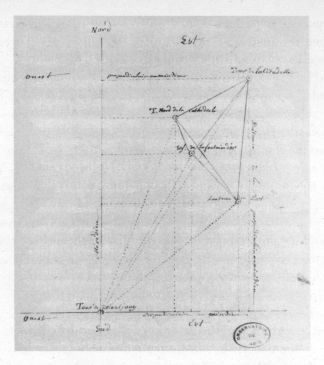

Méchain's Barcelona triangulation of 1794

This is Méchain's own map of the triangulations he made within the city limits of Barcelona in 1794 to verify the distance between the Fontana de Oro and the fortress of Mont-Jouy. The Fontana de Oro is located at the centre of the diamond. Méchain constructed two triangles that included Mont-Jouy and his hotel: one using the north tower of the cathedral and the other using the Lanterna (lighthouse).

(Photo: Observatoire de Paris.)

that his two average results diverged by a horrifying three hundred to four hundred feet. He must have erred in his observations or his calculations. But which? Which set of data could he believe? Most horrible of all: he had already posted a summary of one set of these results, those for Mont-Jouy, to his colleagues in Paris. From this they would want to calculate the length of the metre, the supreme standard for all people, for all time.

It was as if he had set out to fine-tune a Stradivarius and snapped the instrument's neck. His integrity had plunged him into a crisis. His effort to rehabilitate his reputation had only caused him to doubt his own abilities. What had gone wrong?

Under normal circumstance, Méchain would simply have climbed back up Mont-Jouy and taken more stellar observations at the castle. But these were not normal circumstances. His one-day pass to the fortress had been a begrudging exception, not to be repeated for an enemy of the Spanish crown. And day by day, as the Revolutionary armies advanced deeper into Catalonia, the political tensions worsened. The republic promised the people of Catalonia a 'sister' republic of their own; the Spanish crown declared a religious war against atheism. Some residents of Barcelona supported the Revolution; others seethed against French godlessness. It was no time to be a Frenchman in Barcelona.[22]

Moreover, nothing now seemed to prevent the team's departure. Ricardos, who had opposed it, was dead. Tranchot and Esteveny were eager to return to France where their duty lay, as well as their colleagues, friends and families. Méchain, however, faced a terrible dilemma. He had told no one of his error, not even Tranchot. He was free to go, but did he dare to leave his mistake behind? Once he left Spain, how would he ever return to Mont-Jouy? Yet how could he justify staying on in a foreign country – an enemy nation – now that his work there was done? He dared not risk giving the impression that he had decided to emigrate. Even a rumour to that effect might lead the Paris authorities to cut off his salary, imprison his family and bar his return to France for ever.

So, on the advice of his Spanish friends, he secured a passport for neutral Italy – ostensibly because it would not arouse suspicion – bypassing any need to inform General La Unión of his departure. In late May, after two years in Catalonia, Méchain booked passage on a Venetian vessel bound for Genoa, the Italian city nearest the French border. For someone who anticipated the worst, however, Méchain certainly attracted his share of calamities. His pessimism offered him no protection. On 25 May, three days after he had loaded his precious repeating circles on to the vessel in Barcelona harbour, a bolt of lightning struck its mast, bursting the wooden boxes that carried the repeating circles, and charring one instrument's stand. The circles themselves appeared undamaged, but it was a fitting final salute from Catalonia. They sailed on 4 June.[23]

Nothing of these events was known in Paris. There, everyone assumed Méchain had been placed in detention by the Spanish generals (perhaps for having smuggled out the fortress plans). He himself had written home of his 'unjust detention' in Barcelona. He had complained that he was being held 'in irons'. The words were metaphorical, even melodramatic – Méchain had been comfortably lodged all this time at the Fontana de Oro – but his colleagues sent word to General Dugommier that the French astronomer was being held against his will. In mid-June, two weeks after Méchain sailed for Italy, Dugommier wrote an indignant letter to his opposite commander, the devout young General La Unión, demanding that the Frenchman be freed. 'In the name of the French Republic, which protects the savants of all nations, and which knows how to avenge any outrages committed against its own, I seize this occasion to demand that in the name of the arts whose free exercise should be respected at all times and by all nations, you release the citizen Méchain and his two colleagues, charged with the measure of the arc of the meridian, and detained in Barcelona by the orders of your predecessor or by you.' This was not the only lesson the republican general thought he would teach the barbarous monarchist. 'Savants must not be considered soldiers,' Dugommier wrote, 'nor treated as such. The peaceable arts have nothing to do with war. And unless you wish to perpetuate an extraordinary violation of those conventions which govern even the most uncivilized people, you cannot refuse to return him and his two collaborators to liberty and their homeland.'[24] Méchain's mission, he insisted, 'must be respected around the globe'.[25]

General La Unión knew no more of Méchain's whereabouts than did the French. But he knew when his honour had been slurred. Never would he have impeded the advancement of human learning, nor dishonoured his good name by holding an innocent civilian against his will. 'If Méchain were to declare that he had been imprisoned by orders of either the Spanish government or myself, I would pass for an impostor in the eyes of the universe,' he wrote back. And then added his own veiled accusation against the godless French. Like the rest of his countrymen, he announced, he appreciated 'not only Méchain's knowledge, but his moral virtues as well'.[26] Yet just in case those virtues had gone unrecognized, he privately reminded Barcelona's governors to treat Méchain honourably and provide him with financial assistance. Of course by then Méchain had left Spain months before.

That autumn, the siege of Figuères reached its climax. General Dugommier died in battle there on 17 November, killed by an exploding shell as he surveyed his impending victory. 'Dugommier is dead on the field of honour,' the proclamation read. 'He demands vengeance and not tears.' Three days later, General La Unión followed him into the grave, killed by two musket shots during a bloody French assault.[27] The French drove the Spaniards from the crest where Captain Bueno had his tower, forced the surrender of Figuères, and pushed east towards the coast. Their successes, however, quickly got the better of them. Their supply lines unravelled. Desertions mounted. The two nations entered into formal negotiations to end the war, and in July 1795 they signed the Treaty of Bâle, returning the border to the same ambiguous position it had occupied before the war. But by then Méchain had no prospect of returning to Mont-Jouy.

Chapter Five

A CALCULATING PEOPLE

There are certain ideas of uniformity which sometimes seize great minds (as they did Charlemagne's), but which invariably strike the petty. They find in them a kind of perfection which they recognize because it is impossible not to discover it; the same weights and measures in commerce, the same laws in the state, the same religion in all parts. But is uniformity always appropriate without exception?[1]

CHARLES DE SECONDAT DE MONTESQUIEU,
The Spirit of the Laws, 1750

[This] chapter has earned Montesquieu the indulgence of all people of prejudice . . . Ideas of uniformity, of regularity, please all minds, and especially just minds . . . Uniformity of measures can only displease those lawyers who fear to see the number of lawsuits diminished, and those traders who fear a loss of profit from anything which renders commercial transactions easy and simple . . . A good law ought to be good for all men, as a true proposition [in geometry] is true for all men.[2]

M.-J.-A.-N. CARITAT DE CONDORCET,
Observations on 'The Spirit of the Laws', 1793

Delambre had been stopped in his tracks. Méchain had been trapped behind enemy lines. Like a suspension bridge abandoned after its end supports had been raised, the meridian survey had been called off in mid-execution, leaving a span half the length of France unbuilt between them. Not that the leaders of the Revolutionary government cared. They considered the meridian arc a monument to futility. Now that they had the provisional metre in hand, they could leave the ruins of the meridian survey unfinished, a folly of scientific presumption. For them, the challenge was not to push precision to an ever narrower closure, but to bring the advantages of the metric system to the common people. This meant putting metre sticks in the hands of twenty-five million French men and women.

Yet when the date for the obligatory use of the metric system arrived on 1 July 1794 the Revolutionary government had produced fewer than one thousand metre sticks and not a single French citizen was using the new system. Even the petty officials who answered to the dictatorial Committee of Public Safety were still filling out their reports in the old measures, making it impossible for the central government to monitor grain supplies. Prieur de la Côte-d'Or and the other members of the Committee pleaded with their subordinates to conduct the nation's business in the new metric system. They denounced the feudal diversity of measures as a barbarous remnant of the *ancien régime*. They expressed their frustration: why had the people who had pleaded so passionately for metric reform in the *Cahiers de doléances* become suddenly so reluctant to accept the metric system?[3]

This paradox would not have been so surprising had the politicians and savants put aside their wilful disregard for the meaning of measurement in the *ancien régime* and considered the enormity of the change they were demanding. A modern system of measurement allows objects to be described in abstracted, commensurable units that relate to an absolute standard. This is true of the new metric system the French were seeking to establish, as it is of the Anglo-American non-metric measures still in use in America today. In either system a measurement stays fixed, no matter where the object is measured, or which measuring instrument is used. A metre is a metre; as a foot is a foot, a pound is a pound and a kilogram is a kilogram. The dimensions of any other object can be described by reference to these units. The ultimate guarantor of these standards is a national or international agency with precise standards and a staff of inspectors. These inspectors are rarely seen any more because they have built their supervisory role into the measuring instruments we use every day: rulers, scales, graduated cylinders, clocks or gauges. Only in cases of extreme controversy are the inspectors actually obliged to check the calibration. Until that time, we trust the instruments. This form of measurement is adapted to our modern economy, in which buyers and sellers remote from one another in time and space conduct impersonal exchanges, quite certain that their measures are commensurable.

Under the *ancien régime*, by contrast, measurement was inseparable from the object being measured and the customs of the community which performed the measurement. These measurements were not enforced by a remote bureaucracy, but by local people answerable to their neighbours for their fairness. Far from being irrational or unnatural, this hodgepodge of measures made real sense to the peasants, artisans, shopkeepers and consumers who used them every day.

To begin with, each act of measurement in the *ancien régime* referred to a *particular* physical standard, held in local hands and safeguarded by local officials. A town's measure for the length for building materials, for instance, might derive from an iron fathom mortised into the wall of the town's markethall. The local measure for the weight of bread might derive from a master pound preserved in the guildhall of the area's bakers. The district's volume for grain might derive from a master bushel secured in the seigneur's château. And the local volume of wine might derive from a master barrel stored in the cellar of the monastery that owned the vineyard. It was the obligation of local officials – these aldermen, guildmasters, seigneurs and abbots – to enforce these standards, ensuring that exchanges made in the marketplace were fair. In return, they were entitled to extract a small fee for their services.

Not only did the physical standards differ from community to community, but the technique of measurement depended on local custom. One district measured grain heaped high in its bushel;

Ancien régime measures of Laon

These pre-Revolutionary measurement standards from the town of Laon, Méchain's birthplace, are still mortised into the wall under the archway of the town hall. They are among the last ancien régime *measures still* in situ. *From left to right: the 'T' measured the size of barrels; the rectangles are matrices to gauge bricks (above) and roof tiles (below); and the 'I' is an* aune *(an ell, about three feet in length) to measure cloth.*

(Musée de Laon.)

another measured grain after it had been levelled off; still another, after the bushel had been struck to settle its contents. Even the height from which grain was poured into the receptacle was dictated by custom since contents might settle upon handling. A slight nudge might alter the amount of grain in the bushel, a difference of great concern to those who paid taxes in kind or who bought or sold foodstuffs in bulk – that is to say, the vast majority of French men and women. Similarly, the *aune* (the ell), a measure of cloth, generally equalled the width of local looms, so that a square *aune* of fabric could be appraised by folding a quick triangle. Alternatively, the shopkeeper might measure an *aune* by extending the cloth from his nose to his outstretched arm, with a complimentary thumb's worth thrown in 'for good measure'.[4] Quantity in the *ancien régime* was bound up in ritual and custom.

This meant that measurement standards were potentially open to dispute, negotiation and change – albeit with the consent of the local community. Indeed, in many places the quantity that local people called 'a bushel' had actually altered over the years as seigneurs and tenants disputed its 'true' amount (and hence the proper level of taxation and a fair price for basic foodstuffs). As such, local measures served as a living record of the shifting balance of power within the community. Outsiders, of course, did not understand these measures, but local buyers and sellers did – which suggests one of the main advantages of local diversity. They kept outsiders out. Distinctive measures protected small-town traders from big-city merchants, or at least forced the latter to pay the equivalent of a fee before they could enter the local market. Artisanal guilds took charge of their own measures so that they might define their goods in a unique way, identify interlopers and drive them out of business with ruinous lawsuits. This was as true of gunsmiths and milliners then as it is true of the computer industry today. Control over standards is control over the rules of economic life, and *ancien régime* standards were everywhere local. Yet beneath this local diversity lay the deeper meaning of measurement in the *ancien régime*.

Many *ancien régime* measures – especially those that related to the world of production – had at their origin an anthropometric meaning derived from human needs and human interests. This does not mean that they directly reflected the size of the human body, the *pied* (foot) as the size of the king's foot, or as the length of the average human foot. Rather, many *ancien régime* measures reflected the quantity of labour a person could do in a given period of time. Thus, coal in one region of France was measured in a *charge* (a 'load') equal to one-twelfth of a miner's daily output. Arable land was often measured by the *homme* ('man') or *journée* ('day') so as to designate the amount of land a peasant might plough or harvest in one day. Other units expressed the local people's evaluation of worth or quality. Thus, the size of a plot of arable land might also be measured in bushels; that is to say, a plot of land was equal to the number of bushels of grain it took to sow that field. Even in districts where land area was ostensibly measured in a unit like the *arpent*, which referred to a number of square *pieds* (feet), the dimensions of the surface area would actually vary depending on the type of field and the quality of its soil. For instance, pasture land measured in *arpents* was often divided into five distinct *degrés* based on the best use for the field. In some cases, properties described in *arpents* in the official records were in practice divided into *journées* – which could not be compared with one another on the basis of their abstract surface area.[5]

As the economic historian Witold Kula has pointed out, these anthropometric measures expressed features of primary concern to those who worked the soil or produced the goods. After all, a peasant whose plot of land was physically smaller than his neighbour's 'five-bushel' plot, but which took six bushels to sow because it was on a gentle slope and had fertile soil, might well have found that 'six bushels' expressed his stake in the land far more vividly than an abstract surface area. Moreover, these measures did not simply express the value of the land, they guided work rules and set customary limits on the labour a landlord might extract. Thus, when a foreman hired four peasants to pick a vineyard of eight *journées*, the labourers knew not to settle for less than two days' wages each; nor would they do the work with only three peasants on their gang. In this sense, the anthropometric measures of the *ancien régime* acted as a control on productivity and, indeed, masked the very idea that productivity was a value that could be measured.[6]

For just this reason some eighteenth-century landlords had begun to map their property in geometric units rather than in units of labour. They hired surveyors who could 'put all these defective [measures] in good order, so that in each district their content is regulated in either *perches*, *pas* or *pieds* (rods, yards, or feet)'.[7] Armed with the new square units, these landlords hoped to monitor productivity and pocket any gains. This new breed of efficiency-minded landlord-farmer was the great hope of the 'physiocrats', a group of reformers who had acquired much influence with the French royal administration and were also known as 'the economists', being the first to practise that dismal science. The physiocrats touted agricultural reform and free trade as the key to improving standards of living, and they – like economic historians ever since – have expended great effort to determine whether productivity was rising in France.[8] Unfortunately, the question is virtually unanswerable for much of France because the process of translating anthropometric measures into modern measures erases the very information that defined productivity in the *ancien régime*. When England's leading agronomist set out to assess French agriculture in the 1780s, he discovered that he could not rely on the official measures listed in the public records.

> The denomination of French measures, as the reader will see, are almost infinite and without any common standard to which they can be referred . . . The only clue tolerably general that can be in the least relied upon is drawn from the quantity of seed sown . . . [And] inquiries of this kind are not to be made in the bureaux of great cities; books and papers will not afford the information; a man must travel through the country or must always remain ignorant though surrounded by ten thousand volumes.[9]

Even the surveyors hired by 'improving' landlords were daunted by the challenge of transforming land into a factor of production expressible in square units. They warned their employers that for the actual partitioning of fields 'it is best to stick to the report of those who sow the land'.[10] That is because these anthropometric measures of land and other commodities were the outcome of centuries of protracted negotiations among artisans, peasants, traders and seigneurs. Their value had been ritualized and fixed in ways which reflected the relative bargaining power of different members of the local community. As such, *ancien régime* measures had come to express that community's sense of the

proper social equilibrium. And any attempt to substitute a new
kind of measurement was read as a threat to that social balance.

No wonder peasants hated surveyors – and why Delambre and
Méchain met with such mistrust on their route. They too were
surveyors of a sort, come to supplant the anthropometric measures
which were the lifeblood of the peasant economy. They too were
measuring the earth for the purposes of a new partition.

The savants said the new measures would be 'natural' because
they were based on the size of the earth. For these savants, a
metric unit was natural when it could be defined *without* reference
to human interests. The metre, they said, would be independent of
all social negotiation or temporal change, transcending the interests of any particular community or nation. These men invoked
nature as the guarantor that all people would benefit equally
because no person benefited in particular. This spoke to the ideal
of justice as blind. Indeed, this Enlightenment project has often
been read as an attempt to displace personal relations as the foundation of the social order, and in their stead substitute a universal
metric, imported from the natural sciences, by which the social
world might be subject to dispassionate analysis – and schemes for
improvement. But the people of the *ancien régime* also considered
their measures 'natural', in that they had been built into the dimensions of the lived world and expressed their needs, their values and
the history of their shared life. Their anthropometric measures
sanctified man as the measure of all things, and expressed a different notion of justice, one which not only governed the domain of
productive labour, but also the realm of economic exchange.[11]

The *ancien régime* was governed by a 'just price' economy in
which basic foodstuffs were sold at a customary price set by the
local community at a level which most of the people in that community could afford. The just price was enforced by moral
sanction and ultimately by the threat of violence. The theory of
this 'just price' economy had been legitimized moreover, by
medieval scholastic doctrine, although this does not mean that
prices were thought to be divinely sanctioned. The people of the
ancien régime understood that production and consumption would

halt if buyers and sellers were unwilling to trade. To induce production and exchange, then, the just price needed to reflect the
costs of doing business, with these important caveats: that the
authorities intervene in times of dearth, that locals not extort exor-
bitant fees from wayfarers or people in desperate need, and that
sellers not conspire to rig prices.[12]

In such an economy, the diversity of weights and measures
greased the wheels of commerce. In an age where bakers dared
not charge more than the 'just price' for a loaf of bread for fear of
precipitating a riot, bakers who wanted to preserve their livelihood when the cost of flour rose simply baked a smaller loaf. The
same ruse allowed monasteries to circumvent Christian restric-
tions against profits by buying wine in large barrels and then
selling it (for the same price) in smaller barrels. Sometimes this
could lead to accusations of fraud, as when the petitioners of
Notre-Damme-de-Lisque complained in 1788 that their abbot's
tax collector had increased the measure of grain. More probably,
he was simply trying to maintain his own revenue during a time of
rapidly increasing prices.[13]

The workings of this economy were familiar to *ancien régime*
officials. One government agent noted that local grain merchants
profited by buying grain at one measure and selling it (for the
same price) at a lesser measure. But rather than condemn this
practice, he noted that it encouraged commerce in the region,
since attempts to raise prices risked the wrath of the local populace. A provincial assembly warned in 1788 that 'the establishment
of a uniform measure would ruin this genre of commerce, destroying at the same time an infinity of little markets which subsist
only on these differences and, though of no great importance,
supply the needs of nearby consumers'.[14]

In many towns, *ancien régime* officials themselves served as the
'fair mediators' who interposed themselves between buyers and
sellers, setting the just price for essential foodstuffs like bread,
meat, wine and beer. Indeed, superintending the economy in this
way was one of the obligations of a benevolent monarch, and
among the principal justifications for his rule. In setting the just

price, local officials generally took market conditions into account. The price of bread, for instance, was governed by *tarifs*, numerical tables which translated the current market price of wheat into the just price for a four-pound loaf of bread of a specified quality (white bread, brown bread, second-class bread and so on). In major towns, these *tarifs* were drawn up collaboratively by aldermen and bakers who jointly estimated the cost of milling and baking bread, and outfitting a shop, while guaranteeing a modest return for the baker. These regulated prices, however, were 'sticky' in the sense that bakers could not fine-tune their prices to meet daily fluctuations in the cost of wheat. Also, bakers tended to set their prices in round numbers because of a persistent shortage of small coins. Instead of adjusting prices, bakers then altered the weight of their loaves or diluted their ingredients. Such practices were illegal, but even consumers who were aware of them generally tolerated them so long as everyone could still afford a 'pound' of bread. Equity mattered more than efficiency. Yet in times of dearth any attempt to raise prices or to 'short' bread too egregiously could spark violence. Price was not the paramount variable in the *ancien régime* economy, but merely one variable among many, including quantity, quality, the cost of production and local custom.[15]

In short, the old diversity of weights and measures, far from being irrational and unnatural, formed the backbone of the *ancien régime*'s economy. These measures did not simply define a distinct kind of economy, they defined a kind of human being. Today, we assume that the 'market' consists of the aggregate of innumerable one-on-one private exchanges, the sum total of which sets prices. We might call this the market *principle*. The *ancien régime* operated according to the idea of the market as a *place*, which one might imagine as a kind of bazaar or village fair in which buyers and sellers met in public to conduct exchanges under the watchful eye of a third party. That third party – typically an emissary of the king, a town alderman, the local seigneur or the nearby abbot – justified the taxation of these transactions by ensuring that the needy did not go hungry and the producer got a fair return for his troubles. Thus, in addition to providing peasants and artisans with a ready guide to the value of their land and labour, the weights and measures of the *ancien régime* also provided shopkeepers and consumers with some guarantee that their marketplace transactions would be fair.

In this context, the French savants' scheme to reform weights and measures was a revolutionary rupture, far more radical than the sort of translation involved in the switch from, say, Anglo-American units to the metric system. Indeed, the revolutionaries *intended* the metric system to eradicate the assumptions underlying the old just-price economy. Their goal was to make productivity the visible measure of economic progress, and to make price the paramount variable in commercial exchange. They saw the metric reform as a crucial stage in the education of modern *Homo economicus*.

To this end, the Academy of Sciences proposed in 1793 a decimal division for its currency, so that its value too would be based upon the new metric units of weight. The Academy proposed that one *franc* should equal 0.01 grams of gold. 'Thus will all measures, weights and money refer to a unique and foundational base: a quarter of the earth's meridian.'[16] By defining the scale of worth as well as that of quantity, science would provide a secure foundation for a rational economy. On 7 December 1794 the new *franc* was declared equal to the old *livre*, now divisible into 100 *centimes*. This rationalization was the brainchild of the same savants and politicians who had pushed for metric reform: Lavoisier, Condorcet and Prieur de la Côte-d'Or.

Lavoisier was not only the world's premier chemist, he was also one of the *ancien régime*'s 'tax farmers', the financiers who collected the king's taxes – and took a healthy cut for their pains. This position had earned him one of France's great fortunes, as well as the hatred of millions of ordinary French men and women. Despite the source of his income, however, Lavoisier was committed to the physiocrats' policies of laissez-faire and the elimination of the *ancien régime*'s many taxes, both visible and invisible. He had thought long and hard about the optimal way to manage a national economy, and his thinking on this point was closely connected to his understanding of chemistry. His lofty principle that 'matter is neither created nor destroyed, all it knows is transformation' committed his young science to precision measurement. How else could the chemist know whether matter had been conserved or not? If the chemical equation was to be the new mode of thought about the material world, then the finely tuned balance scale would be the proof that such thinking paid off. Novelty, productivity and profit all relied on careful book-keeping. Economic exchange, like chemical transformation, should be measured in universal units so that transactions would be transparent, with buyers and sellers equally informed about the deal they were cutting. Such transactions would also be easier for the centralized state to monitor for fairness and, of course, to tax. Without the decimalization of money, he noted, 'the metric system will have been adopted in vain'.[17]

Condorcet, in addition to his role as Permanent Secretary of the Academy of Sciences, had served as Master of the Royal Mint. Along with his contributions to mathematical social science, he was one of the nation's premier political economists. For Condorcet, economic progress went hand in hand with political progress. He was perhaps history's greatest optimist. His goal was to reconcile freedom, equality and material wellbeing through a programme of universal education and a new social science that would match human laws with social needs. That nature's laws were everywhere the same meant, for Condorcet, that the hodgepodge of human laws must be aligned with universal principles. Reduce the legal code to its essentials and the law would be comprehensible to all literate men and women. This would diminish the unfair advantage that those in authority held over the powerless. Give all citizens equal access to knowledge and they would all have the power to control their own fate. Condorcet had imagined a scheme to classify all knowledge in a decimal system, a forerunner of the Dewey Decimal System. More grandly still, he imagined a language of universal signs to replace all forms of logical thought, much the way algebra expresses mathematics. Such a language would apply to social relations as well as to logical ones. It would 'bring to all objects embraced by human intelligence a rigour and precision that would render knowledge of the truth easy, and error almost impossible'.[18]

Condorcet considered the metric system a first step towards achieving this new universal language for the objects of the material world. In combination with the reform of the French currency, the metric system would bring efficiency to economic relations – and this, in turn, would foster political equality and freedom. '[It] will ensure that in the future all citizens will be self-reliant in all those calculations which bear upon their own interests; because without this independence citizens can neither be equal in rights . . . nor truly free . . .'[19]

As for Prieur de la Côte-d'Or, he had an engineer's appreciation of optimization, plus an administrator's preference for clear protocols – which is another way of saying that he embraced the clichés of the day. Prieur was younger than Lavoisier or Condorcet, and nowhere near their intellectual equal. Under the *ancien régime* he had been a run-of-the-mill military engineer: underemployed, a bit shy, lame in one leg, uncomfortable with public speaking, formerly his mother's darling, in love with a married woman, well trained in mathematics, primed with the ambition to rationalize the world and not much of an original thinker. But from his new position on the Committee of Public Safety, he had the clout to make things happen.

Prieur believed that uniform measures would make France a great nation, smoothly administered from the centre and united through trade. The metric system would transform France into 'a vast market, each part exchanging its surplus'.[20] It would make exchanges 'direct, healthy and rapid', diminishing the 'frictions' which impeded the wheels of commerce. These frictions included anything that masked the true price of an item, such as the variable measures of the *ancien régime*. The price of an item, Prieur argued, necessarily depended on many factors: its scarcity, the work necessary to produce it, the quality of the product. But in the final analysis, price was whatever people agreed it should be. This meant that when people agreed on a price they needed to

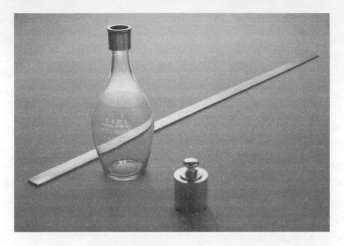

Metre stick, 'cadil', and kilogram weight

This metre stick is the official provisional metre of 1793, which measured 443.44 lignes in the old Paris units. It was constructed of copper by Lenoir. The 'cadil' was the name for the litre until it was changed in 1795. The kilogram weight pictured here was based on the original definition of a gram (then called a 'grave'), which equaled the weight of one cubic centimetre of water at the freezing temperature. For the definitive kilogram of 1799, the gram was based on a cubic centimetre of water at its temperature of maximum density (about 4°C).

(Musée des Arts et Métiers-CNAM, Paris. Photo: CNAM.)

know what they were getting, not be baffled by secret shifts in the quantity being exchanged. Those who claimed that differences in measures aided commerce were just talking about their personal profits. 'The French Republic,' he wrote, 'can no longer tolerate men who earn their living by mystery.'[21] Worse, those who profited from the diversity of measures, said Prieur, corrupted those who tried to conduct honest and transparent exchanges by 'complicating commerce, spoiling good faith, and sowing error and fraud among the nations'. Until commerce was carried out with complete probity, the common people would doubt the advantages of free trade. Only if price were the sole variable in exchange would these exchanges be based on clear understanding between parties.[22]

Instruction in this new form of 'right thinking' about economic matters would come from the metric weights and measures themselves. The people's new rulers would be the rulers they used every day. Rational measures would engender a rational citizenry.

> If we want the people to put some order in their acts and
> subsequently in their ideas, it is necessary that the custom of
> that order be traced for them by all that surrounds them . . .
> We can therefore look upon the metric system as an excellent
> means of education to be introduced into those social
> institutions which conjure up the most disorder and
> confusion. Even the least practised minds will acquire a taste
> for this order once they come to know it. It will be reflected
> by the objects which all citizens have constantly before their
> eyes and in their hands.[23]

Today, many of these ideas are taken for granted and hence unexpressed. But like many things that appear ordinary on the surface, they mask a long history of bitter controversy. Making measurement banal proved to be hard work and would take more than a century of struggle and conflict.

For instance, these reformers all presumed something we are apt to forget: that free trade would have to be fostered by state action. It may be that people everywhere have an innate desire to 'truck and barter', as Adam Smith taught, but the leaders of the new French republic understood that a 'free market' was something quite different and required a new set of social institutions. The proponents of the metric system wanted both a powerful state apparatus *and* a free citizenry empowered to participate in the political and economic life of the nation. To resolve this apparent contradiction, they wished to transform their fellow citizens into a calculating people. The savants, engineers and administrators of eighteenth-century France were already superb calculators who had earned their posts thanks, in large part, to their mathematical merits. They simply wanted the French people to become more like them.

The advocates of the metric system, like today's advocates of globalization, saw their goal as creating at a stroke a new kind of economy and a radical new kind of politics. This is not to say that the savants were innate revolutionaries. The French savants of the eighteenth century had been as fond of their comfortable *ancien régime* lives as the lawyers, financiers and military men who likewise stepped warily into the new age. They had little cause for complaint. Foreign savants who visited Paris before the Revolution often remarked wistfully on the respect their French colleagues received at the hands of the kingdom's great nobles and ministers. The savants were appreciated, and they enjoyed their daily routines. Yet their very routines masked a radical premise. Give scientists a chance to remake the world, and who knows what will be left standing when they are done? What human habit can survive the blade of logic? What social institution can justify its ways to mathematics? What ancient custom can be assayed with precision? The metric system belonged to that radical strain of the French Revolution which sought to destroy all local distinctions to make way for a future in which everything was the same everywhere, much as today's critics of globalization suggest that the Information Age will level all cultural differences around the world. The metric system was to be the new language of the material world. And just as the Revolutionaries, in the name of linguistic unity and rational communication, sought to eliminate the diversity of France's various patois – its many regional languages and dialects – so did the savants dream of extending their metric language to all domains of scientific and public life.[24]

Time was first. The Revolution had marked a new beginning in human history. Where the Gregorian calendar had wed the year's rhythm to Christian holy days, a secular republic needed a calendar based on nature and reason – although pinpointing the

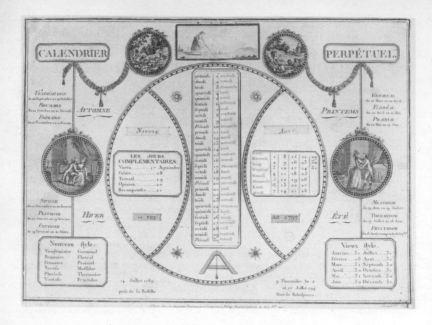

Revolutionary 'calendrier perpetuel'

This calendar of 1797 enabled users to convert between Revolutionary and Gregorian dates. The Revolutionary calendar came into existence in October 1793 and began with the year II. It was abolished early in the year XIV in time to start 1806 on 1 January.

(Photothèque des Musées de la Ville de Paris. Photo: Chevalier.)

exact moment of rupture proved contentious. Was it 1 January 1789 – the beginning of the year which had proved so liberating? Or was it 14 July 1789 – the date the Bastille fell? Both moments of origin, and many others, were proposed. Not until 1793 did the mathematician-turned-politician Gilbert Romme – on the advice of his friend, Jérôme Lalande – settle upon a solution. Year I of the new era would be backdated to the founding of the French republic on 22 September 1792, which happily coincided with the autumn equinox, a most auspicious conjunction of nature and reason. 'Thus, the sun illuminated both poles simultaneously, and in succession the entire globe, on the same day that, for the first

time, in all its purity, the flame of liberty, which must one day illu-
minate all humankind, shone on the French nation.'[25] The
calendar would contain twelve months of thirty days, each poet-
ically named after its season.

vendémiaire	month of the wine harvest	September/October
brumaire	month of fog	October/November
frimaire	month of frost	November/December
nivôse	month of snow	December/January
pluviôse	month of rain	January/February
ventôse	month of wind	February/March
germinal	month of germination	March/April
floréal	month of flowering	April/May
prairial	month of meadows	May/June
messidor	month of the harvest	June/July
thermidor	month of heat	July/August
fructidor	month of fruits	August/September

Each month was then divided into three ten-day weeks known
as *décades*; no more Sundays, no more saints' days. National festi-
vals would commemorate the anniversaries of revolutionary
uprisings, with a climactic festive *sans-culottide* of five days (six in
leap years) to ensure that each year began anew on the autumnal
equinox. No creation of the republic, wrote Lalande, would do
more to break the hold of the priests over their superstitious
dupes. He did admit, however, that ordinary people might find the
ten-day working week a tad on the longish side, and proposed a
mid-week holiday, the *quintidi*, to ensure that the revolutionary
calendar and the Revolution itself would become popular.[26]

And while they were at it, the rationalizers reasoned, why not
divide each day into ten hours, and each hour into one hundred
minutes? A law of 11 brumaire of the year II (1 November 1793)
so decreed. Master watchmakers designed prototype clocks that
pointed to 'V o'clock' at midday, and 'X o'clock' at midnight.
Laplace had the dial of his pocket watch adapted to show decimal
time.[27]

A decimal clock

Between 1794 and 1795 the French government briefly mandated the use of a decimal clock with a day divided into 10 hours of 100 minutes of 100 seconds each. Of all the unpopular changes associated with the metric system, this was the most unpopular. Some forward-thinking individuals, like Laplace, had their ancien régime *pocket watches modified accordingly. A clock in the Palais des Tuileries still kept decimal time as late as 1801. But decimal time was otherwise ignored.*

(Musée des Arts et Métiers-CNAM, Paris. Photo: Pascal Faligot, Seventh Square.)

And why stop at time? Why divide circles into 360° just because the ancient Babylonians had done so? A 400° circle (with a 100° right angle) would not only ease calculation, it would synchronize astronomy and navigation. In a world where the quarter-meridian was ten million metres long, each degree of latitude would then measure a hundred kilometres.[28] This would simplify maps and assist sailors. Already, as a pledge of the metric system's coherence, Etienne Lenoir had ruled his repeating circles for Delambre and Méchain in 400° rather than 360°. The new angular division would require new trigonometric and logarithmic tables. But their production too could be rationalized. By breaking down the complex formulae into a series of simple arithmetical tasks, the savants could portion out the work to semi-skilled 'calculators', as in a factory of mathematical results. Condorcet proposed employing the graduates of the deaf-mute schools because they would be less easily distracted from their labours than other people. In the event, the savants employed out-of-work wig-makers, laid off by the Revolutionary assault on aristocratic hairstyles. This collective human computer – inspired by Adam Smith, and the inspiration for Charles Babbage – prefigured our information economy: universal measures, transparent numbers and the division of mental labour.[29]

Condorcet and Lavoisier were well placed to press for metric reform, at least at first. As Permanent Secretary of the Academy of Sciences, Condorcet spoke for that body. He was also an elected representative to the National Assembly, where he became a chief advocate of equality for women, Jews and blacks. He urged public education for all French children. He believed that virtue and reason were forever conjoined. These views would also garner him enemies, especially when the Jacobin party came to power. Not that the Jacobins disputed those goals exactly, but they despised Condorcet's voluntarist methods of achieving them. When the Committee of Public Safety condemned Condorcet along with the rest of his political allies, he went into hiding. There he composed his great Utopian tract – *Sketch for a Historical Picture of the Progress of the Human Mind* – which he left unfinished when he killed himself rather than face execution in May 1794.

Though he lacked a formal political role in the new republic, Lavoisier had considerable power to promote the metric system. As treasurer of the Academy of Sciences, he controlled the purse strings of the meridian expedition. As the patron of one of Paris' finest salons, he hosted dinner parties on the boulevard de la Madeleine where the scientific elite could hash out policy and win political allies. Lavoisier was a man who knew everyone and who was everywhere respected. He secured an exemption from the military draft for the savants and instrument-makers working on the metric system, including Delambre, Méchain and their assistants. He determined the standard for mass, which he defined as the weight of a cubic centimetre of distilled water at the temperature of melting ice. He was just completing these experiments when the Committee of Public Safety incarcerated him, along with the rest of the tax farmers, in the Porte-Libre prison (the 'Free-Entry' prison).[30]

Lavoisier, who had fretted over the fate of the injured Méchain, now found himself in need of a protector of his own. Borda wrote to the Revolutionary authorities, bravely demanding that Lavoisier be released so that he might resume his labours for the metric system. Where once Lavoisier had cited the on-going metric reform project as the chief rationale for preserving the Academy of Sciences (to no avail), his colleagues now cited his labour for the metric reform as the chief rationale for preserving his life. The Committee of Public Safety responded by purging Borda – along with Delambre, Laplace and several others – from the Commission of Weights and Measures. The signature on the order was that of Prieur de la Côte-d'Or.[31]

For the past few years, Prieur had been a regular guest at Lavoisier's home where the nation's greatest scientific minds had gathered over dinner to thrash out the details of the new metric system. The conversation often turned to politics. Prieur, the youngest person present, and no scientific luminary, often found himself alone in defending the Revolutionary government, of which he was a rising member. Conversation became animated; these were men who spoke their minds. At times, Prieur's views were mocked. It was this personal pique, according to Delambre, that explained Prieur's vendetta against the senior savants. 'As a result he nourished a resentment against Lavoisier and those of his colleagues, such as Borda . . . who showed themselves to be most ardent, lively, or witty in their disputes.'[32] In his own mind, of course, Prieur had only acted to 'regenerate' the Commission of Weights and Measures. He had cancelled the superfluous meridian expedition so that the government might focus on the far more important task of implementing the metric system. The scientific portion of the mission, he wrote, had been 'carried to that point of maturation at which the need for reflective thought separates itself from the need for action'.[33] The time for action had come.

That winter, while Lavoisier languished in prison, the same forces closed in on Delambre's patron, Geoffroy d'Assy, also a wealthy financier associated with the detested taxation system of the *ancien régime*. The neighbourhood council in the Marais – renamed the Section de l'Homme Armé (the Armed Man) – sent two officials to search the d'Assy residence in Paris for evidence of disloyalty. When the house servants explained that the d'Assy family had moved to the country, the officials sealed shut the residence at 1, rue de Paradis. Then, on 25 January 1794 – just as Delambre arrived at the d'Assy country château on his way north from Orléans – d'Assy was arrested there. No explanation was given, but in the light of d'Assy's position in the *ancien régime*, none was needed. A second search of the d'Assy residence in Paris a week later turned up a table lamp engraved with the fleurs-de-lys and made by an artisan who worked 'by appointment to His Majesty'.[34]

It was up to Delambre to remove any further incriminating evidence from the house. After a week spent comforting the family in Bruyères, he drove his custom-built carriage back to Paris, paid a last month of wages to Bellet and his manservant Michel, returned his repeating circle to Lenoir's workshop, and presented himself to the neighbourhood council. He showed them his passport, signed by the Minister of the Interior, declaring his residence to be 1, rue de Paradis. He showed them the certificate attesting to his status as the Republic's Commissioner for the Measure of the Meridian. And he explained to the council that he needed to gain access to the d'Assy residence to recover important astronomical equipment essential to his mission. Needless to say, he did not mention that he had been purged earlier that month from the meridian expedition for 'lacking Revolutionary zeal'.[35]

Delambre's ostensible goal was to retrieve his own papers from his apartment on the third floor. He was accompanied by two officials. On entering his room, he discovered that his secretaire was locked and that he had forgotten the key. This enabled him to make a second trip inside the sealed building a month later. On the second occasion, the officials examined every scrap of paper he removed from the secretaire, including a piece of parchment covered with a Latin scrawl and signed by King George III of England. This document attested to Delambre's membership as a foreign correspondent in the Royal Society of London. The officials seemed uneasy about this document, as well as several other sheets covered with scribbled calculations and drawings; these might be ciphers or secret plans. In the end, however, they let Delambre remove these papers, having found 'nothing suspect'.[36] He later recalled that 'this was a supreme indulgence on their part for a man they thought in correspondence with kings'.[37]

Lavoisier was not so fortunate. He was executed on 8 May 1794 along with the twenty-seven other tax farmers. As one mathematician confided to Delambre: 'It only took them an instant to cut off that head, and it is unlikely that a hundred years will suffice to produce a comparable one.'[38]

By that time Geoffroy d'Assy had spent five months in the Luxembourg prison. Initially, conditions were tolerable; prisoners had the use of a café in the prison's central courtyard. But as war fever intensified, the Committee of Public Safety suppressed dissent by populists and moderates alike. That summer, the Revolutionary Tribunal sentenced d'Assy to death, along with fifty other co-conspirators who had plotted a prison revolt to 're-establish the monarchy and tyrannical power'. A more motley set of conspirators could hardly be imagined: aristocrats, bakers and an entire family. Only an elderly wine merchant and fourteen-year-old boy escaped the death sentence – although the boy's sixteen-year-old brother was executed. Two weeks later, Robespierre himself was guillotined during the counter-revolution of thermidor.[39]

The incarceration and death of Delambre's patron made him the chief protector of the d'Assy family. In June, he returned to the d'Assy home in Paris, armed with a power of attorney from Madame d'Assy, to retrieve various legal documents belonging to the family. In January 1795 he petitioned to recover all his own possessions from the house, including his astronomical equipment, his furniture and a small portrait of himself on the dressing table of Madame d'Assy. He spent the rest of this period far from Paris in the d'Assy country château near Bruyères-le-Châtel, then renamed Bruyères-Libre. When Madame d'Assy's brother died of a fever later that year, Delambre promised to shed tears for him as soon as he had a spare moment. At the time the tragedies were coming too fast.[40]

The Revolution had provided the French savants with a unique chance to rewrite the world's measures. Yet the opportunity carried corresponding risks. Persist in the old ways of treating your subordinates and they might take you down a peg. Place your talents in the service of the state, and the state might call you to account for your findings. Commit an astronomical error and you might wind up in prison.

Most of the purged members of the Commission had sequestered themselves in the countryside during the long hot summer that became known as the Terror. Borda retired to his family estate. Laplace retreated with his wife and two young children to Melun, thirty miles south-east of Paris. But Cassini IV's residence was the Paris Observatory, and he now paid the price of having refused to serve the republic. In the decade before the Revolution, Cassini had hired three men as astronomical aides and housed them in the Observatory grounds, where they might be instructed by Méchain and the rest of the staff. These apprentice astronomers now demanded equality with their boss, as well as the other rights of free men. The eldest, a mild fifty-year-old monk named Nicolas-Antoine Nouet, who also served as the Observatory chaplain, informed Cassini that he wished to marry his personal serving-woman. Cassini was horrified and the two men, once cordial, never spoke again. The second student, a young man of astronomical talent named Jean Perny, returned drunk to the Observatory late one night after a meeting of his Revolutionary club, and banged on his patron's door with the butt of his sword, shouting, 'Cassini the aristocrat must be killed!'[41] He had to be subdued and taken to bed. A few days later he penned an abject letter of apology. The third student, Alexandre Ruelle, a youthful deserter from a dragoon regiment, whom Cassini had harboured and trained until his amnesty came through, became his benefactor's most bitter enemy.

The apprentices' complaint was the perennial complaint of junior scientists in the laboratories of their seniors: they accused Cassini of having appropriated their work to publish it under his own name. They wanted equal credit and equal pay. The 'horrific despotism' of the Observatory director, they claimed, had 'stolen the fruits of their night-time labour'. In truth, Cassini had generally acknowledged his students' contribution in his reports to the Academy, albeit in a patronizing manner.[42]

In this topsy-turvy time, however, the assistants made their accusations stick. With the aid of a sympathetic politician, they reorganized the Observatory along egalitarian lines. Science, after all, was a democratic enterprise, open to all aspirants. No savant should inherit his position like an aristocratic title. The government created four new posts of 'Observatory Professor'. Cassini retained one of these. But rather than give the other three to the country's leading astronomers – Lalande, Delambre and Méchain – the apprentices convinced the authorities that these savants harboured 'aristocratic' sympathies, and had themselves appointed instead.[43] Cassini's salary was halved, and Perny was elected the first director in rotation. Faced with this humiliation, Cassini resigned, ending 120 years of family rule. His resignation only worsened his situation. His students were able to evict him from his apartments in the Observatory. Then the government seized his map of France, the family's great commercial and scientific enterprise, and, when Cassini dared to protest this theft, threw him in prison. Immediately, his student Ruelle, a member of the neighbourhood's Revolutionary council, suggested that his former protector be sent before the Revolutionary Tribunal, a certain death sentence. Mercifully, the council rejected his suggestion.[44]

But topsy-turvy times will flip and flip again. Once Robespierre himself had fallen from power, the student-professors fell out among themselves. Suddenly Ruelle found himself under attack by his own fellow students. Apparently he had committed an error of ten seconds in a solar observation. More damaging still, his results were fraudulent, based on theoretical guesswork rather than on direct observation as he had claimed. For this crime against science – and also, of course, for his affiliation with the radical party now out of favour – Ruelle was himself imprisoned on 22 August. Such were the risks of scientific error in perilous times. To replace him, the other two former students invited Delambre to join them as Observatory Professor.[45]

Thus the winds of revolution shifted in science as they did in politics. Robespierre's republican successors, self-conscious moderates, made a great show of reviving the nation's scientific institutions. In June 1795 they established a new institution, the Bureau of Longitudes, in imitation of Britain's Board of Longitudes, to help France rival that nation's commercial and naval dominance. The Bureau supervised the Observatory of Paris, and was staffed by the nation's finest savants, including Lalande, Laplace, Legendre and Borda, plus Delambre and Méchain. Then they restored the Academy of Sciences as part of the new National Institute, where almost all the (surviving) academicians – Delambre and Méchain included – resumed their old chairs.[46]

Now it was Cassini who was free, while Ruelle languished in prison – though neither found scientific redemption. When Ruelle was finally released, only Lalande would vouch for him, and he finished his career at the national lottery, where one hopes he did not indulge the sort of fabrications that had cost him his career in astronomy.[47] As for Cassini, he spurned the entreaties of Delambre and Lalande that he rejoin the scientific fold. He said he had witnessed too many bitter divisions in the Academy to contemplate a return. He retreated to his country château in Thury with his mother, his five children and nine nuns evicted from a nearby convent. 'My swallows,' he called them. 'I have named it the Republic of Thury,' he wrote, 'where I assure you we lack only republicans.' The savant who had once claimed the meridian expedition as his birthright no longer believed in the metric reform – nor even in science.[48]

'But what of your astronomy?' you ask. I confess, it is nothing to me now . . . 'But,' you ask, 'does not your glory, your reputation, your duty as a savant all speak against this retreat?' My friend, the duty of a father surpasses that of an academician . . . And as for my reputation, my glory, I have sacrificed them, and it has cost me little . . . Obliged to flee the Observatory, I saw the Academy of Sciences delivered to the government of the sans-culottes. And what grieved me most, I saw the savants themselves up in arms, divided against one another, partaking of the delirium and the rage of the Revolutionary horde, adopting their morals, their manners, and even their language . . . How can I recognize myself in the changes they have wrought in our old ways of calculating, our old measures, when we had not ten hours in a day, but twenty-four, and no circles of four hundred degrees . . .? Everything has changed, and I am too old to abandon my old habits and ideas. The year, the months, the almanac, the astronomical tables, all are changed. If Galileo, Newton or Kepler were to descend from heaven and appear at the Academy, they would not comprehend a word in the presentation of Citizen Lalande when he told them that on 20 brumaire, the moon, in a 200 degree opposition to the sun, passed the meridian at five hours . . .[49]

Peasants, shopkeepers and villagers were not the only ones attached to the old numbers of the *ancien régime*. To the numerate, numbers matter. Some old-time savants, like Cassini, considered the metric system an affront to the harmonious values that had once described their universe.

But where Cassini retreated, Citizen Lalande advanced. On 17 May 1795 he became the new director of the Observatory. Throughout the advances and reversals of the Revolution, the great iconoclast remained unbowed. When he was elected to the head of the Collège de France in 1791, his first official act was to admit women to all classes. He ended the announcement of prizes in Latin. He even tried to get professors to teach their own courses. Every evening, rain or shine, he took his long constitutional through the streets of Paris – sometimes walking for five or six miles – handing out alms. With his purple waistcoat and his umbrella (a newfangled invention), he cut a singular figure on the city streets: under five feet tall, unkempt, unwashed, his thick grey hair matted to the back of his aubergine-shaped skull. Yet shameless men can demonstrate great courage. 'I am so constituted,' he reflected, 'as to fear no thing nor person, neither danger nor death.' It may have been his *philosophe*'s vanity, yet he always insisted on speaking the truth. 'I am frank to the point of rudeness; I have never dissimulated, even when the truth might displease.' He had spoken his mind during the *ancien régime*, and he refused to now.[50]

He later admitted that it was probably his notorious reputation as an atheist that saved his life during the Terror. If so, it was the only time his irreligion found favour with the powers that be. 'I don't feel sorry for the nuns who lose their pensions because they refuse to swear allegiance to the state,' he wrote to his daughter. 'It ought to be a joy for them to starve to death for God.'[51] Yet he proved to be an ecumenical saviour. He hid the monarchist Du Pont de Nemours in the dome of his observatory in the Collège des Quatre-Nations, and brought him food and drink there for several weeks at the risk of his own life. Years later, at Lalande's funeral, the founder of the Du Pont Corporation asked God to bless the notorious atheist. Lalande also disguised several condemned priests as astronomers for their protection, telling them not to worry about this deception. 'But of course you are astronomers; who can better claim that title than men who live for heaven?' At a time when it was risky to do so, Lalande also published laudatory eulogies of guillotined savants such as Lavoisier – and then quibbled about their scientific views.[52]

His finest moment was the inaugural Festival of the Supreme

Being, where he helped to celebrate the deity of the ersatz religion Robespierre had hoped to substitute for Christianity. The ceremony took place on 8 June 1794 inside the Panthéon, a year after Delambre had observed the Paris triangles from its cupola. Offered at last a pulpit to preach his atheism and denounce the priestly cabal, Lalande instead seized the occasion to warn against the ferocious patriotism of the times.

> The time has come to declare these important and
> incontestable truths, known to all people, at all times, and in
> every corner of the globe: love of country, love of virtue and
> the reign of reason . . . Love of country is not a patriot's only
> duty; charity is also a duty. We cannot all serve our nation in
> the army, in the state, in the arts and sciences, but we can all
> come to the aid of our brothers . . . It is in this way that
> charity, added to love of country, will make us truly worthy of
> our Revolution, our victories, and the admiration of the entire
> Universe.[53]

But however great his love for his nation and fellow man – and woman! – Lalande's highest priority was always his stars. A few days after his Festival address, he announced he had added 1200 new stars to his catalogue in the past ten days, bringing his total to 21,000. Six months later – during which time Robespierre was deposed and the moderates took power – he added another thousand. He refused to serve on a criminal jury lest it distract him from astronomy. 'There is no sanction,' he informed the authorities, 'which would make me leave my stars; I would do anything rather than defer to your summons.'[54] Sometime in 1796 the family workshop surpassed their initial goal of 30,000 stars and decided to shoot for 50,000. His daughter continued to calculate 'with a courage rare for her age and sex'. Her young son Isaac was placed in child care because he so distracted his mother and grandfather. When Lalande hit 41,000 stars in 1797, he boasted that 'this inventory of the heavens has been my constant project for the past twenty years, and occupies me to the point that I could die without regret, knowing that I have left behind a monument of my passage here on earth'.[55]

All this time, Delambre had quietly been plying his astronomical trade at the d'Assy country château at Bruyères. To be on the safe side, he secured a certificate from the local municipal council attesting to the fact that he was not an émigré and had never been imprisoned. Mostly he stayed out of the public eye. On only one occasion did the Jacobin government call upon his expert advice: Delambre had detected a flaw in the new republican calendar.[56]

In their efforts to keep the autumnal equinox aligned with the birthday of the republic, the calendar's designers had instituted the *franciade*, a leap-year day. But they had failed to see that it would not fall every four years as intended, but occasionally in the fifth year instead. A perfect alignment of republic and nature was not a simple matter. Looking 150 years into the future, for instance, Delambre discovered a year in which it would be impossible to predict whether the equinox would occur before or after midnight on 22 September. He brought these prognostications to Lalande, who informed the calendar's principal author, Romme, who in turn asked Delambre to help resolve the problem. Delambre suggested some possible solutions, although he warned that certain inconsistencies would resurface 36,000 years hence. When Romme presented these modifications to the relevant government committee, they pronounced themselves unconcerned. 'Do you want us to legislate eternity?' one member wanted to know. No, Romme responded, he would be satisfied if the committee agreed to revisit the issue in 36,000 years. And so, to general amusement, it was decreed. Romme himself would not live to see another year, however. He was arrested and executed for his Jacobin sympathies two months later.[57]

By then France's military conquests – thanks to its Jacobin-led armies – had inspired a new sponsor of geodesy. In 1794, General Etienne-Nicolas Calon was appointed director of the Dépôt de la Guerre et de la Marine, thereby uniting the army and navy cartographers under a single command. Calon dreamed of a detailed geographic survey to extend the Cassini map to France's newly conquered territories in the Low Countries, Germany and Italy. He was a perpetual enthusiast, a self-promoting military cartographer, now a brigadier general and a member of the national legislature. More to the point, he had the budget to make things happen. He envisaged a 'museum of geography' assembling forty-five of the nation's leading savants in an effort 'to raise to the highest possible degree the development and glory of the astronomical and geographical sciences'. The spine of this geographical knowledge would be a precise geodetic survey of the meridian.[58]

In pursuit of this goal, Calon decided to consult the learned savant Jean-Baptiste-Joseph Delambre, initiator of the meridian survey. Unfortunately, he did not know in which prison to seek the savant. He asked Lenoir to locate the right facility, and was pleased to learn that Delambre was residing comfortably in a country manor. He invited the savant to Paris to plan a resumption of the expedition, and petitioned the Committee of Public Safety to re-engage both Delambre and Méchain to that end.[59]

Soon afterwards, under the impetus of Representative Prieur de la Côte-d'Or, the National Convention passed the law of 18 germinal III (7 April 1795). This law represented the final evolution of the metric system as we know it today. It provided the final set of prefixes and names that comprise the metric nomenclature. The new law also signalled some retreats from the principle of rationalization. Although the revolutionary calendar was preserved, the decimalization of the hours was abandoned, ostensibly because of the cost of replacing all the nation's clocks, and because the decimalization of time would help only astronomers, not ordinary citizens. Prieur also recognized that the transition to the new measures would have to proceed more 'gently'. To oversee the process he created a Temporary Agency of Weights and Measures under the leadership of the gifted mathematician Legendre. He also decided that the metre would be introduced in Paris first, and set a target date three months in the future so that merchants and customers could prepare themselves. The rest of the country would follow later.[60]

The new law also formally relaunched the meridian expedition. Prieur set aside his preference for a quick, cheap standard, and praised the 'rightly celebrated' savants he had purged from the project eighteen months before.[61] For the metre to become a truly international standard, he now noted, it had to be based on something grander than the fifty-year-old Cassini survey used to define the provisional metre. He urged Delambre and Méchain to recommence their survey 'as soon as possible'.[62] Any further delay was inimical to the public good. He even authorized the savants to approach him directly should they encounter the least obstacle. 'I will endeavour with assiduousness,' he wrote to Delambre, 'to prove to you my zeal for the success of your mission.'[63] Prieur had good reason to demonstrate his goodwill; he was himself under suspicion for having allied himself too closely with the Jacobins during the Terror. Delambre smiled inwardly at this switch. As he noted to Méchain, 'I would have accomplished a lot more last winter had I not been in the bad graces of Robespierre and one of his colleagues – whose name I'll tell you later – and who has since treated me more favourably.'[64]

Delambre had learned a thing or two about state-sponsored research in the intervening years. This time, before setting out on his expedition, he made a few requests. Even a savant could learn to calculate as other men did. 'It is true that until now the astronomers charged with this mission have been unstinting in their efforts and parsimonious in their expenditures, such as one might expect from impecunious savants who spare the Republic every expense as if the costs were their own. They have neither requested nor received any payment for their labour.'[65] Now, however, the expedition leaders deserved a salary like any other citizens employed by the state, plus back-pay for the twenty-one months of geodetic work done before the Academy had been shut. In May 1795, after he and Calon agreed on terms, Delambre rejoined the Commission of Weights and Measures.

On 28 June 1795 Delambre left Paris in his custom-built carriage after an eighteen-month hiatus, accompanied as before by his manservant Michel and the instrument-maker Bellet, plus a new assistant to keep the logbook. The team provisioned their carriage for an extended journey, including thirty pounds of axle grease, a set of ropes and pulleys to hoist the repeating circle into church towers, two crates of astronomical texts, plus tools for repairs: borax, copper, mercury, oil, nails and steel for screws and springs.[66]

Their first night south of the city, the team put up in the d'Assy country château where Delambre was always welcome. Two days later, they arrived in Orléans on the banks of the Loire, where Delambre had been forced to halt his operations eighteen months before. Three days later, they pulled into the cathedral town of Bourges, which was to serve as their base of operations while they worked their way back north towards Orléans. The team took lodgings in the Coeur de Boeuf, an inn just off the square where a liberty tree stood. By sighting from the inn to the nearby Bourges cathedral Delambre verified that the expedition team of 1740 had lodged in the same inn. Thus, trigonometry informed history. The meridian expedition was back in business.[67]

Bourges cathedral is a jewel of Gothic architecture. Above the front portal, the archangel Gabriel weighs the souls of the dead on Judgement Day. Inside, the stained glass scenes of redemption rise in blue panels like an interior heaven. The town had served as an entrepôt for Renaissance finance, the central town of France's Centre. But the sixteenth-century wars of religion had ravaged the region, and Huguenot zealots had decapitated the statues of the apostles in the cathedral. More recently, local revolutionaries had beheaded the copper effigies of the Duc and Duchesse de Berry and turned the building over to the worship of the Supreme Being 'who likewise rules from on high'.[68]

The cathedral tower dominated the countryside for thirty miles around and more. It was 396 steps up the hexagonal stairwell to the platform two hundred feet above the pavement, itself at the summit of the hilltop town. In one corner of the platform, a filigree iron belltower rose twenty feet further still; at its peak turned a weather-vane in the form of a pelican. The selfless metallic bird, a symbol of Christ's devotion, pierced its own breast for blood to feed its young. It would serve as Delambre's signal, the point he would observe from afar. Beyond the low balustrade the rolling plain of patchwork farmland, dotted with small towns, stretched into the vaporous distance. Out there was his destination, if he could ever leave Bourges.[69]

Delambre had left behind the noise and violence of Paris, yet even in this pastoral centre the retreating tide of Revolution threatened to suck him under. No sooner had he arrived in Bourges than the expedition stalled. The cost of travel had risen beyond his means. Inflation had gathered a frightening momentum since the fall of Robespierre. In the earliest days of the Revolution, the legislature had created a paper money called *assignats* to pay off the national debt (itself one of the principal causes of the Revolution), backing its value with the sale of confiscated church lands and the property of émigrés. This paper currency had always been treated with scepticism in the countryside, and the war had set off a first round of price increases. The Committee of Public Safety had tried to contain the escalation with wage and price controls, but the moderates had decided to lift the controls and print more money. The value of the *assignats* began to plummet with alarming speed. Delambre's expense book documents the accelerating price of food, lodgings and transportation as he travelled away from Paris. At each successive post, the cost of rented horses doubled and doubled again. The first stage out of Paris cost 92 *francs*; the last stage to arrive in Bourges a week later cost 804 *francs*. A few months later, the cost of a stage had doubled again to 1400 *francs*. As the price of ink, paper and basic foodstuffs soared, so did the cost of repairing church towers, erecting scaffolding and building observation stations. Even the gratuity for the stableboys had risen by a factor of ten. Within a few weeks Delambre had spent his entire budget for the campaign season. He pleaded with General Calon for more funds. Without hard currency, he would be stuck in Bourges for the entire summer.[70]

Nearly a month passed before Calon came through with the money. The Treasury would only release *assignats*, even though provincials would accept nothing but hard currency. To compensate, Calon raised the men's salary (reduced to a pittance by inflation), and awarded them military ranks: Delambre, Méchain and Tranchot became captains. This entitled them to food rations. The challenge was to get the peasants and innkeepers to accept the army's coupons.[71]

Money was not the only obstacle. Geography also presented a challenge. The mournful region between Orléans and Bourges – *triste* Sologne – was one of France's most level. Where the expedition had once battled suspicious peasants and northern fogs, they now confronted a marshy terrain almost impossible to survey. The green ponds, tall grasses and patchy forests offered few views into the distance. The rare church steeples were hard to make out through the mists: the steeple at Salbris, used by the surveyors of 1740, had been incinerated by a lightning strike; elsewhere, the Revolution had taken its toll. Delambre expressed his disappointment. 'The sans-culottes have destroyed half the steeples in the Bourges region for "daring insolently to rise above the height of their humble cottages"'. It took Delambre three trips up and down the region to select a chain of workable triangles.[72]

The villages of the Sologne are as isolated today as they were before the region's swamps were drained. Ponds collect seepage, ditches surround farms and ruler-straight roads run between twin rows of plane trees. The whine of an occasional motorbike only emphasizes the quiet. The churches are locked year round and are badly in need of repair. The area is depopulated, the number of priests dwindling. In the eighteenth century, the region was already known as the nation's Siberia. Sandy soil, barren heaths and thick swamps made even subsistence agriculture burdensome. 'There is no soil more unyielding and painful to labour in all the world,' said the locals. Peasants rarely owned their own land. The cattle were sickly, the sheep weak. The ponds bred 'Sologne fever' (probably malaria) which afflicted the villages each autumn. And on top of this, the people were beset by a plethora of taxes.[73]

Such a climate bred suspicion. Certain families were said to have the power to bring storms upon their enemies. Sorcerers gathered before dawn at ponds like Boisgibault to beat the waters with great sticks and shout horrible cries. 'It was enough for them to seek to change the weather and the blue sky would cloud over and the thunder rumble.' To clear away the fogs and evil vapours required holy incantation and the continual ringing of church bells, called the *dindon*. The curés had to stuff cotton in their ears to withstand the continual pealing.[74]

The church in Vouzon dates from the sixteenth century and the bell in the square tower used by Delambre still tolls the hours, although the rest of the edifice burned down in the 1880s. At Souesme the church tower has been rebuilt since Delambre's day, and is covered with scaffolding in anticipation of further repairs. The octagonal Sainte-Montaine belfry still stands, dwarfed by a venerable chestnut tree. Luckily for Delambre, when he visited the site in November the tree was bare of foliage and he could make out the surrounding signals between the branches.[75]

Where no steeple was available Delambre paid locals to build him an observation tower. At Oizon, he had a twenty-two-foot signal built in the form of a pyramid, which he covered with planks of wood painted white. At Ennorde, he had a twenty-four-foot pyramid built, which he then had transported to the middle of a hummocky wheat field. At Morogues, north-east of Bourges, he had a twenty-five-foot signal erected beside a road as it crested a small hillock. Delambre and Bellet climbed the scaffolding and perched themselves on bales of hay to sight the surrounding stations. At Méri-es-bois, just north of Bourges, an elderly villager led them to the ruins of the signal-post used in the survey of 1740, and assured Delambre that he remembered Cassini's passage through town fifty-five years earlier, but that some local lads had torn down the signal last year as a sign of 'feudalism'.[76] All this activity attracted unwanted attention.

> Even those with the best notion of who we were, took us to be
> prisoners of war being transported from place to place.
> Others, on seeing the crates for our circles, took us for
> charlatan salesmen and refused to give us lodging. That's
> what happened in Vouzon. At Souesme, we were also refused
> a place at the inn; but that was because they knew who we

were, and knew we only had *assignats* to pay them. Without
the help of the municipal council, who promised to
compensate with grain anyone who sold us bread, we would
not have been able to procure any food. Even so we passed
several days with nothing to eat but bread . . . Not only that,
but an epidemic was then sweeping through Vouzon, and one
of my collaborators fell sufficiently ill that we had to leave
him behind when we left the town for Chaumont.[77]

Triste Sologne took Delambre several months to measure, and
proved the least accurate sector of the entire survey because the
triangles could not be evenly spaced. By working well past the
optimal season for geodetic measurement, however, he managed
to complete the chain of triangles between Bourges and Orléans
by late November. He planned to use the winter – otherwise
impractical for geodesy – to conduct his astronomical measurements at Dunkerque and determine the latitude of the northern
extremity of the meridian arc, the counterpoint to the measurements Méchain had already conducted at Mont-Jouy.[78]
    So before leaving Orléans for Dunkerque Delambre composed
a letter to his distant colleague. This letter renewed their correspondence after a two-year silence. Delambre was pleased to be
back in touch with his colleague: he was eager to trade information about data, funding and personal matters too. There was one
thing in particular he wished to know. As Méchain had already
performed the latitude measurements at the southern end of the
arc in Mont-Jouy – and with such precision – might he advise
Delambre as to which stars he had observed, what methods he
had used to measure their height and what precautions he had
taken against the possibility of error? This information would
ensure that his own results could be most fruitfully compared
with the superb results Méchain had already achieved.[79]
    No innocent request could have touched so sensitive a nerve in
so sensitive a man as Pierre-François-André Méchain.

Chapter Six

FEAR OF FRANCE

Must it be ever thus – that the source of happiness must also
be the fountain of our misery? The rich and ardent feeling
which filled my heart with a love of Nature, overwhelmed me
with a torrent of delight, and brought all paradise before me,
has now become an insupportable torment – a demon which
perpetually pursues me.[1]

JOHANN WOLFGANG VON GOETHE,
The Sorrows of Young Werther

Upon leaving Barcelona harbour Méchain's precious repeating
circles had been blasted by a bolt of lightning, and no sooner had
he sailed within sight of neutral Genoa than an English frigate
blockaded his vessel and forced her eighty miles south to the port
of Livorno, where he and his fellow expedition members were
placed in quarantine while customs officials threatened to
impound his instruments.

Was ever a natural philosopher so beset by misfortune? Storms
of fate, both natural and man-made, had forced him from the
narrow meridian where his duty lay, and if the calamities of war
and the frailties of his body were not enough to drive any man,
even a natural philosopher, to hopeless despair, these multiple
woes were compounded by the knowledge of his own mistake,
greater because it was self-inflicted, greater still because it was
unrectified and greatest of all because it was secret. A stoic natural

philosopher might have coped with headstrong revolutionaries, capricious generals, mischievous nature and violent machinery; but the error for which he blamed himself was more than enough to plunge a sentimental natural philosopher – such as he – into a most profound melancholy.

Méchain was a stranger in Livorno, without acquaintances there. He could expect no help from his wife and colleagues in distant Paris; any letter to them would have to travel over the Alps, across battle lines, and through the thick of revolutionary chaos. So, from the local lazaret – the compound where goods and visitors waited out their obligatory ten-day quarantine – he wrote to the director of the astronomical observatory in the nearby university town of Pisa, ten miles north in Tuscany.

Méchain could claim no prior acquaintance with Giuseppe Slop de Cadenburg. But the Pisan had long been a regular correspondent of Méchain's *maître* Lalande and had also assisted the French navy with its Mediterranean maps, Méchain's task for much of the past twenty-two years. They also shared a fascination with comets. So although they had neither met nor corresponded, the two men knew one another by reputation, always a savant's most precious possession. As a scientific colleague, Méchain importuned Monsieur Slop to use his credit with the local authorities to get his instruments through customs. In return, he would visit Pisa upon his release to express his gratitude in person, and also to demonstrate the new astronomical repeating circle, should the Pisan wish to examine the most advanced scientific instrument of the day. The device was not contraband, and as his mission served the general good, it ought to be protected by all nations. 'When war divides people,' Méchain wrote to Slop, 'science and the love of the arts must reunite them.'[2]

Such an eloquent plea and collegial connection was sufficient to unite two astronomers wherever they met. Slop sent an associate in Livorno to assist 'the famous Méchain', introducing him to top officials and negotiating the various payments – which some might call bribes – to clear his instruments through customs. A grateful Méchain arrived at the astronomer's house in Pisa on

22 June 1794, the day of the summer solstice. He stayed for three weeks.[3]

During this respite Méchain poured out his overburdened soul. Here at last was a sympathetic ear, as well as a knowledgeable astronomer. Méchain called Slop a kind of father, though he was only four years older. Virtuous, honourable, generous, good, venerable, worthy, with a noble simplicity of heart and a bounty of wise counsel . . . there was no end to the virtues Méchain ascribed to Slop, who practised astronomy in Galileo's former haunts. Slop had married a lively woman of English descent, Elizabeth Dodsworth, and they had raised their three children in an atmosphere of tolerance and free-thinking. Francesco, the eldest, had been arrested the year before for dabbling in revolutionary politics. Now he was back in Pisa, supposedly working on astronomy.[4]

His stay with Slop's family reminded Méchain how much he missed his own virtuous wife and children. Yet this recollection was a bitter one, for in the same breath he remembered how he had abandoned them to all the dangers of Revolutionary Paris. The contrast, he confessed to Slop, was painful. 'I will tell my wife and our children that if they desire happiness they should pray that heaven aid my efforts to resemble a husband and father as respectable as you, so tenderly dear to his own family and to all those who have the advantage of knowing him.'[5]

Méchain told Slop everything: his honourable motives for undertaking the mission, his unexpected misadventures along the way, his modest triumphs thus far, his growing frustrations with his assistants, even his tormenting doubts about the latitude measurements of Mont-Jouy and Barcelona. As a fellow astronomer, Slop understood the niceties of celestial observation – and the myriad ways that science might go wrong – yet he would not judge the results with the same critical eye as Méchain's French colleagues, so concerned for the exactitude of the grand mission upon which all their reputations, careers and perhaps even their lives depended. It was safe to confide in Slop. 'You are the only one to whom I can speak so intimately,' Méchain would later write, 'my only friend, the most worthy, the most virtuous, the most respectable.'[6] To be on the safe side, he also swore Slop to secrecy, and later asked him to burn the wrenching letters in which 'I opened my heart to you, as to a father'.[7] It helped too that Méchain could be confident these letters would not be opened by the agents of the French Revolutionary state, nor find their way across the Alps to his colleagues in Paris. When read through the inevitable fog of self-deception, these letters offer a remarkable window into a man consumed by scientific doubt. In the days and weeks to come, he would confess his error at Mont-Jouy to Slop and to no one else.

Méchain's immediate goal remained Genoa. He considered making the voyage by boat, but ultimately set out instead with his team on the overland route, a two-day ride by saddle horse. The coastal road skirted the edge of mountains suspended above the sparkling Mediterranean. The terrain was rough, and the instruments travelled separately by coach, arriving undamaged in Genoa just one day after Méchain, Tranchot and Esteveny rode into town on 11 July. That very evening, Méchain dined with the French ambassador and dispatched letters to his colleagues in Paris to ask for instructions, to Slop to tell him of his safe arrival, and to his wife to inform her of his whereabouts. He told them he planned to return to France as soon as possible and that everyone had assured him that the mail-boat to Nice was perfectly safe. Given his recent mishap at sea, however, he preferred to wait for official instructions. He had not heard from Paris in a year. He knew nothing of the status of the meridian project, of his own status, or of that of his family. No one back home even knew he had left Spain. Letters took at least a week to travel between Paris and Genoa and sometimes much longer; many never survived the trip at all, producing that tangle of unreliable and delayed knowledge which is the stuff of tragedy or farce, depending on the circumstances – or a gnawing anxiety, if one is that way inclined. Anticipating a delay, Méchain went down to the customs house to extract clothing and linen from his trunks, and settled into the Albergo del Leon d'Oro (the Inn of the Golden Lion).[8]

Genoa had once disputed dominion of the Mediterranean with

Venice, and its vessels and bankers still traded from Gibraltar to the Levant. The spectacular semi-circular harbour, with berths for scores of the largest vessels, formed an amphitheatre at the foot of the Apennines. On a rocky promontory at one end of the harbour rose the spindly Lanterna lighthouse, like a Renaissance minaret, four hundred feet tall. Most of the well-fortified town of one hundred thousand lived clustered near the harbour under blue-slate roofs that mirrored the colour of the sea. Strewn on the steep hills behind the town were sumptuous palaces and terraced gardens of orange trees. The prosperous city boasted an exquisite opera house and theatre, plus plenty of street pageantry, from the annual San Giovanni procession to appease the floods of spring to the nightly promenade of elegant men and women along the ramparts. The nobles dressed all in black, with a short cape and no sword, and ruled the patrician republic as a tightly knit oligarchy. For the past century, however, the republic had been plagued by internal bickering, and French and Austrian armies had occupied the city in turn, obliging the patricians to trade the troublesome colony of Corsica to the French in order to preserve Genoa's independence – this, two years before Napoleon's birth on the island in 1769. Genoa still prided itself on its autonomy, and maintained a guarded neutrality as the French Revolution unleashed warfare across Europe. This neutrality pleased no one, of course. For the past year the English navy had intermittently blockaded the city. For his part, the French ambassador had been trying to incite the lesser nobles to challenge the ruling oligarchy, luring disaffected artisans to the French cause and printing Jacobin propaganda in his basement. Meanwhile, not far to the north-west of Genoa, the French army had attacked the Austrian and Piedmontese armies, according to the plan of a young general named Bonaparte.[9]

Indeed, three days after Méchain reached Genoa, Napoleon Bonaparte arrived there too. He had come on a diplomatic mission to urge the Genovese to ally themselves with France – or else. He stayed a week, covertly casing the town's fortifications and assessing Italian politics. The upstart national republic was threatening to supplant the venerable city republic.[10]

Genoa harbour

'Veduta di Genova', by Ippolito Caffi.

(Photo: Galleria Internazionale d'Arte Moderna di Ca' Pesaro, Venice.)

While Méchain and his team waited for officials in Paris to decide their fate, the mail-boat only brought word of fateful upheavals in the capital. On 9 August – ten days after the event – report of Robespierre's fall reached Genoa. For the moment, Paris was calm. Even this calm, however, could be variously interpreted. Tranchot hoped the radical cause would be regenerated now that Robespierre had been deposed. The engineer wrote to Francesco Slop, Guiseppe's radical son, that despite the suppression of the Jacobins, 'the same political order rules [in Paris] as it did before all these heads fell . . . once again bringing despair to the aristocrats here who carry their heads so high, but who will soon resume their usual bowing and scraping'.[11] All across Europe, Tranchot boasted, French armies continued to advance. 'Bois-le-Duc is ours,' he noted, 'as is Düsseldorf with its suburbs incinerated . . . also the fortress of Coblenz, where the town will be razed the moment it falls into our power.' Closer to Genoa, the French army was bombarding Cúneo just eighty miles to the west.[12]

Tranchot did more than sympathize with the radicals. During his stay in Genoa, he operated in concert with the French ambassador and cultivated young Francesco Slop for the French cause. The two men shared the hope that the Revolution would come to Italy. Tranchot advised his young recruit to bide his time.

I feel, as you do, the disgust of living out my life in a world for which you are not born. But you know as well as I that our revolution is not yet at that point at which a man should abandon the wellbeing he enjoys in his native country. That happy moment will come, however, and we may even hope that it will not be far off, for the basis for the new government seems to be gradually settling into a form that will inspire respect.[13]

Beneath this faith in the Revolution's promise lurked a deeper pain. Tranchot had reason to believe that much of his family had been imprisoned or killed in the Revolutionary violence of his native Lorraine, a fact that must never be mentioned, Méchain told Slop, because his subordinate would 'think it ill-intentioned of me to have let it be known'.[14] Relations between the two men had deteriorated since their departure from Spain. Tranchot's health was as robust as ever, and he was impatient for Méchain to order the expedition back into action. It was tedious to wait for official instructions from Paris. He had had enough of Méchain's fastidious perfectionism, his endless second-guessing, and his refusal to allow anyone else to conduct observations, perform calculations or even look in the expedition logbook.[15]

No word came from Paris that entire summer. Every Saturday the mail-boat arrived with only the public newspapers in its satchel. As the weeks passed, Méchain grew anxious. 'For a long time now I have been destined to languish in a state of uncertainty, destined to be tormented by the cruellest worries as to the fate of my family and the fate which awaits me.'[16] These fears took many shapes. Had the meridian expedition been cancelled? Might a jealous rival have denounced him to the Committee of Public Safety? It would not be difficult to make him appear culpable in their eyes: his long absence from France, his sojourn in countries hostile to the republic, rumours that he had been offered a stipend and a position by the Spanish crown, perhaps even whispers that his results were imperfect. No wonder he secretly asked Slop whether he might, if the worst came to the worst, seek refuge in some 'obscure corner' of Pisa while he searched out a more permanent home for himself in exile.[17]

In mid-August, he finally heard from his wife. She informed him that the meridian mission had been suspended 'at least until spring', and that only the Committee of Public Safety had the authority to revive it. On the one hand, nothing now prevented him from returning home. On the other, he feared she was concealing even worse news from him. 'Perhaps she only embellishes the state of affairs to calm me down for a few moments.'[18] He poured out his doubts to Slop, professing his innocence, swearing his love for his family and cursing his wretched luck.

> But forgive me; when I write to you I feel myself to be by your side, for it is the only moment when calm reigns in my heart. Oh, you know all too well what so cruelly torments and afflicts me. A virtuous, tender and sensitive soul like yours feels all too well why I tremble as I approach my family, whereas under any other circumstances I would fly to them with transports of joy. It is the ardent desire to see them again which has made me break all my ties with them and separate myself for an indefinite period. Might not my return bring them new and more fearsome alarms; might I not trouble the little peace they now enjoy? How many frightening incidents and events will my return not present? May God not visit upon my family any of the harm which I might cause them.[19]

He bared his soul to Slop. 'You are now reading, Monsieur, into the very depths of my heart. You see there the sources of that fear which so strongly agitated me when I was with you.'[20] He doubted whether he was worthy of the friendship and counsel of so benevolent a man as Slop. He begged indulgence for his weakness. His health, he noted, had been deteriorating rapidly these past several months.

Méchain's spirits were not improved by the greetings sent by his colleagues on the Commission of Weights and Measures. At the end of August they finally sent him a copy of the law of 1 August 1793, which had established the metric system and which set the metre provisionally at 443.44 *lignes*. Moreover, they informed him that Delambre had been purged from the Commission, with no one assigned to take his place. Méchain drew his own conclusion: the meridian expedition, he decided, had been 'definitively abandoned'. In which case, he had to wonder what purpose had been served by a mission 'for which I have so tormented myself'. The little good he had hoped to accomplish had turned to ashes in his mouth, and his agony over the latitude of Mont-Jouy had become a bitter farce. 'For it matters little now whether the latitude and longitude of Mont-Jouy be a quarter-minute too great or too little.'[21] The new system of measures had already been established on the basis of the old measures of the meridian.

> All of which throws my spirit into the most extreme disgust. The ambition to be useful and win a little glory, which once animated me, turns out to have been a vain fantasy. What interest will my feeble labours inspire in the Commission or in the government when they no longer serve any useful purpose? Why then should I remain zealous to continue this mission? My attachment to my family? Oh, why have they set me on this course which has led me God knows where? Oh, if only I had chosen another path, the one which presented itself to me most naturally, I would today be at peace and sheltered from all reproach or suspicion. My work would be interesting to both [my colleagues and the government], and all would pity me. But I wanted to get the best out of the project, the best for my colleagues, and for myself, and it may be that with the purest of motives I will have brought about the unhappiness of my family and myself without having contributed in the least to the success of the mission, and

without any hope of recognition. Well, enough of this; the more I think on it, the less bright the future seems. My lot is cast, I wait upon the event, and console myself with my conscience, secure in the knowledge that my actions have always flowed from the best of intentions.[22]

Yet the apparent cancellation of the mission, while it made all his labours meaningless, also brought a sense of relief: his measurements no longer mattered. If his efforts were to be buried in an unmarked grave, so too would their flaws. Indeed, the cancellation of the mission prompted Méchain to calculate how little difference his Catalan extension would have made to the estimate of the provisional metre. Now that the new standard had been set – albeit provisionally – what difference would his small addition make? 'As you can see I add little, and any errors I committed in my small sector cannot have much influence, whether they cancel each other out or whether they are enormous; so that I am now trying to contain and calm the deadly disgust which is killing me.'[23]

Then, the very next week, came the good news that returned him to the depths of despair, plus some truly black news that nearly finished him off. His supposition had been premature: the meridian project was being revived. He heard it from his wife and from Lalande. General Calon had been placed at the head of the department of military topography, with Méchain appointed chief of naval cartography at an annual salary of 6000 *livres*, two months' of which had already been paid to his wife. Tranchot had received a subordinate position in the new organization. Official news of this appointment arrived from Calon himself two weeks later, with this request: that Méchain return immediately to Paris so that they might discuss the future of the meridian project together.[24]

As a practical matter, this promotion could only improve his family's condition in the capital – where, his wife informed him, their provisions were running low – and it clearly gratified Méchain's sense of self-worth. Yet he was not sure that a 33 per cent pay hike would compensate for his added responsibilities. In his former humble job, he had only had to answer for the exactitude of his own work. Did he really wish to assume responsibility for the results of an entire department, conducted by men whose exactitude he could not vouch for? And how could he be sure this arrangement would outlast the next topsy-turvy twist of the Revolution? On top of this, Calon was ordering him to return to Paris, where he would surely have to turn over his data. 'Oh, how well I see and feel why each man trembles for his own fate and for the interests which are dear to him, and why no one dares act.'[25]

Worse, the revival of the meridian project meant that his mistake mattered again, that the discrepancy in the Mont-Jouy data was once more an affront to the accuracy of the metre, 'the most important mission with which man has ever been charged'. So many reversals seemed to have upended Méchain's mind. He admitted to Slop that he could barely think straight any more. 'In my last letter to you I importuned you with my guessed-at results, and you saw the dying efforts of a downed combatant who still fights for a victory and a success which long ago eluded him. In picking myself up, shamefaced and dispirited, I can barely remember what I was running from or what I said.'[26]

Most unsettling of all was the news that accompanied the announcement of the meridian project's revival. Lalande confirmed that Lavoisier, Condorcet and several other colleagues had been guillotined. Worse, Lalande informed him that the Terror had struck even closer to home – in the grounds of the Observatory, where his family still lived. Alexandre Ruelle, the young apprentice astronomer whom Méchain had taken under his wing, sheltered from the police and trained for eight years, had denounced Cassini to the Revolutionary police, condemning his former protector to prison, and then capped his hideous betrayal by demanding that Cassini, Lalande and Méchain be sent before the Revolutionary Tribunal and thence, inevitably, to the guillotine. Had Méchain been living in Paris at the time, the consequences would have been too awful to contemplate. At least Ruelle himself was now in prison – for his foul betrayal, for backing the radical party out of favour and also for having committed a scientific error. Yet even after telling Méchain all this, Lalande still expected him to return to France that month.[27]

Méchain also heard from other voices, friends who had sought refuge from the Revolution overseas. They urged him to join them in exile, to give up all hope of rescuing his wife and children, to cease to serve a country which had bloodied itself so horrifically.

[My friends] tell me I should count myself lucky if the Committee of Public Safety abandons me. They say that if the Committee spares my life, yet strips me of all my resources, exposing me to a thousand perils, that my wife, children and I will come to seek death as the way out of our misfortunes. They tell me that my return to Paris would do my family more harm than good. And they urge me, with great ardour, to follow them into exile. But my family, my duty and my honour also call me, and I have always hearkened to their voices. Why should I repudiate them now . . .? Why do such dire auguries traverse the seas to find me? Am I really so guilty?[28]

The native son of verdant Picardy, dazed by the light of the summer Riviera, sat by the writing desk in his darkened hotel room and considered his dilemma from every angle. Emigration was a capital offence. Even the rumour that he was contemplating emigration might cost him his position, the livelihood his family depended on, and any hope of ever returning to France. Yet how could he return there now? He had been sent out on his mission with the best technology in the world. He had made himself personally responsible for the observations. And he had performed the calculations according to the most reliable methods. Yet the results were inconsistent. Méchain could not bring himself to blame the instruments or the calculations. He blamed himself. And if he blamed himself, then surely others would do so too. In a rare display of understatement he informed Slop: 'I have been through some tough moments since I left you [in Pisa], but it is all my fault. I have placed myself in the hands of chance when I should have stuck to the path of certainty. One must suffer and

not complain, or risk suffering still more.' He told himself to put
the past behind him and think only of the present and future. Yet
he could not help but worry and he could not help but complain.[29]

In early October the French ambassador to Genoa was recalled to
Paris to account for his Jacobin sympathies, and his replacement,
Ambassador Villars, arrived with funds and passports to speed the
team's return. The time had come to leave Genoa. Tranchot went
down to the warehouse to pack the instruments for their voyage.
Calon asked that the instruments travel across the Alps by mule,
for safety's sake.[30]

Méchain publicly announced his intention to return. Indeed, he
wrote to the prominent Milanese astronomer Barbera Oriani
(whom Méchain had met a few years before in Paris), to tell him
he would be leaving Genoa 'on the thirteenth or perhaps the fif-
teenth of this month'.[31] He asked Oriani to supply him with his
most recent observations to carry across the frontier to Lalande.
For his part, Méchain appended a gift-offering, one of those gifts
by which one savant elicits the free concession of scientific information from another: a summary of his own astronomical findings
in Catalonia, including various eclipses, various stars and a summary of the latitude data for Barcelona.

Oriani's response was to rush down from Milan to visit
Méchain before he left Italy. In honour of this visit, the
Frenchman retrieved one of his repeating circles from the depot
(even as Tranchot was packing their trunks) to demonstrate its
marvellous capacities. They set up the apparatus on the terrace of
the Hôtel du Grand Cerf – 'one of the best hotels in town, mag-
nificently situated right across from the sea' – and together took
latitude measurements through the mid-October nights.[32]

Oriani was smitten with the circle, and when his colleagues in
Milan learned of its capabilities they were eager to procure one of
these marvels for themselves. The Milanese had been conducting
their own geodetic survey, and Oriani suggested that he and
Méchain connect the French and Italian grids via Genoa. 'It is
desirable that so beautiful an enterprise be undertaken.'[33]

Méchain wrote to Calon for his approval. To Méchain's surprise Calon not only seconded the project, he even promised the Milanese a repeating circle of their own as soon as Lenoir could fashion one. The project would, after all, link the maps of France and Italy just as the French army was closing in on Turin to the immediate north. But there was to be no reprieve for Méchain; Calon still insisted that he report to Paris. 'For you are not destined to return to Italy, but to extend the measure of the meridian in concert with Delambre.'[34]

So Méchain concocted another plan to justify his sojourn in Genoa. (Never underestimate the ability of a scientist to generate new and interesting scientific questions when the need arises.) As Méchain pointed out, Genoa was near the 45th parallel, halfway between the equator and the pole. Bordeaux was not the only site suitable for a pendulum experiment to determine the length of the metre. If Calon would just send him the Observatory's platinum-bob pendulum, Méchain would save the Commission the trouble of transporting a scientific team to Bordeaux. Or, alternatively, he might conduct observations at Genoa to supply new refraction corrections (and secretly resolve the discrepancy in his own Barcelona data).[35]

In the end, the new French ambassador rescued Méchain by taking matters out of his hands. A month into his job, Ambassador Villars had become familiar with Méchain's vacillating ways, and Méchain had come to consider Villars 'a friend of the arts and sciences'. The ambassador certainly proved a good friend to Méchain.[36] Villars was savvy in the ways of officialdom. He advised Méchain to request further instructions from Paris – no one could ever fault a public servant for seeking to clarify his orders – and in the interim he would refuse to issue Méchain a passport, thereby taking responsibility for the delay upon his own head. All Méchain had to do was to write the request. On Monday morning, Villars told him that he expected the official request by two that afternoon, before the mail-boat set out. When Méchain arrived at the post office at half past three – 'I was still hesitating,' he admitted to Slop – the outgoing mail was already in the satchel and Villars and the carrier were waiting impatiently. Villars snatched the letter out of Méchain's hand, and stuffed it in the satchel. 'Your shillyshallying is futile,' he told Méchain. 'Your worries have no basis. Calm yourself down and spend the rest of the afternoon with me.'[37]

Villars' manoeuvre worked. Tranchot stopped packing, and Méchain spent the rest of the winter on the Italian Riviera. 'Our ambassador here,' he informed Oriani, 'has required me to defer my departure.'[38] In the meantime, he even took in some theatre. While his travelling companions favoured melodrama, Méchain preferred the classical theatre, in which 'order and tranquillity reign, virtue is honoured, and the happiness of all is every day more assured'.[39] Some days, he even felt a degree of calm in his heart, only to suffer horrible presentiments of ruin at night. Folded inside a letter to Slop, he slipped a note labelled 'For your eyes only'. 'I do not believe I have given any cause for reproach,' he wrote, 'but in the present circumstances, who can be sure he is safe from reproach, envy, enmity, and jealousy?'[40] He had heard rumours that his enemies in Paris were conspiring to thwart his plans – although his wife seemed to be shielding him from these intrigues. 'You see that I am well informed about events six hundred miles away and that my fears are not all chimeras.'[41]

Today we have our own clinical terms for such psychological states. Méchain was depressive, we would say; he was paranoid, obsessive, passive-aggressive. No doubt this is true. Yet even feelings have their history, and Méchain was a man of the eighteenth century, a man who suffered from the exquisite malady known as melancholy. Melancholy was a complex affliction fuelled by a disequilibrium of body and mind. It preyed on solitude and could encompass many moods. A voluptuous melancholy delighted in tombs and barren landscapes; nothing could be more delightful for an elegiac poet. A bitter misanthropic melancholy, such as beset Candide when he encountered the world's cruelty, extinguished all hope in the future. And an anguished melancholy, oppressed by regrets too great to bear, might drive the mind to madness, or suicide.

Méchain exhibited all the melancholic symptoms featured in the nosology of Dr Phillipe Pinel, director at the Bicêtre asylum and Méchain's colleague at the Academy: taciturnity, gloomy suspicion, monomania and love of solitude. But if Méchain suffered from paralysing doubt, it was because he faced an awful dilemma. If he trusted no one, it was because he did not trust himself. If he worried about conspiracies, it was because he himself was keeping a secret.[42]

That winter, Méchain did not conduct the pendulum experiment. Nor did he measure geodetic triangles with the Milanese. A nameless colleague back in Paris was preventing Calon from sending him the pendulum, or so Méchain believed. As for the triangles linking Italy and France, Lenoir had yet to fashion another circle. At least Méchain had a neat solution for the latter problem. He asked Calon if he might sell one of his own circles to Oriani and take Lenoir's new circle when it was ready. This would equip his team with hard currency and obviate the need to transport the circle home. To his surprise Calon again approved, and a week later Méchain was able to offer Oriani his choice of repeating circles: the one ruled in the traditional 360° scale, or the one ruled in the new 400° scale. Oriani chose the 360° circle and the two savants settled, somewhat awkwardly, on a price of 1200 *livres*.[43]

Méchain did conduct a few astronomical observations that winter from the belltower of San Lorenzo cathedral, and at the famous Lanterna. One purpose of these observations was to test his conjecture about the refraction correction. The results were inconclusive. He also learned that Oriani intended to publish his Barcelona results, despite Méchain's warning that they did not seem to follow the usual rule of refraction.[44]

In late December Esteveny, the instrument-maker, decided to return to his family and business in Paris via the mail-boat to Nice. The voyage was a month-long fiasco. A sudden storm forced all the passengers to toss their belongings into the sea, and Esteveny arrived in France with only the shirt on his back. Then, as he disembarked, the French authorities arrested him as a returning émigré. Freed on the say-so of the captain, he was hampered at every stage of his overland trip by a lack of hard currency. When reports of his unhappy voyage filtered back to Méchain, it only confirmed the latter in his caution. Just imagine what would have happened had Méchain entrusted Esteveny with his precious instruments or, worse, with irreplaceable data![45]

Méchain still had Tranchot, though the two men had grown increasingly estranged. They no longer lodged at the same inn, nor spent time in one another's company. Both men came from the lower strata of the *ancien régime*, yet neither had found his place in the new. Tranchot was a military engineer, a man of action, robust and confident. He was a bachelor who had yet to draw upon his expedition salary. He wrote short frank letters in a neat square hand. Yet he aspired to the trappings of polite society. During his twenty years in Corsica, he had amassed a small collection of minerals and fossils, and during his sojourn in Italy he added petrified fish, feldspar and crystals 'very sought after by connoisseurs in Paris'.[46]

Each man operated according to his own exacting conscience. Tranchot had his military patriotism, and Méchain his passionate rectitude. Perhaps this was why they found one another so exasperating. Tranchot considered Méchain's delays a dereliction of duty, and his vacillation pathetic. When Méchain announced for the umpteenth time that they would soon be leaving Genoa, the engineer wryly commented, 'According to Monsieur Méchain we will be leaving after Easter, but I don't yet know if that is really true.'[47]

For his part, Méchain swore he bore no enmity towards his adjutant, despite his suspicion that Tranchot was plotting to usurp his leadership. Tranchot's impatience was a continual reproach, his competence a continual challenge.

[G]iven what I have seen since our departure from Barcelona, I would be deluding myself to count any longer on his friendship, affection or trust. Since arriving here he has expressed himself all too clearly, and I am not so stupid as to expect any resumption of his previous feelings. But honesty and probity have always been my rules of conduct with him, as with others.[48]

Yet Méchain had to wonder whether Tranchot knew his secret. And whether he might give it away.

As spring approached, and the season for geodetic campaigning drew near, Calon tried again to lure Méchain back to France. He tried to make the prospect appealing to Méchain's honour. Many of France's most illustrious savants had signed on to his new geographic enterprise, he noted: Lalande, Delambre, Laplace, all his old colleagues. 'You will not find yourself a stranger in the presence of those who share your labour,' he said – not realizing, of course, that these were the very people Méchain feared would unmask his error.[49]

Not until the law of 18 germinal III (7 April 1795) formally revived the meridian survey and restored Delambre to the mission did Méchain pack in earnest to leave Genoa – though even at this last minute he claimed intrigues in the capital jeopardized the expedition. Again he began making excuses not to return. This time, however, he could no longer postpone his departure. To fail to leave now would cost him his position and his family's sole means of support. In his last letter to Slop he even summoned a stoic courage: 'But the die is cast, and I will attempt the adventure.'[50] In late April, he boarded the mail-boat for Marseille with his one remaining circle and his one remaining assistant.

Even as Méchain left Genoa, the Revolutionary republic moved against the patrician city. Tranchot had been tracking the war's progress. In November, he boasted of French victories which would oblige the enemy to recognize the sovereignty of the French people. Nearer to Genoa, plans were afoot to land twenty thousand French soldiers behind Austrian lines. In mid-March, Tranchot climbed the hills above town to watch the French fleet engage the English navy. The guns sounded from four in the morning until three-thirty in the afternoon, and Tranchot initially reported a French victory, one which he hoped would oblige the Genovese patricians to abandon their cowardly neutrality. Then, when he discovered that the battle had actually forced the French to retreat to Toulon, he dismissed this as yet another betrayal of

France from within. Oh when, Tranchot wondered, would France find a hero to right its reversals?[51]

Within a year Bonaparte led his armies across Italy in one of the most dramatic military campaigns of modern history. In April 1796 his army occupied Genoa, the city he had judged ripe for the picking during his visit of 1794. In May 1796 Milan opened its gates to the conqueror, only to have the city pillaged by French troops – though Napoleon assured the astronomer Oriani that men of science would gain from his conquest: 'All men of genius, all those who have achieved distinction in the Republic of Science, are French, no matter what their native land.' In June 1796 French troops occupied Livorno and Pisa, where Slop's son went to work as an agent of the French republic.[52]

When Méchain and Tranchot arrived in Marseille by mail-boat in the spring of 1795, Méchain's colleagues had every reason to think he would proceed immediately to Paris to consult with them before resuming his mission. Paris was only a week's ride north by carriage. Or perhaps he would prefer to head directly to Perpignan to pick up his triangulations where he had left off. Perpignan was only a few days' ride to the west. Yet Méchain holed up at Marseille for the next five months. During those summer months, the best season for geodetic observations, Delambre measured the entire sector from Orléans to Bourges.[53]

Méchain's dithering infuriated Tranchot, exasperated Calon and baffled his colleagues. To prompt him to resume the mission, Calon sent Esteveny down to rejoin him in Marseille, as well as two new assistants to replace Tranchot. The French army was desperate for surveyors, and Tranchot was one of the nation's most experienced military cartographers. Calon wanted him to triangulate the mountainous frontier along the Swiss-Italian border where Napoleon was preparing his invasion.

Privately, Méchain admitted that Tranchot no longer wished to serve under him. 'I know he does not want to be under my direct authority, and may God grant his wish!'[54] Yet he did not want to forgo Tranchot's services either. Nor can he have been eager to have his assistant – the one person on earth who might have guessed his secret – far out of sight. So when his new assistants arrived in Marseille, he refused to release the engineer. He wrote to Lalande that Calon wanted to deprive him of his most capable aide. 'We got on well together while on mission; he assisted me marvellously in my astronomical observations.'[55] He wrote to Delambre that without Tranchot his progress would be slow. Admittedly, a second observer might seem superfluous now that he had only one repeating circle. But Calon himself had approved the sale of the other circle to the Milanese, and it was not Méchain's fault that Lenoir had yet to supply a replacement. In mid-August, Calon relented. 'I must at last defer to your wish that I not separate you and Tranchot. You may use him as you judge most convenient.'[56]

This suited Méchain. With Tranchot under his direct authority, and with only one circle available, he controlled both the observations and the data.

Tranchot, however, could hardly have been more disappointed. Instead of commanding his own survey, he was once again under the authority of the vacillating, exasperating savant. He obeyed Calon's orders – it was wartime and he was an officer – but he insisted that the expedition get underway. He requested permission to rent a coach and driver at 10 *francs* a day to carry the team overland to Perpignan. Yet, after approving this plan, Méchain balked at boarding the carriage and instead booked a passage on a naval vessel travelling to the port of Sette. (Four years later, their expedition done, the driver of the Marseille coach turned up in Paris with the signed contract to collect his 15,000 *francs* in accumulated daily fees, and Delambre was obliged to bargain his bill down to the cost of a new coach instead.) In late August 1795, Méchain and his team sailed from Marseille to the port of Sette. From Sette the team was rowed in longboats on to the beach at Canet. And from there soldiers were commandeered to haul their equipment four miles across the flatlands to Perpignan. After a hiatus of two and a half years, Méchain had at last resumed the measurement of the meridian.[57]

Chapter Seven

CONVERGENCE

And every Space that a Man views around his dwelling-place
Standing on his own roof, or in his garden on a mount
Of twenty-five cubits in height, such space is his Universe:
And on its verge the Sun rises & sets, the Clouds bow
To meet the flat Earth & the Sea in such an order'd Space:
The Starry heavens reach no further but here bend and set
On all sides & the two Poles turn on their valves of gold;
And if he move his dwelling-place, his heavens also move
Where'er he goes, & all his neighbourhood bewail his loss.
Such are the Spaces called Earth & such its dimension.[1]

WILLIAM BLAKE, *Milton: Book the First*

When Delambre and Méchain resumed their mission in the summer of 1795, the red-stone cathedral town of Rodez, where they had promised to meet, lay roughly halfway between them, some two hundred miles apart. Delambre, who had come further, still had the slightly longer route to run. He had put the northern lowlands behind him, though to the extreme north the latitude at Dunkerque remained unmeasured. To the south, Méchain had come through Catalonia, though he had yet to join the Pyrénées peaks to the Cassini triangles of France. As the two savants turned at last to face each other, they looked across *la France profonde*, the ancient provinces of Auvergne, Rouergue, Languedoc and Roussillon, a mountainous high plateau dominated by the Massif

Central, a series of ranges studded with domed volcanoes and laced with icy rivers.

The Revolution had carved these lands into quasi-geometric administrative *départements* and uprooted the *ancien régime*'s aristocratic and religious rulers, yet the great geographic centre of France still lived by the ancient rhythms: subsistence farming, up-country summer pastures and market days in the villages. From this high plateau, Paris was a remote metropolitan rumour. The two centres – the political and the geographical – were joined in a perpetual struggle to define France. The political centre, with its universalistic pretensions, strained after the linear progress of science and empire; while the geographical centre, with its pride in its particularity, laboured just to get by. The spokesmen of the political centre condescended to the geographical centre and urged the rustics to emulate them; while the inhabitants of the geographical centre, apart from a few official emissaries, did their best to ignore Paris and its fantastical plans – such as the absurd proposal for a new metric system and the impossible mission to measure the world. From their point of view, Delambre and Méchain were just a new kind of emissary sent from the political centre to traverse the geographical centre and, in the name of number, tame it.

Before resuming his mission, however, Méchain wanted to coordinate with his northern colleague. He wrote to Delambre to enquire how his colleague kept his logbook. Did he record his data in order of observation, or in an order fit for calculation? Did he record every observation, or just the summary values? Did he group the data by station, or by triangle? How many times did he observe each angle? How did he construct his signals? 'I ask you all these questions,' Méchain explained, 'so that I may follow the same sequence as you, and so that we may present our results uniformly. You have already measured a great many triangles; I have completed far fewer. So it will not cost me as much to redo all my logbooks in the manner you have chosen.'[2] And while he was at it, might he know how Delambre handled expenses? Did he pay for his assistants? Prices were rising in the south of France; the government's *assignats* had lost almost all their value.

The world's pre-eminent scientific nation had equipped a mission to measure the size of the earth with the greatest precision in human history; yet the expedition leaders, two of the world's most meticulous astronomers, had not coordinated their methods of recording data. The savants of the eighteenth century might wish to subject the world to the dominion of uniformity, but they were wary of placing themselves under the same rule.

By the time Delambre received this letter (via Paris), he was triangulating his way through the autumn marshes of Sologne. He gladly detailed his methods for his colleague. He always recorded all his observations in the exact order he took them, in a logbook kept in ink with each page numbered. Only then did his assistant copy the data into another notebook in a sequence more convenient for calculation. He always noted who performed the observation, the instrument he used, as well as the time, the weather and any other relevant circumstances, including a hand-drawn sketch of the site, with all its features labelled. He did so with the conviction that, as an emissary of the state, he was leading a mission of national importance. 'The Commission will ultimately decide what to publish; in the meantime, I suppress nothing.'[3]

That went for his expenses too, especially now that the cost of food, lodging and transport had reached astronomical levels. His assistants' salaries were insufficient to put food in their bellies, and Delambre paid all their costs out of general funds. So far the administration had refused to reimburse his receipts, and his own salary was several months in arrears. But Calon had lately promised to compensate him. Méchain could likewise rely on a sympathetic hearing if he encountered any problems. In return, Delambre had one question of his own. As he was about to travel to Dunkerque to determine the northern latitude, might Méchain tell him which stars he had observed at Mont-Jouy, and what steps he had taken to assure the precision of his measurements?

Luckless man on a pile of *assignats*

This stack of assignats, a paper money issued during the early years of the French Revolution, shows bills ranging in nominal value from 50 francs to 10,000 francs. The assignats lost value with staggering speed during the hyperinflation of the years 1794–7. They were never popular outside the major towns.

(Photothèque des Musées de la Ville de Paris. Photo by Briant.)

No innocent question could have prompted such contradictory reactions from Méchain. There was no subject more painful – and none he was more anxious to discuss. Méchain's response was longer than most scientific articles. It was certainly longer than any article he had ever published. It covered nine pages in his crabbed script and took him twelve days to compose; it was begun in Perpignan and finished in Estagel, halfway up the mountains.

He began by expressing gratitude for Delambre's guidance. 'Your information offers me the most instructive lessons, from which I will endeavour to profit. I only regret that the circumstances, conditions and my own inadequate know-how have made it impossible for us to work in concert; but henceforth, guided by you, as well as by the canvas mapped out by Cassini in 1740, I will endeavour to follow a more reliable method.' He then described the precautions he had taken at Mont-Jouy to obtain the most precise results possible – while hinting simultaneously that he had doubts about their trustworthiness. He was concerned, he confessed, about the correction he had used for refraction. His discordant data for Mizar were troubling. Why did his results for that one star diverge from the average value three times as much as the others? He had even contemplated a return to Barcelona to double-check his data. He only wished that Delambre had already completed his Dunkerque observations so that 'the comparison of results from the same stars would be complete, and the judgement on my case rendered'.[4]

He neglected to mention the contradictory measurements he had taken the following winter at the Fontana de Oro.

Courtesy can be formulaic: an opening flourish to honour an esteemed colleague, a closing embrace 'with all my heart'. In between come the data, the conjectures, the counterhypotheses and the criticisms of the work of third parties. Scientific life in the *ancien régime* had been guided by such formulae, with variations as subtle as the formula for the refraction of light: gradations of respect from token to veneration, and shades of affection from feigned to heartfelt. Seniority, scientific eminence and social standing bent the spectrum one way or another, as did friendship, camaraderie and the rivalry of the schools. The new age called for sterner stuff. Manly citizens only had time for frank exchange. Overnight, the formal *vous* became the republican *tu*. Everyone was equal, at least for the time being. Yet the old forms were already reasserting themselves, and some people had never wholly given them up. Méchain and Delambre never addressed one another as 'tu'.

Méchain's elaborate courtesy came in many hues. Courtesy is a camouflage, even when sincere. Avowals of respect and affection are pleas for reciprocity. Even an embrace 'with all my heart' warily seeks its return. Méchain rarely published or addressed public sessions of the Academy, except to report comet sightings or to contribute to the celestial ephemerides, the annual tables of celestial events. Already, in a career one-third as long, Delambre had published three times as much, as if his late scientific bloom had finally given him something to trumpet. Méchain recoiled from the cold finality of the printed page, with its anonymous audience. He preferred the personal letter, crafted specifically for its one sympathetic recipient, a human engagement with a like-minded mind. He corresponded with savants around the world, sharing data and disappointments with men who sympathized with the self-abnegation that went into the making of new knowledge: Bugge of Copenhagen, Zach of Gotha, Maskelyne of Greenwich, and now his compatriot, Delambre of Paris.[5]

Méchain and Delambre were not friends, not yet. They were cherished colleagues who had once observed the stars together at their *maître*'s behest, but who had not seen one another for three years. They were former fellows of the Academy (Méchain with ten years' seniority) who had been sent in opposite directions to measure the world, and were now republican comrades in a world destroyed and remade since their mission began. They were skilled savants who had been asked to assemble thousands of pages of numerical data in locations up and down the length of France, all the while pitted against one another to gather that data with the greatest dispatch and precision. They were collaborators who would soon be expected to boil that mass of unorganized data down to a single quantity, the length of one metre.

There was no guidebook through this high plateau of cooperation and competition, just as there is no primer for friendship or betrayal. To guide them, Delambre and Méchain had only the codes of etiquette they had absorbed from the *ancien régime* and the new egalitarian codes they were learning as they went. By combining the two, they would have to fashion a new integrity.

Méchain's self-deprecation tapped a vein of sympathy in Delambre, who responded (as he was meant to respond) with praise for Méchain's talents and virtues. He offered words of reassurance.

> Why do you speak of the 'judgement of my case?' If there is a judgement to be rendered, it is with regard to Bradley's tables of refraction. You will be the judge and I think he will lose his suit. Nothing could be more precise than your observations. I would give much to be assured of making ones as good, and I care little whether they conform to a theory widely known to be inadequate, and which have recently been criticized with a vigour that is far from your excessive modesty.[6]

Méchain needed to put more faith in his own exacting observations and precision instruments, Delambre admonished. He then showed Méchain how he might reanalyse his summary Mont-Jouy data, demonstrating how the results varied with different assumptions about the relationship of refraction to temperature, altitude and angle. None of these assumptions made the data for Mizar comply with the rest, but Delambre promised to look into the matter further when he conducted his own observations at Dunkerque. In the meantime, 'I make so bold as to urge you to calm yourself about your observations. I consider them definitive. There is no need for you to return to Barcelona . . . I hardly flatter myself to think I will do as well.'[7]

Finally, Delambre offered comfort of a more practical nature. He conceded every advantage to Méchain as to salary and position. Times were tough and funds were scarce, but Delambre reassured his colleague: 'I am a bachelor; your situation is quite different. Any preferment should fall to you for many reasons, not to mention your seniority and lengthy labours.' Most of all, he hoped Méchain would return to Paris that winter, so they might together mourn the losses they shared: Lavoisier, Condorcet and the other savants executed by the Jacobin régime, now, fortunately, far from power. But even if they did not meet in Paris, he looked forward to the day that he and his esteemed colleague finally joined their triangles together in Rodez, for 'that day will mark an epoch in both our lives'.[8]

On his way to Dunkerque to measure its latitude – his last task in the north – Delambre passed through Paris. He dined with Lalande and Calon, and participated in the inaugural assembly of the new Academy of Sciences, where he heard the metric system extolled as the centrepiece of the Academy's mission. The morning after, he continued on his way, passing through Amiens in time to celebrate Christmas Eve (3 nivôse of the year IV), and arriving in Dunkerque five days later.[9]

For the next three months Delambre took the latitude measurements that were the northern counterpart to Méchain's measurements at Mont-Jouy. These were the most delicate observations of the entire mission because any error here would translate directly into an error in the final length of the metre. The equivalent observations had absorbed Méchain's attention for two full winters. Yet Lalande suggested that Delambre expend no more than a week on them. Four nights, his *maître* advised, would be enough to close to within one second of the correct latitude, 'and you should satisfy yourself with that'.[10] The old astronomer did not share his former students' fetish for precision. From his point of view, these superfine measurements wasted time and effort. The standard metre could be established by legal fiat and, if scientific window-dressing were needed, a few adequate measurements would suffice. Either way his pupils ought to return to the real business of astronomy, cataloguing the heavens.

Instead, Delambre took pains to match Méchain's precautions. He set up his observatory in the attic of the Intendance, a military building within sight of the belfry. He pierced a hole in the ceiling (with permission) and slept one storey below. The only drawback to this cosy astronomical arrangement was the unsteady floorboards. Even though he calculated that their maximum disturbance caused a deviation of only 0.001 seconds, he built a wooden platform to stand on, just to ensure that his movements would not perturb the instrument.[11]

Precision is painstaking work. It demands meticulous precautions, stratagems planned like war. Delambre used astronomical theory to prepare his observations. He verified the verticality of his circle by three different methods. He drew up formulae to correct his data for refraction and temperature. He estimated in advance the best precision he could expect. And only then did he begin his sightings of Polaris, a star particularly suitable for assessing latitude because its proximity to the pole meant that its angular height as it crossed the celestial meridian would, with only minimal correction, supply the angular distance of the observer from the equator – or in other words his latitude.

His thirty-eight observations of Polaris as it transited the celestial meridian below the pole gave him a latitude of 51°2'16.66", which shifted by a minuscule 0.06 seconds when he removed his least reliable data. The two hundred results for its transit above were trickier, due to the cloud cover, and differed by one full second from the earlier results. But when he excluded the less reliable data, the difference narrowed to within 0.5 seconds. And when he summed up all his Polaris data, his total differed from his best data by only 0.25 seconds (or some twenty-five feet). It was another demonstration of the repeating circle's precision, as well as a testament to Delambre's preparation, skill and integrity.

Throughout these latitude measurements, Delambre followed the same procedure he had followed for his geodesic measurements. He always recorded all his data, and he and Bellet signed the bottom of every page of his logbook, so that anyone who examined his records would see that nothing had been altered or suppressed. He had decided he did not have the authority to reject data unilaterally. As he later explained: 'Once an observation was taken, I considered it a sacred thing: whether good or poor I faithfully recorded it.'[12]

Delambre then set out to verify his Polaris results with the nearby star known as Kochab, also located in northerly Ursa

Minor and hence likewise suitable for latitude measurements. Unfortunately, he found the dim star difficult to pinpoint. 'It was never completely in focus for me,' he admitted, even though he had Bellet file down the scope to increase the magnification.[13] This had happened during the geodetic sightings as well. At such times, Delambre relied on his assistant, who observed with 'much zeal and exactitude'.[14] As he noted occasionally in his logbook: 'Bellet thinks he can see the signal; I, who could not see it, took absolutely no part in these observations.'[15] The results for Kochab were mixed: those taken on the lower transit were of especially poor quality because of the cloud cover, and came in three seconds short of the Polaris data – a rather discordant result. Those for the upper transit, however, agreed to within a stunning 0.02 seconds.

At this point, Delambre might have left Dunkerque with a clear conscience. His overall results agreed to within one second. January had been temperate; February was cold with clear skies; but in March the cloud cover worsened. The weather over the English Channel could never rival the skies of Catalonia. Yet when budget woes stranded Delambre for an extra three weeks in Dunkerque, he decided to take supplementary observations. Imagine his horror when the new results radically diverged from the old. For nearly a month, Delambre felt the anguish which so tormented Méchain – until he identified the problem as two loose screws on the lower scope of the circle. Once it had been repaired, the new results matched his old ones. He left Dunkerque, satisfied, on 29 March.

Delambre had promised Méchain a full report of his latitude measurements, but so far all Méchain had received was a preliminary report from Lalande which spoke of Polaris and Kochab. This left Méchain bewildered. 'No doubt,' he wrote to Delambre in May, 'you have since observed the others.' He was counting on Delambre's results to help him solve the refraction problem and fix the outer limits of his error. He requested a full report 'if you can spare me a few moments'. In the formula of courtesy, this was a plea. 'Adieu, my dear colleague, I depend upon your friendship and indulgence, which I claim with insistence, while I await your response as a token of them both. I embrace you with all my heart, wishing you perfect health.'[16]

Delambre responded to this plea with a lengthy letter, the core of which he had already read to a public session of the Academy of Sciences. As he explained to Méchain (and to the Academy), he had not observed the four additional stars Méchain had sighted at Barcelona because each presented disqualifying difficulties. The star known as Capricornus only came into position during daylight; it might be observed in Barcelona, but not in grey Dunkerque. The star Mizar, which had caused Méchain so much anxiety, passed too low on the horizon to be observed with confidence on the North Sea coast. And so on. Delambre declared himself satisfied with only the two stars. 'Indeed, these are perhaps the only stars a savant ought to observe if he wishes to seek the certainty he needs, rather than to foster doubts.'[17]

A public paper for the Academy is not the same as a letter to a colleague. So Delambre appended a personal note for Méchain. He lauded Méchain's superior skills at observation, his ability to pick out faint stars that eluded others. 'I could never pretend to compete with you.' As for Méchain's doubts about refraction, he assured his colleague that the problem was not worrisome. None of Méchain's data suggested that the correction for refraction should differ at different latitudes. All their colleagues – Borda included – agreed that Bradley's tables were inaccurate, and that Méchain's Mont-Jouy data should be considered definitive, 'the most perfect that could be hoped for'. By general acclamation, he told Méchain, his Paris colleagues had declared the astronomical portion of their mission complete. 'That task is done,' he said.[18]

Two savants defer to one other. Each denies he is competing. Each concedes the other's superiority. Yet the same phrases in different mouths have different meanings. Delambre's formulae rang with self-assurance. Méchain's formulae quivered with self-doubt. The formulae of polite society – like the formulae of nature – can convey a great variety of meanings.

Delambre appended a final postscript. He informed Méchain that he would be dining two nights hence in the company of

Madame Méchain, and hoped that he would likewise embrace his colleague soon. Méchain was missed in Paris, and would find a warm welcome there. The new Academy was just like the old one; the new Bureau of Longitudes was staffed by their dearest colleagues. The government – now called the Directory – was still dealing with the aftershocks of Revolution and hoped to stabilize the currency by creating a new paper money called *mandats*. Delambre himself had arrived in the capital just in time to witness the suppression of the 'conspiracy of equals', a revolt led by Gracchus Babeuf, his old political sparring partner from Amiens. Currently he was assembling funds to resume his triangles to the south of Bourges. By mid-summer he expected to be back on mission, advancing rapidly towards Méchain.

All this time, Méchain had been stalled in the mountains outside Perpignan, struggling to advance a few triangles north. The region was in turmoil. French soldiers, back from their Spanish conquests (bargained away by the diplomats), were being billeted in every spare room in town. Prices were soaring. The 24,000 *assignats* supplied by Calon fetched less than 800 *francs* in hard currency; a month later they were worth half as much. Not even Méchain's parsimony could keep pace. His team members could barely afford their daily pound of bread and half-pound of meat. Wine and any other 'succour' had to be paid for out of their own pockets. Conditions were even worse in the mountains, where his assistants were setting up signals for observation. Villagers would not accept *assignats* at any price. No one would rent them horses or mules. The equipment had to be carried on foot, and porters were demanding wages of 100 *francs* a day. The workmen who built the signals expected still more. The bill for the two masonry pyramids that Méchain proposed as markers for the ends of the baseline near Perpignan came to 24,000 *francs*, his entire budget for the year.[19]

To help out, Tranchot offered to transfer his three years of back pay to the expedition budget. This calculated generosity meant that Tranchot's daily expenditures had to come out of general funds. To cope with inflation, the Bureau of Longitudes multiplied its pay scale eighteenfold that winter. But even Méchain's astronomical salary of 144,000 *francs* (in *assignats*) did not match the inflation rate. If the republican government had wanted to teach its people that price was the paramount variable, they could hardly have picked a more painful lesson plan. It was the world's first experience of hyperinflation. Méchain noted that he and his collaborators would have died of hunger in the mountains without the hard currency he had put away from the sale of his repeating circle, 'and that's no jeremiad, but the truth'.[20]

Tranchot's first task was to re-establish the frontier signals, destroyed by two years of warfare. He revisited Puig de l'Estelle, where he had been ambushed by *miquelets* two years before. He replaced the signals at Mont Bugarach and Forceral. Meanwhile, Méchain assured General Calon of his determination. 'Never believe, Citizen, that I am disposed to abandon the mission; I will exhaust every resource.'[21]

Among those resources was his physical stamina, diminished by his accident, but unexpectedly robust. Méchain decided not to reclaim his customized carriage, stored in Perpignan these past three years. The terrain was too rough for carriage travel, and horses were prohibitively expensive. He headed into the mountains on foot, alone. The assistants Calon had sent him were more trouble than help. With only one circle, he had nothing for them to do; so he sent them back to Paris – all except Tranchot. He did have one guiding light, however: Cassini's triangles of 1740. The problem was that Méchain was trying to operate at a level of precision which was itself a kind of uncharted country.[22]

The first stations around Perpignan were relatively benign. The summit of Mont Forceral, a barren cone-shaped hill to the immediate west of the town, overlooked a spectacular panorama of dusty vineyards, salt lagoons and the Mediterranean coast. Méchain slept under the stars there because he could not afford a guardian to watch the circle at night.[23] Mont d'Espira, to the immediate north, stands in the foothills of the Corbières range, where the terrain became more forbidding – a landscape of stony valleys dominated by ruined castles. In the interior of this range stands an isolated, twin-peaked mountain, used in Cassini's 1740 survey, and which Méchain would use as well: Mont Bugarach, known in local dialect as the Pech de Bugarach.

The Pech de Bugarach nearly broke Méchain's spirit. This enormous limestone rock was considered sacred by the inhabitants of the valley. The town at its foot sheltered eight hundred inhabitants, a general store, three water mills and a nearby mine which produced jet, a dense black coal that the locals fashioned into jewellery. Méchain had hoped to camp on the summit, but the twelve-foot-wide peak could not accommodate both his signal and his tent. So he lodged in a farm on the mountain's flank and tackled the two-hour climb every day, scrambling up the slope on his hands and knees, hanging on to shrubs and dwarf trees for balance, then sidestepping his way warily across the bare rock of the final ascent, while loose pebbles underfoot rattled into the abyss below. One misstep and all was lost. He could cite a 'thousand examples' of men who had plunged to their death, he informed Lalande. But at least he climbed unencumbered. 'I trembled with fear for the men who carried the case for the circle and the timbers for the signal.'[24]

Nothing would induce these men to repeat the performance. They also refused to guard the instruments overnight, or to keep watch over them during the day. The shark's-tooth summit was scoured by fearsome winds. The new signal was the third on the site. A violent storm a month earlier had destroyed the signal Tranchot had built to replace the one he had built two years before. The place was precarious. Villagers spoke of the *sinagries*, spirits whose malevolent glance was enough to strike men dead.

Even today, a zone of mystery surrounds the mountain. The name Bugarach derives from Arabic, and means either 'father of all rocks' or 'the banished father', signifying a high place where exiles are sent to die. The mountain still attracts mystics, who believe it to be the navel of the world, or a place extraterrestrials will one day land or a crypt for ancient gods and suppressed human memories. It is a stiff climb to the top. From the farmhouse where Méchain lodged, a climber in a hurry needs nearly two hours to reach the peak. Above the pasture lands, the trail takes a series of switchback turns through a steep scrub of beech trees and mineral springs. The earth is clinging and heavy. After the vegetation clears, the trail passes through a cleft in the double-toothed peak for the final ascent to the summit. At the top, if the atmosphere is still clear, one can see the cool blue curtain of the Pyrénées drawn across the far southern horizon and, nearer in, a half-dozen ruined fortresses at strategic intervals along the receding ridges, their broken ramparts an extension of the broken hills. To the north, guarding the broad green trough that runs from the Atlantic to the Mediterranean, the citadel of Carcassonne, Méchain's destination, is barely visible. Yet such clear days are an exception. In the old Occitan dialect of the local shepherds:

Quand Lauro porto cinto   When Mount Lauro wears his belt
et Bugarach mantelino   and Mount Bugarach his cloak
Aben la pleijo sur l'esquino.[25] Then it rains on the slopes.

Imagine what it was like, Méchain asked Lalande, to set out on a clear morning and reach the top only to find that the surrounding signals had been obscured. Clouds advanced like armies between the ridges, or flew in at a high altitude, smothering the entire region and lingering for days. Méchain slept in the farmhouse, and every day he had to climb the mountain to check his circle, left unguarded on the summit. Every morning he reassembled it for observations. Every evening he packed it back in its case and secured it under an oilcloth weighted down by stones, with only the *sinagries* to watch over it at night.

Then, when the weather cleared, Méchain saw a full panorama, including a view of the trail of his ascent. He could trace it back with his eye as it traversed the steep ridge, slipped through the rocky saddle, then emerged on the other side, before plunging into the tangled brush. It was the trail of his immediate past – and his future – the only way up or down the mountain. From the peak, time stretched out before him, like the view towards Carcassonne. Already it was late October. Each series on the repeating circle took an hour, and he needed a dozen series. The wind was cold. The clouds were closing in – then, just like that, his future was obscured.

This was not the sort of life he had envisaged when he took up astronomy. Astronomers are generally sedentary folk. They keep odd hours, to be sure: midnight vigils and pre-dawn discoveries. But after a certain age they stay put and trade information by post. In the ceremonious days of the *ancien régime* Méchain had held a sinecure as *capitaine-concierge* of the Royal Observatory. Until his survey across the English Channel in 1788, he had never left the dense soil of northern France. The *ancien régime* had been the fixed backdrop against which time advanced, like a fine chronometer. Every night he had strolled across the midnight gardens of the Observatory to the starry roof, and every dawn he had returned to his neat little house. But the Revolution had broken time, reset the clocks and torn down the calendars, filling the days with events so rapid he could no longer hear the beat of the pendulum clock. The Revolution had cast him out into the periphery, where time had slowed to a crawl and his days were filled with repetitive ado. Now he slept in decrepit inns in provincial towns: the pork roasted to a chip, the servants always late, no parlour to write in, chairs that defied all notion of rest, doors that gave windy music as well as entrance, whitewashed walls and tapestries so old as to be a 'fit [nest] for moths and spiders'.[26] When he was lucky he slept in the manor houses of the local gentry or, even better, in the homes of amateur savants, such as the Arago household in Estagel near Perpignan. When he was unlucky, he slept in the straw of an up-country cowshed, without candlelight to verify his calculations, or in a windy tent on an icy summit.

Yet he lived in an age when voyagers had approached the ends of the world, returning with sperm whale oil from Antarctic shores, breadfruit from Tahitian Edens and visions of the Northern Lights. What glory was there then in traversing France, a nation which tens of thousands of ordinary French men and women crossed every year, only to bring back some numbers? One might just as well ask what challenge Thoreau faced in his two years at Walden Pond while tens of thousands of ordinary Americans braved a savage frontier a thousand miles west. For the men of Paris, their nation's interior was a foreign land, as exotic in its way as any Andean highland. The people of the central provinces did not speak French, but a gamut of Occitan dialects more closely allied to Catalan or Provençal. The village mayor might understand French, but he would not speak it. The local measures enclosed each village in its own economy. The challenge was not *whether* France could be traversed, but *how*, by *whom* and *to what end*. Shepherds might herd their flocks to up-country slopes. Brigands might hide in the mountains. Only a man of science would climb a peak to prove a point.[27]

Ten days later, when Bugarach was done, Méchain moved to Mont Alaric, just as a storm knocked down the adjacent signal at Tauch. Tranchot was sent to repair it. Then it was Tauch's turn. By the time he had finished with Tauch it was late December and bitterly cold. He had climbed each mountain-top station fifteen or twenty times, and each had required some eighty miles of travel through snow-covered fields over ice an inch thick, while violent north-west winds blew. The grease in the repeating circle clotted in the cold. His fingers were too numb to tighten the screws. And despite his efforts he had only extended his arc from Perpignan to Carcassonne: laying down three pitiful triangles in six months.[28]

This might have been an opportunity for him to return to Paris for some much needed rest with his family. Certainly none of his colleagues would have begrudged him a break during the geodetic off-season. But Méchain had once again decided to winter in the south. He told Calon he wanted to choose a site for his baseline, a place where he might best measure one of the sides of his triangles from end to end. For this, he needed a straight stretch of level terrain at least five miles long, whose terminal points could be triangulated from a nearby station. In 1740 Cassini had used the beach near Perpignan, but Méchain considered the shifting sands

there too unstable. Instead, he picked a segment of the Grande
Route that ran from Perpignan to the fortress of Salses. Tranchot
directed the army engineers to build a masonry pyramid at each
terminus, and Méchain linked them by triangulation to his station
at Mont d'Espira. He also wanted to conduct the actual baseline
measurement right then and there – Tranchot even assayed a preliminary run with a surveyor's chain – but his colleagues in Paris
insisted that he finish his geodetic angle measurements first. The
baseline measurement would have to be postponed until the special equipment could be prepared in Paris.[29]

Méchain took this refusal to mean that his colleagues did not
trust him. His spirits, after a temporary respite, had darkened. In
March he admitted to Lalande that 'my strength no longer
matches my courage'.[30] The cold weather had aggravated the old
injury to his arm, but he himself admitted that the accident's long-
term effect had been more mental than physical. He was
despondent, and did not quite know why. 'All that remains now is
to cure my head. I will make every effort. I still have hopes that I
will conquer the apathy and lethargy which alienate me from my
true self, and which chill my spirit the moment I am at rest or
alone with myself, enervating the few faculties I ever had.'[31] His
melancholy had begun to overwhelm him. He felt paralysed.

Then, in the next breath, he wrote to Delambre to offer to triangulate north of Rodez on to his partner's side of the arc. The
extra work would not strain him, he said, but simply allow him to
make up for his inadequacies and balance their respective contributions. Delambre never answered this request. Spring came and
went. The summer slipped away. Méchain remained in Perpignan
taking latitude measurements.[32]

Friendly bureaucrats there let him set up his observatory in the
courtyard garden of the district office, where he also constructed
a sundial. These measurements were doubly superfluous. The triangles, the baseline measurements and the latitudes of Dunkerque
and Barcelona were all that was needed to calculate the metre. But
recently Borda, Laplace and the other physicists on the
Commission of Weights and Measures had decided that a series of

intermediate latitude measurements along the arc would fine-tune their knowledge of the curvature of the earth, and hence improve the final extrapolation from the French meridian to the whole. They asked the two expedition leaders to gather this additional latitude data at three other sites: in Paris, within Delambre's arc; in Evaux, at the halfway point of the meridian arc; and in Carcassonne, within Méchain's arc. And they urged Méchain to join Delambre in Evaux to take the supplementary latitude measurements together there.[33]

Méchain declined. He thought it best that Delambre measure alone at Evaux. That would ensure, he said, the accuracy of the latitude. 'I know all too well that my results are far inferior to yours with regard to their exactitude!'[34]

This remark, conveyed to Borda, prompted a sharp rebuke. This was no way for a self-respecting man of science to talk. When Méchain denigrated himself, he denigrated the mission. Borda set aside the usual courtesies and spoke frankly to the astronomer. The commander's advanced age, his eminence and his regard for Méchain gave him the right. 'I have good cause to be angry with you,' he wrote. 'Where did you get the notion that Delambre's results, either for the latitude or the triangulation, are better than yours? Why do you deprecate your own work – or rather, the Commission's work – when everyone else finds it excellent?'[35] The discordant data for Mizar simply proved that Bradley's old refraction tables were inadequate. If Méchain's results did not match Cassini's findings of 1740, then it was 'tough luck for them'. 'You were not sent out to find the same results as your predecessors,' he reminded Méchain, 'but to find the truth.'[36] Armed as he was with superior instruments, and thanks to the precautions he had taken in the light of his mission's supreme importance, his new results would undoubtedly differ from the old. If not, the entire mission would hardly have been worth the effort. In the meantime, Borda would use Méchain's results to derive a new formula for refraction.

For his part, Delambre advised Méchain to leave the resolution of his problems to the superior theoretical skills of Borda and

Laplace. He could take pride in the fact that the world's pre-eminent physicists would be constructing their theory from his superb Catalan data, 'the best and most certain which exist'. Méchain's integrity and meticulousness were legendary. Yet Delambre reminded Méchain that he must not try to accomplish the impossible. 'No matter how much effort we put in, it will always be difficult to surpass one second of precision. You and I have reached that goal, I think, by avoiding exaggerated claims and contenting ourselves with the art of the possible.'[37] Minor errors were inevitable in any operation of this magnitude, and would hardly affect the overall results. 'Anyone with the least notion of the difficulties we have faced will take them into account when they consider the exactitude we have achieved.'[38]

These words of solace do not seem to have given Méchain much comfort. Having let an entire geodetic season pass without closing a single triangle, Méchain shifted his operations to Carcassonne that winter, so that he would be ready to advance on Rodez come the spring. But at the same time, he admitted to Lalande, he was on the verge of collapse. 'After having withstood so many tests, my courage has given out, just when there are no more major difficulties to conquer . . .' Weighing upon his mind was an unflattering comparison with his distant partner. While he stalled in the south, he noted, 'Delambre works his way towards us with the swiftness of an eagle, having already traversed almost the entire length of France'.[39]

In the summer and autumn of 1796, Delambre measured seven stations south from Bourges towards Evaux, each station posing its own particular problem, and each requiring its own solution. For instance, at Morlac, the first station south of Bourges, the belfry, which had once risen forty feet above the twelfth-century church, had been dismantled by the 'hammer of the revolutionar-ies'.[40] Delambre offered to split the cost of a replacement tower with the villagers (population 748), but they refused. So Delambre offered instead to cap the church with a cheaper eight-een-foot-tall wooden pyramid, the cost of which he would split with them and which would also keep the congregation dry when the rains came. They spurned this offer too. So, somewhat peeved, he cut a deal with a timber merchant, who helped him build the tower at a reduced fee in exchange for the right to reclaim the wood once the readings were done. When the merchant arrived to remove the church roof a month later, however, the villagers balked. A trial ensued, and, five years later, the church tower was still in need of repairs. Then at Arpheuille, the last station in this sector, Delambre's measurements were skewed by a large oak tree which cast a distracting shadow against the church belfry; even so, he refrained from pruning it because it also shaded the peasants' Sunday dances.[41]

Delambre arrived in Evaux – the halfway point of the arc – on 24 November 1796, and took lodgings at the Auberge du Cheval Blanc. The inn also rented him space in an attic granary by the main town gate. There he pierced a hole in the roof to fashion an observatory, much as he had at Dunkerque. After a run of clear skies and 210 observations of Polaris, the weather turned dreary. Snow fell overnight. During the next two months only two nights were suitable for astronomy.[42]

His initial results disagreed sharply with those obtained by Cassini in 1740. This discrepancy, which would have sent Méchain into paroxysms of anxiety, led Delambre to recheck Cassini's work. He discovered an error in Cassini's methods. He spent his days updating his calculations, refining his formulae and triple-checking his results. All of these he supplied regularly to Méchain, at Méchain's request. No doubt he sampled the thermal baths as well.[43]

Evaux lies at the northern edge of the Massif Central where a hot spring emerges from the hillside at 60°C, flush with minerals. The town's Roman bathhouse was destroyed by fire in the third century CE. During the Middle Ages, the baths served as a pilgrimage site, the waters being said to heal injured limbs and chronic diseases. By the eighteenth century there were three bathhouses, superintended by a consulting doctor. After months of hard travel, Delambre would have found Evaux a welcome respite.

Evaux is a fine place to rest, but no place to be stranded. Delambre was determined to finish the expedition in the coming season. 'No sacrifice is too great for me to finish this summer.'[44] He hinted at some 'additional reasons' for returning to Paris, though he did not specify what they were. We may speculate, however, that he wished to resume his scientific career, interrupted so soon after it had begun. And he presumably wanted to spend more time with a charming Parisian widow he had recently come to know.[45]

In December he began soliciting an advance from Calon for the next campaign season, yet another round in the scientist's endless scramble for grant renewals. Calon promised to do what he could, but the government had shifted strategies on inflation. While *assignats* and *mandats* were being phased out, hard currency was still difficult to come by. Delambre might supplement his pay with army rations; as a cavalry captain, he was entitled to 9 *livres* a day, plus double rations for himself and his horse. The problem was that these rations could only be redeemed near the war front, and, as the paymaster kept reminding him, there was no place in France further from the front than Evaux. Worse, Calon himself was beginning to lose influence. He no longer sat in the legislature, and he had been accused of mismanaging his budget. He requested detailed accounts from Méchain and Delambre so that he might in turn justify their expenditures to his superiors. 'Some of the employees I've had to fire are vile worms who seek to blacken even the most honourable conduct.'[46] By the middle of spring, Calon himself had lost his post. To make up for the loss of their military patron, the Paris savants called in their political chits. Lalande lobbied Lazare Carnot, a former engineer, now one of the nation's executive directors. Lalande also offered to chat up Lavoisier's widow, 'who they say is very rich'.[47]

The truth, Delambre wrote Méchain, was that he had salted away a back-up fund of 2000 *francs* in hard currency, money he had saved out of his salary and his small annuity. He also hoped to divert his honorarium as an Academy member. It would only be enough to fund one signal, but that was one less signal that had to be covered out of the budget. This slush fund had to be kept secret, of course. 'It is best if they think us more beggarly than we are.' But if the only way to start his expedition in the spring was to finance it out of his own pocket, he would do so. And in the end, that is what he had to do.[48]

On 1 April 1797, after four months in the bathhouse town, Delambre began his final push towards Rodez. He had thirteen more signals to plant across the high Auvergne plateau, eleven triangles left to close. As he advanced south, each station posed a greater challenge, like the stations of the cross. In his eagerness, he may have set out prematurely. The early spring weather was horrendous; almost every day it rained. Some days he sheltered in an inn. On other days he was caught in the downpour. To the east, through the gloom, he could see a row of brooding beehive volcanoes, their black humps flecked with snow. The greatest of these was the Puy de Dôme. Over a century earlier, the great Blaise Pascal had sent his brother-in-law to the summit with a crude barometer to prove that the atmosphere was finite. Over a thousand years ago, the Romans had worshipped Mercury in a magnificent temple by the crater. Over ten thousand years ago this entire plateau had been born in a volcanic surge – although it had only been ten years since savants had first suggested as much, and their theory was still controversial.[49]

Years, decades, centuries, millennia . . . Day by day men crawled like ants across the corrugated arc of the world, peering ahead to the next ridge, seeking to uncover the processes which had shaped the earth. Where they had once expected perfection, they were beginning to learn how eccentric our planet is, suspended as it is between accident and necessity. And in the interstices between geological time and their daily labour, human history unfurled, likewise poised between accident and necessity. The planting season had begun. The earth was damp, the rivers flush, the air heavy with rain. The rich black soil fed lush green pastures. Cattle, sheep and horses grazed by the road. Everywhere nature had been shaped for human ends, in turn shaping human choice. Even the Gothic churches had been constructed of the black lava stone.

From a distance, through the gloom, Delambre had difficulty picking out the clock tower of Herment, a walled medieval town of 527 souls perched on top of the steep summit of a conical hill. Each of the triple towers of Herment's church had been demolished and rebuilt: destroyed by Huguenots, rebuilt by the Catholics, destroyed by revolutionaries, and now rebuilt by science. When Delambre arrived, the fifty-three-foot clock tower was a dark skeleton. He filled the interior of tower with bales of hay to make the tower solid and so visible from afar. But when he tried to drape the frame with a white signal cloth, the locals balked. White was the colour of the royalist flag – and the region's administrators were battling a reactionary resurgence. The townspeople did not want to be mistaken for counter-revolutionaries. Only a few weeks before, a crowd of hooligans had absconded with a baptismal font that they considered sacrilegious and heckled the *curé* for having sworn allegiance to the republic. To appease the patriots, Delambre sewed a red strip of cloth to one edge of the sheet and a blue strip to the other, transforming his signal into a makeshift Revolutionary tricolour flag. This satisfied the townsfolk long enough for him to conduct his measurements and leave town. The day after he left, a rowdy royalist crowd circled the church and forced the *curé* to take part in their procession. 'The enemies of law, government and order have not surrendered their foolish hope of sowing chaos here,' complained one local administrator.[50]

At his next station Delambre needed to call on administrative help. No sooner had he set his signal on the strange grey organ-pipe cliffs above the town of Bort-les-Orgues than a torrential storm brought a mudslide down from the hills, filling the streets with a three-foot-deep sludge of earth and stones. Residents blamed the flood on the bizarre signal on the mountain top and demanded that it be torn down. This self-important town along the Dordogne had long suffered from inundations, despite its annual sacrifice to the river. Early each spring, on the eve of Ash Wednesday, a procession of boys in white robes marched through town, bearing torches and singing a dirge-like chant as they carted an effigy of an old man on a tumbrel: 'Farewell, old man, you must go, I remain! Farewell, farewell, farewell!' When they reached the river, the oldest living man present set the effigy on fire and tossed it, still burning, into the river. The rite dated back to the Celts, who sacrificed their eldest travellers at rivers too wide for them to cross. By the eighteenth century, the ceremony had been supplemented by official demands for flood mitigation. Municipal leaders dissuaded the townspeople from destroying Delambre's signal; it helped, too, that the signal was well out of reach beyond the high cliffs.[51]

His next station, the peak of Puy Violent, was the highest point along the entire meridian arc: six thousand feet above sea level. There Delambre had the choice of lodging either in the nearby Renaissance town of Salers – an onyx outpost of judicial palaces, slate-roofed inns and black-stone battlements, but an arduous three-hour climb to the mountain top – or in a cowshed an hour from the peak. As it was mid-August, Delambre thought it would be safe to save himself the daily hike, and he decided to lodge in the cowshed.

For the ten days of my labour, I slept in my clothes on bales of hay, living on milk and cheese. I could almost never sight two stations simultaneously because a thick fog obscured the horizon. In the long intervals while I waited ten or twelve hours for a view from the summit, I was scorched by the sun, chilled by the wind and drenched by the rain, all in succession. But nothing irritated me more than the inaction.[52]

Though Puy Violent was not named for its weather, it might as well have been. It rose beside its sister peak like the cusp of a molar. To the east – to Delambre's back – a range of higher mountains, scarred by barren cirques, hemmed in the view. To the west, the panorama opened out over a green skirt of lava that plunged towards convergent river valleys draining into an invisible Atlantic two hundred miles away. Geologists had recently come to suspect that this entire region had once been the site of a single vast volcano. When the clouds cleared, Delambre could look out over its well-worn shards. Through the scope of his circle, he could see the black battlements of Salers on its basalt prominence across the valley, plus all his surrounding stations: his signal above the organ-pipe cliffs of Bort, the church tower at La Bastide, and his next destination, the signal upon the ruined castle wall at Montsalvy. It was a silent perch under a long low sky, without a human being in sight, only hawks angling on the wind and cattle grazing on the slopes. Today's red-coated Salers cattle are the result of late nineteenth-century breeding, but their eighteenth-century forebears were already famous for their cheese, the region's chief export. On summer days they grazed on the high slopes; every evening, to protect them from wolves, the shepherds drove them down to the shed where Delambre joined them for the night.[53]

The up-country people were handsome, with blue eyes and dark brown hair. They mistook Delambre for a sorcerer. Who else would have paid a team of men to cart four twenty-two-foot timbers to the peak of Puy Violent and assemble them in a pyramid? When a cow refused to give milk, when a plough broke in the field, when a journey proved unlucky, the sorcerer's evil eye was to blame.[54]

As part of his mission, the government had asked Delambre to assess the common people's view of the metric system as he travelled through the countryside. Delambre found that the vast majority of the common people had never heard of the new measures. The workmen who built his signals were illiterate, innumerate and did not speak French. This did not mean they were inarticulate or unskilled; Delambre found them quite adept at constructing his bizarre pyramids. For a sympathetic hearing, however, he addressed himself to the region's 'enlightened citizens': its magistrates, state officials and educated men. These citizens looked forward to the new era. During the *ancien régime*, the province of the Auvergne had been ruled by a jumble of legal codes; half a village might owe allegiance to Roman law and the other half to the common law. This legal tangle nourished the region's multiplicity of measures, turning every marketplace into an arena of 'frauds, deceptions, cons and thefts'. Or so the region's enlightened citizens believed.[55]

In the past two hundred years, customs have changed, people have changed, animals have changed, even the terrain and weather have changed – all, paradoxically, while seeming to have stayed essentially unchanged. The population of these central regions has been stable for two hundred years, although there has been a steady emigration towards the towns and no one under the age of sixty habitually speaks Occitan. The Salers cattle have multiplied, bred to a glossy muscularity, but the wolves are gone. Cantal cheese is still the mainstay of the region's economy (after tourism), and is now exported around the world. The winters are less bitter than they were in the eighteenth century, though the great storm of 1999 levelled three hundred million trees. Even the modern highway still follows the route laid by the *ancien régime* engineers, though it is now paved in asphalt. On the road out of Salers on his way to Montsalvy – known today as the D920, then as the Route de l'Intendant – Delambre was caught in a powerful rain storm. It felt, he said, like travelling inside a cloud accompanied by continual thunder and lightning.[56]

At Montsalvy on 12 August, Delambre sighted Rodez through his repeating circle for the first time. The horizon was hazy, and he had trouble distinguishing the target. But the next day he saw it sharply outlined against the blue: the serene red head of the Virgin Mary rising from her pedestal on top of the cathedral. The statue would link his chain of stations and Méchain's. Delambre's observations from the hilltop just north of Montsalvy are commemorated today by a finely wrought orientation table, which points out features as near as Rodez and as far as the Pic de Nore, Méchain's station north of Carcassonne.

Delambre's final goal lay just ahead. Rodez sat on a hillock in the warm basin of Rouergue, and, as he descended, the temperature rose to meet him. The soil lightened to a crumbly orange, the yellow houses turned their open windows towards the sun. Suddenly, he could no longer sense the Atlantic behind him, but

The tower of Rodez Cathedral

*The Renaissance belfry of Rodez Cathedral served as the liaison between
Delambre and Méchain's triangulations. The head of the statue of the
Virgin Mary, which they used as their common signal, is the most elevated
point in the centre of the tower.*

(Photo: Roman Stansberry.)

instead smell the Mediterranean ahead: the fruit trees, the corn
husks, the olive trees, the dry dust of the south. Lizards, some a
foot long, skittered behind rocks. Delambre had entered the Midi.
Half the Massif Central was yet to come, with pine-blue ranges
and deep gorges that carried cool air down from the mountains.
But it was the south none the less. Delambre had only two stations
left, two stations to link his triangles with those of Méchain.[57]

He expected to hear from his colleague any day now, either at Rodez or at neighbouring Rieupeyroux. The two savants had been out of contact since spring, presumably because Méchain was also converging on Rodez, travelling from town to town and out of reach of the mails. On 23 August, while observing at Rieupeyroux, Delambre sighted one of Méchain's signals to the immediate south. It was a good sign. It meant that he and Méchain were closing in simultaneously on Rodez. What a glorious finish that would be, the perfect resolution to their six-year mission of competition and cooperation. Hurrying to complete his measurements, he and Bellet set out the next day to drive the last twenty-five miles to Rodez. On the road, they met a traveller journeying alone in the opposite direction. It was Tranchot, on his way to look for them. His mission was complete, he said. Méchain had given him meticulous instructions to build a chain of signals from Carcassonne to Rodez, so that Méchain might follow behind, taking the geodetic measures with his repeating circle. It was Tranchot's signal at La Gaste that Delambre had sighted a few days earlier. Yet Méchain was nowhere in sight.[58]

Two days later, on 9 fructidor of the year V of the Revolution – otherwise known as 26 August 1797 – Delambre arrived in Rodez and wrote in his logbook this epigraph from Virgil's *Aeneid*:

> *Hic labor extremus, longarum haec meta viarum.*
>
> This is the final labour and the end of long travels.[59]

Then he, Bellet and Tranchot climbed the 397 steps of the cathedral tower to observe the surrounding stations. Statues of the four archangels watched from the corners. On a central pedestal, higher than any other point for fifty miles around, stood the statue of the Virgin Mary. After a lightning strike in 1588, the bronze figure had been replaced by a statue in the same red stone as the rest of the tower. Some Revolutionaries now wanted to replace her with a statue of liberty. Others insisted that the entire tower be razed. But the local Revolutionary Society had voted instead to reconsecrate the cathedral as a Temple of Reason. Like the basilica at Saint-Denis, the past had been preserved to serve new ends.[60]

The wind was blowing hard; the horizon was clear. They completed the sightings in two days, then packed to return to Paris. Delambre was almost done.[61]

Chapter Eight

TRIANGULATION

'Tis the same, he would say, throughout the whole circle of
the sciences; – the great, the established points of them, are
not to be broke in upon. – The laws of nature will defend
themselves; – but error – (he would add, looking earnestly at
my mother) – error, Sir, creeps in thro' the minute holes and
small crevices which human nature leaves unguarded.[1]

LAURENCE STERNE,
The Life and Opinions of Tristram Shandy

Where was Méchain?

That spring and summer of 1797, as Delambre triangulated
from Evaux to Rodez, and Tranchot unfurled his chain of signals
from Carcassonne to Rodez, no one heard from the leader of the
southern expedition. In early August, as he approached Rodez,
Delambre had become sufficiently worried to contact Madame
Méchain at the Paris Observatory. Perhaps she knew her husband's whereabouts.

While Méchain worked his way slowly through the remote
mountains of southern France, his wife had moved from her little
home on the edge of the Observatory grounds into the 'big house'
of the Observatory, into the grand apartment once occupied by
four generations of Cassinis. She had been offered the apartment
while her husband was in Marseille. But the practical Thérèse
wanted to make a few improvements before moving in, as the

rooms had been damaged during the Revolution. It must have been a glorious day when the Méchain family entered their new home. Their eldest son, Jérôme-Isaac, now seventeen years old and named after his godfather Lalande, intended to follow his father's profession of astronomer. Already he had been hired as an Observatory assistant. The younger boy also showed scientific promise. The daughter was perhaps the brightest of the three. Who knew? With the Cassinis in eclipse, perhaps the Méchain clan would rule the Observatory for the next few generations. Of course, their father first had to found an honourable line.[2]

Thérèse Méchain expressed surprise that her husband had not been in contact with his esteemed colleagues. His last letter to her, dated 21 July, had indicated that he was about to begin his triangles in the Montagnes Noires, north of Carcassonne, and that he hoped to complete his mission by summer's end. Now the summer was over, and the race – if that is what it was – was over. The only question now was whether Méchain could salvage his honour.[3]

When a letter from Méchain finally did arrive, it was winter and Delambre was back in Paris, preparing to measure the northern baseline near Melun. The letter was dated 10 November 1797, and had been mailed from the town of Pradelles, where, by his own admission, Méchain had made little progress. The problem was the weather. He had not been able to squeeze in more than two hours of observations in the past two months. And while waiting for the skies to clear he had been agonizing over his Barcelona results. He revisited his perennial obsession with Mizar. 'It has caused me nothing but despair and disappointment; I regret that I ever observed that star.'[4] He also now worried about the looming moment when he would have to present his data to his fellow savants. He had done some preliminary calculations and they had revealed some shameful comparisons. When Delambre's geodetic and astronomical data were combined, the values for Dunkerque and Evaux matched to within one second. When he performed the same operation with his own data, he found an inconsistency of nearly five seconds in his values for Barcelona and Carcassonne. His analysis was admittedly premature, perhaps even forbidden. (Were the savants allowed to have a sneak preview of the final result while the expedition was still in progress?) But he *had* made the calculations, and they had convinced him that he *must* return to Barcelona that winter. For this, he needed the assent of both the French and Spanish governments, and he feared that Borda would not approve. Might Delambre approach the old physicist-commander on his behalf?

It was the same old obsession, with an ominous new tone. No matter what the cost, Méchain informed Delambre, he was determined to complete his mission. 'In this situation, I have chosen to remain in the horrific exile I have long bewailed, far from my other duties, far from all I hold dear, and far from my own best interests. I will make every sacrifice, renounce everything, rather than return to Paris without having finished my portion of the labour . . . and if I am not allowed to complete it, I will never return.' Méchain saw only two options. 'Either I will soon recover the strength and energy I should never have lost, or I will soon cease to exist.'[5]

To Delambre, this smacked of a suicide note. And where, in the name of Cassini, was Pradelles? Delambre could not locate the town on any map. Presumably it lay somewhere in the Montagnes Noires of the Languedoc, which suggested that Méchain had not completed even a single station for the second year in a row.

Delambre decided to consult Borda. He sent the commander excerpts of Méchain's letter as a sign of the 'moral state' of their colleague. 'I don't like it when he says, *I will recover my energy or I will soon cease to exist.*' Personally, he would have liked to see Méchain return to Paris for the winter so that the two expedition leaders might compare data and check over each other's calculations. Delambre had so far shared all his data with Méchain, yet Méchain had refused to reciprocate. 'Not only do I really wish to see his data, but I think this precaution necessary. If Méchain returns to Spain, who knows if he'll ever come back, or if his data will ever be recovered?' It was essential that they find some way to 'heal his mind and return him to his right senses, to his family, to astronomy, and to his colleagues'. To that end, Delambre urged that they enlist Madame Méchain.[6]

All that winter, while Delambre soaked in the hot springs at Evaux, Méchain had been holed up in the southern citadel of Carcassonne. It was one of the coldest winters on record; even the Canal du Midi had frozen over. The citadel had been fortified since Roman times and repaired many times since. Méchain spent the season taking latitude measurements from the tower of the Saint-Vincent church in the *basse-ville*, a 'modern' town of the thirteenth century, with clean straight streets, decent sanitation, a hospital, a courthouse and a theatre. It also housed the cloth merchants and professionals who had prospered since the Canal du Midi had connected the Atlantic and the Mediterranean.

Amateur astronomers had long haunted the Saint-Vincent tower. During the Revolution, the church had been put to more mundane uses. A factory for artillery carriages had been established in the nave, with forges in the side chapels. Restored to its traditional functions in late 1795, the tower once again became the site of astronomical observations. In Carcassonne, Méchain found a pair of amateur star-gazers eager to assist him.

Raymond de Rolland and Gabriel Fabre were local magistrates who shared a passion for the heavens. Like Méchain, both men were approaching fifty, their careers and families up-ended by the Revolution. Rolland, the son of a wealthy manufacturer, had become the region's chief judge and a partisan of 1789, only to lose his post in the judicial reforms of that year. Fabre was the principal author of Carcassonne's *Cahier de doléance*, which had called for uniformity of measures and many other enlightened laws. He now presided over a criminal court, where he had a reputation for combining legal analysis with sympathy for the unfortunate. His personal motto came from Seneca, the Stoic philosopher and astronomer: *Res est sacra miser*, or 'A suffering man is a sacred being.'[7]

These men welcomed Méchain into their homes with honour and sympathy. His mission made sense to them. They admired him. Their hearts went out to him. It was a glorious opportunity for two amateurs to assist one of astronomy's elite practitioners. Méchain considered his stay in Carcassonne the most gratifying period of his life. His personal comfort and sympathetic new friends also combined to make it the most miserable.[8]

In mid-April, shortly after Delambre set out from Evaux, Méchain wrote to his northern colleague to inform him that he would be heading into the Montagnes Noires the following month, or as soon as the snows melted. On clear days, he could make out his next station, the forbidding Pic de Nore, the highest peak in the range. He made a final plea to Delambre that he be allowed to extend his triangles north of Rodez to atone for his shameful performance of the previous season. But it was too late; Delambre was already on mission.[9]

Then, instead of heading north, Méchain stayed in Carcassonne all that spring and summer. He was comfortably lodged near Fabre's home, not far from the civic theatre. He often dined at the house of Rolland, whose wife was a charming hostess. Among these sympathetic friends, Méchain was overwhelmed by guilt.

Precision is an obsession. Why else would anyone sharpen the blade of knowledge to its ultimate fineness? Precision is a quest on which travellers, as Zeno foretold, journey halfway to their destination, and then halfway again and again and again, never reaching finality. We generally expect our heroes to possess virtues we might envy: courage, generosity, insight, honesty. The heroism that Méchain was expected to display was more prosaic: the ability to focus his attention for numberless days and nights on a repetitive task while he worked towards an ever-receding goal. Scientific knowledge is a prize which recedes as we advance. They are also heroic who practise self-abnegation. The price they pay is equally prosaic: anxiety, the abyss of self-doubt and a certain fastidiousness. Precision is an obsession, and the sharp edge of Méchain's exactitude was cutting him up inside. He was terrified of being caught, of being accused, of being blamed. His results 'oppressed him day and night'. Endlessly he replayed the events of the past.[10]

He confessed his anxiety, but would not reveal the data. He confessed his pain, but not its cause. He appeased his conscience with self-reproach, without quite admitting what he had to reproach himself for. Every letter he wrote to Delambre alluded to the Barcelona latitude without explaining exactly what was wrong. He was not ready to turn over his logbooks, he said, not even to General Calon, his administrative patron and military superior, despite the General's promise that they would not be shown to others. Still Méchain refused. He needed more time to correct his corrections, he said.[11]

Error was the great enemy of enlightenment, the loathsome *Infâme* the *philosophes* had journeyed forth to slay. In that battle, mathematical science was their most fearsome weapon. For four millennia, astronomy had been the supreme quantitative science, gathering more and more of the world under its domain. Ptolemy's geocentric model had been an orrery of mathematical refinement. Galileo, Kepler and Newton had showed that God's geometric perfection existed on earth as well as heaven. And now Laplace and the other eighteenth-century savants were using mathematical analysis to show how the very dust of creation had formed our solar system, while they were simultaneously engaged in an epic struggle to direct their great mathematical weapon against the corrupt society around them. The metric system was another extension of this programme to bring mathematical rigour down from the heavens so as to reorder the most mundane of earthly affairs.

Yet none of these great minds had ever treated their own data as rigorously as they treated the movements of the heavens or the shape of the earth. They averaged results, they looked for discrepancies, and they tossed out the data they considered unworthy of nature's perfection. The question they asked one another was not *which* data to trust, but *whose*. An honourable savant made himself personally responsible for the consistency of nature's data, without defining what that consistency consisted of. What counts as an error? Who is to say when you have made a mistake? How close is close enough? Neither Méchain nor his colleagues could have answered these questions with any degree of confidence. They were completely innocent of statistical method.

In their more reflective moments, the savants did acknowledge that the paths of error were multiple, that investigators could hardly express the degree of their error without first having access to the truth, and that the acquisition of truth was a voyage through a labyrinth. For six years now, Méchain had been wandering in a maze of mountains and regrets. At the core of his agony lay a tight cluster of doubts. Whom could he trust? Could he trust Delambre? If he told his colleague his secret, would his colleague betray him? Did his colleague trust *him*? Could he trust himself? With whom did his allegiance lie?

With his acute self-awareness, Méchain posed this question directly to Delambre. 'Can I always be assured that when I write to you I am addressing a friend and absolutely him alone?'[12] Sometimes, he made every pretence of trusting his colleague. 'I throw myself into the arms of a friend and can only hope he will not fold his arms against me; I confide in him and in him alone.' Other times, he pleaded for compassion, leniency and forgiveness. 'If you treat me with rigour, I won't know whom to turn to; my position is abominable.'[13] Still other times, he begged Delambre to conceal everything that passed between them. 'If you still have any friendship for me, you will throw this letter in the fire.'[14]

That letter survived – albeit under seal.

Pradelles was and is a hillside hamlet on the southern slope of the Pic de Nore. The village was even marked on Cassini's map. It had 561 inhabitants and twelve times as many sheep. Before the Revolution, it had filed a complaining *Cahier de doléance*, like tens of thousands of other French villages, demanding fairer taxes, more reliable justice, the end of road tolls, regular meetings of the National Assembly, and 'individual and civil liberty for every citizen'. In short, it was a typical mountain village in rural France.[15]

While in Pradelles Méchain lodged with the richest man in town, Joseph-Louis de Lavalette, sieur de Fabas, a young former nobleman, whose Pailhès manor house is now a resort hotel with a dining room and a swimming pool. Several caches of Fabas's treasure of gold and silver coins have been unearthed over the centuries. In all that time the town's population has hardly varied. The mountain that dominates the village is today itself dominated by a factory-sized meteorological station, whose red-and-white striped tower rises another 150 feet into the atmosphere like a giant barber's pole. It is a felicitous perch for those who wish to peer into the distance or predict the future – and only a ten-minute drive up the paved road, or a stiff one-hour hike from Pradelles. From the summit, Méchain could see the citadel of Carcassonne to the south, the round blue bowl of the Mediterranean to the east, and the forest peaks he had yet to measure up north.

Not that clear days are common in the Montagnes Noires. By the time Méchain arrived in October, the summit was already inhospitable. 'The unhappy Pic de Nore is redoubtable for its cold and fogs,' he wrote to his Carcassonne friends.[16] He was still in Pradelles in late November when three feet of snow fell on the mountain in one night. It was the biggest storm of the decade. He considered calling a halt. 'I will cede the terrain to the snow, the icy winds and the wolves, which are not uncommon in these parts.'[17] The long trough from the Atlantic to the Mediterranean funnelled alternating winds against the mountain side; the western *cers* brought icy chills and the eastern *autun* induced 'aches and nervous afflictions . . . sapping strength and vitality'.[18] Méchain finally retreated to Carcassonne in mid-January, after climbing the mountain more than thirty times. In two years of campaigning it was the only station he had measured. Méchain seemed to have lost interest in finishing the mission.

At that same moment, the Paris Academy of Sciences made a momentous decision for the future of the metric system. The assembled academicians decided to convoke a meeting of the world's most capable savants to review the meridian data and prepare the final determination of the metre. The idea was to give a global imprimatur to the metric system, to demonstrate that it was not merely a French reform, but was truly 'for all people, for all time'. The decision was made in January 1798; the meeting was scheduled for September. This meant that the expedition data had to be assembled for examination in nine months at the latest. It was to be the first international scientific conference in history.

For the conference to succeed – for the provisional metre to be rendered 'definitive' – it was essential that Méchain complete his triangles that year, while Delambre measured the two baselines – the one in the north near Paris and the other one that Méchain had already prepared in the south. As spring rolled around, Méchain again made promises. He did not need Tranchot to complete the survey. His friends had found a local lad named Marc Agoustenc to assist him. This year he would start at Rodez, and work his way back south towards Carcassonne. That way he could measure the temperate tablelands around Rodez in the spring, and then return to the frigid Montagnes Noires in the summer. As of June, however, he had yet to leave Carcassonne.

After years of quest, time had suddenly contracted. Delambre hurried to measure the northern baseline, a straight stretch of the king's highway near Melun, known today as the N6. In the company of Laplace and the nation's chief engineer, he had scouted the terrain and requested the construction of two masonry markers, each containing a copper plug that would define the exact terminal point. Above each stone foundation, he now erected a sixty-one-foot wooden tower. Yet even from this high platform, the beautiful twin rows of plane trees along the highway blocked his line of sight. Over the next six weeks he had workmen prune back some six hundred trees, while he triangulated the two ends of the baseline from the roof of the farmhouse at nearby Malvoisine. Six years earlier the proprietors there had let him raise their chimney so that he might sight the d'Assy country château, thereby weaving his own life story into the making of the impersonal metre. Since then his patron had been executed, and the world turned upside down. Now back on the same farmhouse roof, Delambre and Bellet took their final geodesic measurements.[19]

At eleven o'clock on the morning of 24 April, the team laid down the first of their high-precision rulers to measure the baseline. Each of their four rulers was two *toises* (twelve feet) long and a marvel of artifice. Lenoir's workshop had fashioned them out of pure platinum, the newest and most expensive metal on earth. Borda had then calibrated each one against a one-second pendulum and set it in a wooden sleeve alongside a strip of copper so that the relative expansion of the two metals could be read with microscopic precision. The routine was this: Bellet laid the rulers, Tranchot checked their alignment and level, Delambre read the temperature gauge, and each man recorded the results in his own separate logbook. An additional logbook was kept by a tall grey-eyed young man of seventeen named Achille-César-Charles de Pommard, the son of Delambre's widow-companion. After the fourth ruler was placed, the first was shifted to its end, and the process resumed. It took them all of the first day to advance 528 feet. At night they marked their point of progress by driving a lead-topped stake into a hole in the road and marking the extremity of the final ruler with a plumb line. They then covered the hole with heavy planks to shield the marker from passing carriages. It took them forty-one days, working from dawn till dusk, to traverse the six miles.[20]

Eminent visitors came to watch their tedious crawl across the earth. Lalande rode down from Paris for the afternoon. A party of savants arrived on 3 June to celebrate the final measurements: among them Louis-Antoine de Bougainville, a seventy-year-old circumnavigator who had been the first European to discover Tahiti, and the youthful German geographer Alexander von Humboldt, on the threshold of a world tour that would make him the most famous explorer of the age. Both were impressed with Delambre's approach. 'Delambre's personal character inspires as much confidence as the excellence of his instruments,' Humboldt wrote. 'To complete such a task in the face of so many physical, moral and political obstacles, it is essential that the expedition leader have this calm temperament, this tranquil joy, this perseverance.'[21]

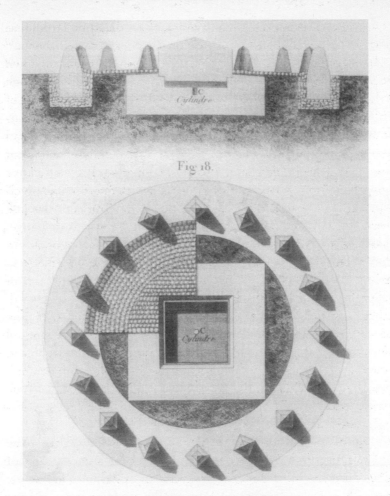

Marker for the baseline of Melun

Masonry pyramids were placed at the terminal points of the Melun baseline, a six-mile stretch of the Paris–Melun highway now known as the N6. The pyramids were completed in 1798, in time for Delambre's measurements. They were replaced in the 1880s, when the baseline was remeasured. They have since been destroyed, after being damaged in a road accident.

(Delambre, *Base*, 3, plate vi. Photo: Roman Stansberry.)

High-precision instruments mean nothing unless you trust those who wield them. Humboldt had secured one of Lenoir's precious repeating circles for his own world tour. His first stop would be Barcelona, where he hoped to take latitude measurements and send Delambre his data. Trust, but verify.

The next order of business was the southern baseline at Perpignan. This was originally to have been Méchain's baseline, linked to his chain of triangles, and so Delambre invited his colleague to join the measurement party. Méchain refused. Three years ago the Commission had not trusted him to measure the baseline on his own, and had insisted that he finish his triangles first. Well, he had yet to finish them. And there was another reason. He refused to have anything to do with Tranchot. The fact that the Commission had allowed Tranchot to assist Delambre proved that his colleagues trusted Tranchot more than Méchain. Let Tranchot have all the glory then. Méchain should have surrendered command of the southern expedition to him long ago. Instead, he had botched his mission, failed his colleagues and besmirched his reputation. Shame was all he felt. Painful, acute and deserved shame. 'After all that has happened,' he wailed, 'I can no longer show myself anywhere and my only wish is to be annihilated.'[22]

Madame Méchain agreed to help bring her husband to his senses. As he had refused to come and visit her in Paris, she would track him down in southern France. And once she found him, she assured Delambre, she would convince him to join his colleague at Perpignan for the measurement of the baseline.[23] She would stay with him until his mission was accomplished, assisting him with his observations, as she had done in the halcyon days before the Revolution. In return, she had only one request: scientific Paris was full of gossips; no one must know that Pierre-François-André Méchain, member of the Academy and joint-commissioner for the measurement of the world, had been unable to complete his mission without the help of a woman, not even if that woman were his wife. She wrote to Delambre in the strictest confidence.

Paris, 30 May 1798

Monsieur,

You ask me to induce my husband to put the final touches on
the important work with which you are conjointly charged.
No one takes a greater interest in this than I, and I have long
considered joining him myself, so that I might bring him
words of consolation and peace. Until recently, diverse
circumstances have prevented me from carrying out this plan.
I now leave immediately for Rodez. I have notified my
husband of my trip without awaiting his reply, so as to give
him no occasion to dissuade me. As I suspect he is no longer
in Rodez, I have asked him to indicate a place where we
might meet. Do not think that I will waste his time. On the
contrary, my goal is to accelerate the measure of the triangles.

I have told him emphatically not to accommodate me by
proposing a rendezvous in a town appropriate to a lady. I will
not waste even a quarter-hour of his time because he does not
have the time to waste. I have told him that I will gladly meet
him on the mountain tops, sleep in a tent or a stable, and live on
cheese and milk; that with him, I will be content anywhere. I
have told him that we will work together by day, and let the
nights suffice for conversation. I am hopeful that the esteem
and absolute trust he places in me will allow me to dissipate the
unwholesome thoughts which devour his spirit, and which,
against his will, distract him from his purpose. When I am done
with him, he will be ready to be delivered into your hands.
Perhaps I will wait until you have joined us before handing him
over to you, so that we may together regenerate his spirits. You
may judge for yourself, Monsieur, if the signs of your friendship
for me have won for you my esteem and my gratitude.

This, regrettably, is all that it is in my power to do, my final
effort for the good of the service, for the interests of my
husband, and for glory. Needless to say, all of this must remain
between you, me and Monsieur Borda, who entirely approves
of this plan. For all the world, I beg you to keep this all a

secret. I have announced that I am going on a visit to the country and no one knows the purpose of my voyage, so as to give no one grounds to say, 'She has gone to fetch her husband.'

I've not heard from him since the letter of 16 floréal [5 May 1798], in which he says he is leaving for Rodez. I've waited until the last minute to find out whether he has resumed his triangles. As soon as I have joined him, I will inform you of the exact state of affairs. I will also ask him why he has not sent the document releasing Tranchot from his employ. All this will soon come to an end.

I have the honour to be, with feelings of the highest esteem, your very humble servant,

Madame Méchain.[24]

One month later, on 7 July 1798, in the red-cathedral town of Rodez, Monsieur and Madame Méchain met for the first time in six years.

What did they talk about, Monsieur and Madame Méchain, upon seeing each other for the first time in six years? We do not know. There is no record of their conversation, just as there is no trace of the scores of letters they sent one another during his travels. What we *want* to know, of course, is whether he told her what was torturing him. Again, we do not know. All we know is that he was a man given to confession, and he had already confessed his error to at least one person (Giuseppe Slop). We also know that his wife had sufficient astronomical knowledge to grasp the import of his error. And we may presume that, having travelled so far to lure him out of exile, heal his affliction and talk some sense into him, she would have felt entitled to some kind of explanation. So it is quite possible that he told her. Indeed, Madame Méchain may have had sufficient astronomical knowledge to put Méchain's error in perspective. Such knowledge, after all, did not depend on subtle mathematics, but on a grasp of the perils of observational science and the practical goals of the meridian mission. And Madame Méchain was a practical woman . . .

Why hadn't he come home once in six years? It was not that far to travel. Paris was only a week away by carriage. Delambre had been back a dozen times, and although his northern sector lay closer to Paris, he had also travelled as far as Rodez, returned home and then set out again for Perpignan – whereas in six long years Méchain had not found two spare weeks to visit his wife or children (who hardly knew him) or to confer with his colleagues; not even during the dark winters, when all geodesy stopped. Think how much easier it would have been for him to visit her than for her to visit him.

And now that she had come all that distance, all he wanted to talk about was how they were measuring *his* baseline without him; how Tranchot was plotting a revolution to usurp his leadership of the southern expedition. What harm had Tranchot ever done him, anyway? The engineer was an honourable man, an able geodeser who had laboured hard on this expedition. After a year of working together, Delambre had only words of praise for him. According to Delambre, Tranchot was hard-working, exact in his measurements, intelligent, never presumptuous; in short, an ideal collaborator. Moreover, Tranchot had never spoken a single word against Méchain in Delambre's presence. To be sure, he lacked Méchain's education and his skills in calculation. All the more reason to be generous with him. Had he insulted Méchain? Raised a hand against him? (Méchain had called Tranchot a violent man, and claimed that he had once threatened him. And for his part, Tranchot had admitted that once or twice, in Genoa and Marseille, he had expressed, perhaps 'intemperately', his frustration with Méchain's delays and Méchain's tight-fisted control over the mission's purse strings.) Well, if Tranchot had insisted that Méchain resume his mission, he was right to do so. Any further sulking would forever keep Méchain from the recognition he deserved.[25]

As for this error Méchain kept worrying about: who was to say whose fault that was? It could be a problem with the instrument (no matter what Borda said). Or it could be a flaw in the formulae or the correction tables. Or it might even have some natural cause. In a sense, it was presumptuous of Méchain to take all the blame. The burden of the mission – like its success – was great enough to be shared by all. No one's work is perfect. And a relatively minor and inadvertent error was nothing to be ashamed of. His colleagues would not hold it against him, not even Laplace. They were not as judgemental as they sometimes seemed. They had no intention of replacing Méchain, nor of assigning someone else to finish his sector for him – provided he finished it. They were simply desperate to see the job done before the foreign savants arrived in Paris, and desperate to please the politicians who had committed such huge sums to the mission.

Or perhaps Méchain's refusal to return to Paris and his self-reproach were just a roundabout way of casting the blame on others. Yes. By accusing himself of so small (and inadvertent) a sin, he underscored the guilt of all those who had accepted the Revolution, as if those who lived in Paris were somehow complicit in the crimes of those dark years. Well, this was not acceptable. The Méchain family had survived the awful year of 1789, when both of Thérèse's parents had died and the mob had invaded their home in the Observatory. Méchain himself had missed the horrors of 1794 when so many honourable men and women went to prison or death, and thank God he missed it, because who knew whether he would have been among them. But in the end, things had not turned out so badly for the Méchain clan. Instead of their little house, they now lived in the Observatory apartment. (And Cassini was perfectly content on his estate in Thury; he said himself that he wanted nothing more to do with science.) Now that inflation had settled down, Méchain's salary was worth more than Thérèse's inheritance ever was. And their new rights and liberties were nothing to belittle. So why did Méchain continue to hide in these remote provincial towns as if Paris were tainted, or as if he could somehow turn back the clock?[26]

She could see he was suffering, that he was exhausted by his travels, by the tedium and terrible burden of his guilt. But he had to think of his children's future, even if he could not contemplate his own. He always analysed everything from every angle, turning things over in his mind, working himself into a state. It was all so

counterproductive. As he himself wrote to Delambre, their sufferings were nothing alongside the horrible fate of 'the millions of
people who would have given their whole fortune to be in our
shoes, not having suffered for an instant the dire needs which have
befallen so many of our fellow citizens'.[27] Consider their sufferings
then, complete the mission, and come home . . .

Of course, we do not know what he told her, or whether he told
her his secret – *all* his secrets, for what man spends six years away
from his wife and does not amass a multitude of secrets? (And
what of her secrets – a wife who was conspiring with her husband's colleague to 'hand him over'?)

All we know we know by triangulation. All our knowledge of
nature, of history, of one another, or even (some might say) of ourselves, comes to us by mediation. Delambre and Méchain
investigated the shape of the earth by measuring angles along a portion of its surface. We investigate their relationship by taking
angular measurements of their letters to one another, their letters to
their *maître*, their letters to third parties. They learned about one
another through a similar angular measurement: a colleague triangulates the relationship between a husband and a wife; a wife
triangulates the relationship between her husband and his colleague;
and a husband triangulates the relationship between his colleague
and his wife. We learn about ourselves by comparing ourselves with
others, sometimes with others from the eighteenth century.

All we know is that Madame Méchain spent the next five
weeks by her husband's side, and that when they parted in the
gloomy village of Rieupeyroux, he still refused to join Delambre at
Perpignan. Her husband had rebuked her, she said, even lied to
her. Or so she informed Delambre when she wrote to him, as
promised, on her way home via Carcassonne.

Carcassonne, 1 September 1798

Citizen,

Having completely failed in my mission, and with a heart
pierced by a thousand griefs, I return to Paris. As some

business obliged me to pass through Carcassonne, Citizen
Fabre has shown me your recent letter, which has disturbed
me yet again. I will make every effort to relieve your worry. I
informed you in May that I would soon be leaving Paris to
take my place beside my husband. I promised you then to
send you news as soon as I had joined him. A thousand fatal
accidents conspired to prevent us from meeting in Rodez until
7 July. Since that time, I have begged him in vain to write to
you and to agree to measure the baseline of Perpignan with
you. Not wishing to upset me, he always answered vaguely.
For the first time, my husband has dissimulated with me. Not
knowing his intentions, what could I tell you? I realized I was
wronging you, but the situation dictated my actions. I only
know that Citizen Tranchot, who knows perfectly well what
he is doing, informed all my husband's acquaintances here
that he alone would measure the baseline, not Méchain.

I told my husband that only a bar-room savant would think
a T— could replace a M—, that everyone knew who was the
best man for the job, and that the matter would have to be
resolved by the Academy. Above all, I urged him not to throw
away the fruits of all his years of suffering and sacrifice for
something so idiotic. I decided to force the issue. I insisted
that I would not leave his side until he had finished all his
triangles and had been reunited with you. He thereupon
renounced the baseline irrevocably, and vowed that he would
concede all the glory to those whom fortune had favoured. He
said that he would never show up unless Tranchot had been
set aside, that in any case his presence in Perpignan would
hardly be agreeable, and that he would rather be sent to his
death than go to Perpignan. I lacked the will to oppose him
further. But forgive me if even now I lack the courage to
address any longer a subject that is killing me.

In the end, I managed to make him promise to inform you
himself of his resolve. I wrote to Borda and explained my
situation, no longer able to withstand the horrific blows which
assail my heart. I know my husband to be a man of talent and

virtue. I affirm and attest that his abilities and faculties have not been alienated in any way, that they remain unchanged, only his heart has been ulcerated by a man who has sworn to bring him down, to bring an entire family down. In my mind's eye, I have seen my husband covered with glory and public acclaim. You put it well, Monsieur, when you say that a moment which should be the happiest of his life may yet consign us to oblivion. I do not complain. I accuse no one, least of all my husband. It is the extreme sensitivity of his soul which has ruined him. He is more unhappy than blamable.

I have obtained from Citizen Méchain a promise that he will not break off his labour until his triangles are complete and the results sent to you and Borda. It's all I could do. My family calls me back to Paris, and I cannot stay here until the mission is done. When I left him on the first of this month, the stations of Rodez, Rieupeyroux and Lagaste were done. Puy St Georges, Montredon, [Combatjou] and Montalet are, I believe, all that remain. You can currently write to him at Lacaune in the Tarn.

Citizen Fabre would be glad to grant you the rendezvous which you wish to have with him. I hope with all my heart that it takes place. I will urge my husband to join you here.

I have the honour to be, with the most distinguished sentiments, your fellow citizen,

Madame Méchain.[28]

She had failed, she said, to convince her husband to join Delambre at Perpignan. She had failed, she said, to stay by his side until his triangles were done. Most tragically, she had failed, she said, to bring him to his senses. Méchain remained bitter and melancholic. Yet she had succeeded – despite her gloomy prognostication – in the central task. News of her impending arrival had shamed him into leaving Carcassonne to meet her at Rodez. And once she arrived at Rodez, she had restored him to his mission. Méchain was triangulating again. Though her name nowhere

appears in the expedition logbooks – an invisibility she undoubtedly sought – it is quite likely that husband and wife observed
together the angles at the cathedral of Rodez and the chapel of
Rieupeyroux, the two stations which linked Méchain's southern
chain with the northern chain of Delambre.[29]

So did Madame Méchain dissimulate with Delambre? Did she
hide, even from him, her role in her husband's revival? It cannot
have been easy for her to work behind her husband's back, in
league with men he considered his persecutors. She told
Delambre that Méchain had refused to come to Perpignan
because of Tranchot. And Méchain, when he wrote to Delambre,
told him much the same story. But in his letters to his friends in
Carcassonne, Méchain suggested that this was not the entire
truth. Méchain considered the Perpignan baseline 'his'. To join his
colleague there now, he said, would be to submit himself to 'the
authority and supervision of Citizen Delambre'.[30]

For his part, Delambre promised Madame Méchain that he
would bring her husband back to Paris, come what may.[31]

The International Conference was to begin two months hence and
Méchain had five stations left to measure, which was just about
possible. At the Pic Combatjou site, he observed from a fertile hilltop, then as now under cultivation. At the Montredon site, he
observed from a ruined castle all overgrown with trees, now a rendezvous for teenagers. At Puy Saint-Georges, he observed from
the ruins of a medieval abbey, with its freestanding Gothic arches
that still open on to the air – although a quaint orientation table
placed on the site in 1907 thoughtfully directs your attention
towards Paris (535 kilometres), Tokyo (14,050 kilometres),
Madagascar (9285 kilometres) and New York (7350 kilometres).

Three stations down, two to go. All well and good, except that
it was already mid-September, the savants had begun to arrive in
Paris and the two remaining stations – Montalet and Saint-Pons –
were located in the rough Montagnes Noires of the Languedoc
where 'malevolent persons' had pulled down the signals Tranchot
had erected the previous year.[32] All through the season he had

been doing battle with the locals and the weather. One signal had been sawn off, another burned to the ground, another dismantled for nails, another toppled by storms. In one town, a village joker told the peasants that a nearby signal was a new kind of guillotine, so the peasants tore that one down too. Even some local officials seemed to fear that the signals might serve as a secret aid to the republic's enemies. For his part, Méchain said he feared the 'fanaticism' of the people.[33]

His fears were not groundless. Thirty-five years before, a young cartographer working for Cassini III had been surveying from a church tower in the Montagnes Noires, not fifty miles away, when he had been hauled down from his ladder and all but hacked to death by a crowd who claimed that his 'sorcery' was sowing death among the villagers. He managed to escape, blood streaming from his head and hands, but the town officials were too intimidated to help him, as were the few strangers he met on the road. Only at nightfall did he manage to stagger into a neighbouring town and find asylum in an inn run by the widow Jullia. There, a doctor and surgeon tended his wounds. This story was well known to all geodesers and illustrated, sadly, a continuing hazard of the job. It was not so much that the young cartographer was a sorcerer, but that his sorcery was the sorcery of numbers. When surveyors came to measure the earth, peasants had reason to be fearful. In the judicial investigation into the attack, one villager explained that the cartographer was 'a sorcerer who had come to harm them and was the agent of taxes, and had come to increase the income tax, ruining them and causing them to die of hunger'. The court obliged the village to pay reparations and sent the ringleaders to prison. Local priests were told to command their flocks to leave cartographers in peace. But suspicion persisted. And the rough countryside, long famous for its bandits, now sheltered many new fugitives: refractory priests, unrepentant royalists, army deserters, draft-dodgers and rebels of all sorts, all reviled as bandits.[34]

As an agent of progress in this region, Méchain was obliged to call on official protection to ensure his safety. After local ruffians tore down his signal at Montalet for the fourth time, he asked that a garrison of militia post a seven-man guard at the site. The jagged Roc de Montalet rises from the pine forests like a shattered cathedral. Blueberry thickets grow at its perimeter. A plaque there commemorates Méchain's visit, and proudly records his troubles with the locals. Méchain spent ten days at Montalet, sleeping in a tent at the edge of the rock and scratching out daily letters, his fingers numb with cold. Behind his armed guard, he had slipped back into a dark melancholy. He poured out his soul on the page. 'I renounce it all,' he told his friends in Carcassonne, 'and the instant I complete my assignment I will abandon everything to seek, if I can find it, some refuge of obscurity and peace, the sole balm my lacerated and broken soul can bear.'[35] He feared he was losing his mind. How do you measure the world when the earth is turning beneath your feet? As he told Delambre, 'I have spent all my time in the cruellest anxiety, unable to concentrate on what I am doing, continually reproaching myself for the past because the present is unbearable and because I tremble for the future.'[36]

At the time, Delambre was waiting on the other side of the Montagnes Noires, no more than sixty miles away as the crow flies. He had arrived in Perpignan with his team in late July to prepare the baseline. Tranchot was of great assistance here, having laid the markers along the Grande Route back in 1796 and having already gauged the distance with a surveyor's chain. The Grande Route ran just to the west of an ancient Roman road, the Domitia, which two thousand years ago served as Hannibal's invasion route. It had been shifted slightly by the medieval Catalan rulers of Perpignan. Then in the middle of the eighteenth century the *ancien régime* engineers had transformed it into one of the king's magnificent highways, laying it straight and true from Perpignan to Salses, between dry vineyards on the left and the salt lagoons on the right. And just as Delambre arrived the republican engineers had begun reinforcing the road in anticipation of heavier traffic. In another hundred years, modern engineers would pave it with macadam and then asphalt. Today, it is known as the

N9 and it still runs straight and true, except where it swerves to accommodate a shopping mall.[37]

Geodesy may be a natural science, one which measures the size and shape of the earth. But geodesy is also a science which depends on human history and human works. To measure a baseline for their triangles, the geodesers needed a straight stretch of terrain, and what could be straighter than a Roman road, adjusted by medieval surveyors, rectified by *ancien régime* engineers and finalized by rational republicans?

On 6 August Delambre, Tranchot, Bellet and Pommard laid the first ruler. Because they could not afford overnight guards for the rulers (they were, after all, made of the world's most valuable metal), Delambre had retrieved Méchain's carriage from storage to ferry the equipment and the team to and from the work site every day. Their goal was the fortress of Salses, an impregnable ochre bastion which blended into the stony red terrain. An English traveller had called this 'the most barren country on earth'.[38] Suffocating heat alternated with the winds of the sirocco. The rulers had to be shielded from the sun to prevent them from overheating, lest they expand. Fiery gusts of desiccated air pushed them out of alignment. A flash rain storm forced the savants to take shelter. Then, towards evening on the thirty-sixth day, a pack of wild dogs charged their camp and scattered the rulers, over-turning an entire day's work. While his assistants laboured in the angry sun, Delambre sat in the shaded carriage, reworking the calculations. He had to correct for deficiencies in the thermometers on rulers 1 and 2, for a slight kink in the line of the road, for the bridge over the Agly river and for the forty-eight-foot increase in altitude from one terminus to the other. The price of precision is continual vigilance.[39]

The southern baseline took two days longer to survey than the northern baseline. But the results corroborated one another to a remarkable degree – proof, the savants said, of the care with which they had been measured. On 19 September the tedious operation completed, Tranchot and Bellet began packing the instruments into Méchain's carriage for their return to Paris.

'Only one obstacle holds me back in spite of myself . . .' Delambre wrote, 'Méchain.'[40]

Lalande had predicted as much. 'Our poor Méchain cannot finish,' he wrote to Delambre, 'and it is up to you to repair the damage done by his illness, or else two months from now we will be no further along than we were last year.' With the conference about to begin, there was no time to spare their colleague's feelings. Lalande told Delambre to take over and complete Méchain's triangles for him. 'You keep telling me you don't want to distress him, but when you deny he is ill it's as if some nutcase were to go around denying that you were an astronomer.'[41]

It was true. Méchain seemed to have taken the only honourable course for a savant who had failed in a mission whose purpose was perfection. He had had a nervous breakdown. His letters had become inchoate, inconsistent, obsessive. Every day he offered another reason why he had yet to finish, why he could not come down from the mountains, why he refused to return to Paris – ever. 'The truth is,' he wrote, 'that anyone who does not have cause to shed tears for the loss of those dear to them, fearing the loss of their own life and liberty, could hardly be sorry to leave this theatre of misery, except for those sick souls who crowd around the guillotine.'[42]

Delambre travelled to Narbonne, then to Carcassonne, to be as near his colleague as possible. The signal at Saint-Pons, Méchain's final observation point, was a remote mountain station, an arduous three-hour trek up from the ancient abbey town of the same name. But it was only a day's ride away. Delambre offered to come and assist his colleague with his measurements if he so desired. He could be by his side by evening.[43]

By return post, Méchain warned Delambre to stay away. Delambre should not waste his time by coming to Saint-Pons when the foreign savants were waiting for him in Paris. 'As it is,' he said, 'there is barely enough time for you to make the journey in good time. I will send you my results there as soon as I am done.'[44]

Delambre neither returned to Paris nor charged up the

mountain to finish the triangles. Instead he politely requested
that the two men meet. He would wait for his colleague to finish,
he said.[45]

He was working as fast as he could, Méchain replied. 'But I
cannot see through clouds, nor stand up to hurricanes which carry
everything before them.'[46] He had arrived at the Saint-Pons station in early October, and taken lodgings in the abandoned
Moulinet manor house, a half-hour walk to his signal. Moulinet
had belonged to the archbishopric before the Revolution. The setting is spectacular. The mountain slopes are densely forested: the
fallen leaves form a moist October bed around the mossy rocks.
Fox, deer and wild boar flash through the woods. When the view
is clear you can see a fourth of territorial France from the peak:
from the ragged blue-and-white bands of the Pyrénées to the
south, to the black-and-green ridges of the Auvergne to the north.
It would be a lovely place to make one's hermitage, if the affairs of
the world allowed it.

The days slipped by: ten days, twenty days, thirty days, forty
days . . . Still Delambre persisted. He persisted when Méchain
promised on 4 October to send all his data the next day, and was
still sending summaries a week later. He persisted when Méchain
promised on 13 October to finish tomorrow, and then wrote on 19
October to say he was still wrapping up. He persisted when
Méchain wrote on 22 October to say he would be in Carcassonne
the day after tomorrow, and then again on 28 October to say the
muleteer had cancelled the trip because a three-day storm had
flooded the roads. He persisted, although the foreign savants were
waiting and their French hosts had their reputations on the line.
He persisted, because Méchain had the data without which a
definitive metre was impossible.[47]

Méchain was running out of excuses. 'It would be an infinite
pleasure to meet you,' he wrote, 'though I should fear the occasion.'[48] He kept advising Delambre to leave. You are missing
your moment of glory, he reminded his colleague, your chance
to present the fruits of your seven-year labour. Which was why,
in the end, Méchain had to come down. He could risk his own

reputation, but he could not risk Delambre's, not when Delambre seemed so willing to risk his own moment of glory just to ensure that Méchain shared it. Delambre had let Méchain complete his mission alone – and in that sense, they had done it together. It was a remarkable act of friendship. Delambre's restraint brought Méchain down the mountain.

They met in Carcassonne in early November 1798 at the home of Gabriel Fabre, the criminal judge whose motto came from Seneca: 'A suffering man is a sacred being.' In Fabre's view, there was no man who deserved his fate less than Pierre-François-André Méchain, 'but it does not always lie within us to master our feelings'.[49] And Méchain was certainly a man of feeling.

For three arduous days, Delambre laboured to convince his colleague to return with him to Paris. Méchain tried every evasion. 'I am unshaken in my resolve not to return to Paris this winter,' he said.[50] He had repeated this as many times, in as many ways as he could think of. 'I will not change my resolution for anything in the world.'[51] He came up with half a dozen alternatives. He might spend the winter in Rodez collecting more latitude data. He could seclude himself in a mountain hermitage and polish up his calculations. 'In the springtime, we will see if my existence is still of some use somewhere.'[52] Perhaps he would return to Barcelona to verify his latitude data. Nor had he given up on the idea of extending his triangles as far as the Balearic Islands. To his Carcassonne friends he admitted that he now regretted turning down the job offers from abroad, and that he might yet 'seek his fortune elsewhere'.[53] Under no condition would he return to Paris. What kind of greeting could he expect there except 'reproaches, disdain and contempt'? His shameful behaviour was already public knowledge throughout France; must it now be paraded before all the savants of the world? 'I will not expose myself to this final humiliation,' he said.[54] Let Tranchot take all the credit. (He now referred to his former assistant as 'my director'.) He would accept any punishment the Academy doled out. He deserved no less. As for his family, his return could only aggravate their problems, 'burdened as I am with myself'.[55]

Yet once he had come down the mountain, what choice did he really have? Delambre would not return without the southern data, Méchain would not let go of his data, and Méchain could not allow Delambre to delay his own departure any longer. Ergo: Méchain had to go.

If the metre was a social convention, then the social conventions would have to be observed. If two savants had been sent on a mission to measure the world, then two savants must return – together. Science is a collective enterprise: its highest achievement is to 'make a contribution'. If Delambre wanted to claim his contribution, he needed to let Méchain make his. He needed to bring Méchain back to Paris with him – along with his data. Delambre also had one final card up his sleeve, and on the third day he played it. He showed Méchain a letter from the Bureau of Longitudes importuning him to return and promising him the directorship of the Observatory when he got back.[56]

They left in early November. Delambre had waited fifty days in all.

Although they had lied about Méchain's date of departure from Paris, there was no way to disguise the date of his return. The foreign savants had been waiting for two months. It was Lalande who trumpeted the news to his colleagues on 14 November. He had just received a message by post: Delambre and Méchain had arrived at the d'Assy country home in Bruyères-le-Châtel. Tomorrow they would be in the capital. The last triangle had been closed.[57]

Chapter Nine

THE EMPIRE OF SCIENCE

The most magnificent prizes are reserved
For those whom mathematics serve.
For triangles connected at enormous cost,
Never to be renounced, whether true or false.
. . .
Have they decreed new weights and measures?
Subjected old folks to the latest tortures?
To hoist a pint, or cut a yard of cloth,
Or adjust the hands on the family clock
Was the arc of the meridian really worth it?
We can cut our fabric without measuring the earth;
And if our calculations are not free of error,
To break old habits is still false rigour.[1]

LOUIS-SÉBASTIEN MERCIER,
Satires on Astronomers, 1803

Méchain returned to a hero's welcome. He scarcely had time to clean himself up before he and Delambre were ushered into a formal banquet hosted by the President of the Directory, the Minister of the Interior, the Minister of Foreign Affairs and the entire Academy of Sciences, all of whom were lined up to offer their collective (if belated) greetings to their prodigal expedition leaders and the visiting savants, obliging Méchain to accept their heartfelt congratulations, 'which they presumed I deserved,' he remarked, 'for the completion of my mission'. It almost pained him to admit it, but his colleagues even offered him 'a demonstration of their most tender friendship, expressing satisfaction with the accomplishments of the past and confidence for the future'.[2] And this was not all. In the days that followed, they confirmed his elevation to the directorship of the Observatory, the highest honour in French astronomy. They elected him temporary president of the Bureau of Longitudes. They crowned his head so high with laurels that he dared not look behind him.

'The first days are always glorious and festive,' Méchain confided to his friends in Carcassonne. 'Those that follow will be the days of trial . . . Will I then be able to fulfil the expectations for which they honour me now?'[3]

In his absence, Paris had changed. If its buildings were recognizable, their inner purpose had been reconsecrated. The same might be said for its inhabitants. The Panthéon had become a national mausoleum; the old nobility had given way to a new notability; and the Méchain family had moved out of the little house on the edge of the gardens and into the redecorated Cassini apartments in the main Observatory, where Méchain now ruled in the Cassinis' place. His children had grown up. His youngest boy had been six when he left; he was now thirteen and wanted to be an astronomer like his father. His oldest son had departed on a geodesic expedition of his own, working as an astronomical aide on Napoleon's expedition to Egypt. While Méchain had been travelling far from home for seven years, those he had left behind had travelled further still.

After months of delay, the world's first international scientific conference could begin. The pre-eminent savants of all the nations of Western Europe – all those nations, that is, allied to France by conquest or watchful neutrality – had gathered in Paris to settle together the true length of the metre. These were not men who could be easily fooled, and who would try?[4]

It was Laplace – the most eminent of them all – who had first proposed the conference. It would guarantee, he said, the metric system's universality. Let the final determination of the metre be set by an International Commission, and it would dissipate any lingering 'jealousies' caused by the decision to base the metre on a meridian that ran through France alone.[5] Let the foreign savants consider the metric system their own work, and they would ensure its spread to foreign lands. Privately, of course, Laplace assured Delambre that the gathering was a 'mere formality'. As the basic parameters of the metric system had all been set in advance, the foreign savants would come to Paris simply to rubber-stamp the preordained results.[6]

Not all the French expected their guests to be so docile. Commander Borda, the prime mover behind the meridian expedition, objected to the conference. If the meridian expedition had produced a metre based in nature, why did it need the imprimatur of the savants of all Europe? The truth did not care who spoke on its behalf.

But Laplace's proposal found two powerful backers. Minister Talleyrand, the perennial master of French foreign policy, was still committed to metric reform as a tool of international diplomacy, although France was now in a position to command rather than beseech. Where Talleyrand had once proposed that Britain and France cooperate on the new measures, his Foreign Office now invited only savants having 'at heart not only the progress of the arts and sciences, but also the glory of the nations prepared to collaborate in this undertaking'. The British were pointedly excluded.[7]

Laplace's other ally was the most junior member of the Academy of Sciences. Ordinarily, a young academician would not dare to intervene in such a momentous controversy between his seniors only one month after his election, but Napoleon Bonaparte was an extraordinary academician on several counts. For one thing, he had never published a scientific paper. His main claim to scientific fame was the fact that he had been Laplace's examination pupil at artillery school. He had no pretensions to original invention or research. Rather, Laplace had advanced his candidacy (over the marvellous Lenoir, among others) in the hope of allying the Academy to France's rising political star.[8]

For his part, the general had political ambitions, and science was part of his campaign. He did not merely cultivate the sciences, he cultivated the savants. He returned from his Italian conquests bearing Renaissance art and the latest theorems. He strode into Academy meetings to the applause of men and the cheers of women. He was no wilting Méchain, cowering from praise. In the throng after the meeting, Delambre expressed surprise that the general was back in town again so soon. 'I am indeed, and will dine with you tomorrow if you so desire.'[9] Over dinner, he spread his napkin on the table to illustrate a diagram of a new geometric proof from Italy. 'My dear General,' fawned Laplace, 'we have come to expect everything from you, except a lesson in mathematics.'[10] He was the universal man: blending thought and action, science and romance, inspiration and planning. He was as delighted as a child with his election to the Academy, and intervened immediately in its affairs. An international scientific conference on the metre dovetailed nicely with his own vision of a Europe unified under French leadership.

Invitations had gone out in June 1798 to savants from the Netherlands, Denmark, Switzerland, Spain and the Italian republics; those nations, in other words, which would form the nucleus of the League of Armed Neutrality directed against Britain. None of the savants of Britain, America or the German states was invited.

From the beginning, the French had expected America – their sister republic – to be the first country to join the metric system. They had been delighted when Jefferson dropped his preference for a pendulum standard at the 38th parallel (near Monticello), in place of a standard at the 45th (near Bangor, Maine), clearing the way for trilateral Franco-British-American cooperation. In 1792 a committee of the United States Senate even recommended this pendulum standard as the national unit of length. But when the French savants switched to a meridian standard that traversed

France alone, Jefferson became convinced that the French show of internationalism was a sham. Congress put off any consideration of the legislation.[11]

The French did not give up so easily on America, however. Soon after the passage of the metric law of 1793, they dispatched the naturalist-explorer Joseph Dombey to convey the new (provisional) standards to the United States in the form of a copper metre stick and a kilogram weight. In January 1794 Dombey set sail from Le Havre on the American vessel, the *Soon*. Unfortunately, a storm drove him to the Caribbean, to the fractious French colony of Guadeloupe. From there, his mission went from bad to worse. Local plantation owners imprisoned Dombey as an emissary of the radical Jacobin government. Released upon threat of violence by those loyal to Paris, he disguised himself as a Spanish sailor and boarded a Swedish schooner, only to be captured by British corsairs and escorted to the prison-island of Montserrat. There he died of illness that April.[12]

Miraculously, Dombey's papers and the precious copper metre bar and kilogram weight arrived safely in the United States (where they are still preserved in the Museum of the National Institute of Standards and Technology) and the French ambassador took up Dombey's mission with enthusiasm. Ambassador Fauchet said he was delighted to learn of the metric reform and expressed his confidence that 'an enlightened and free people would receive with pleasure one of the discoveries of the human mind, the most beautiful in theory, and the most useful in application'.[13] By this he meant the French people. He also hoped the adoption of the metric system in America would 'cement the political and commercial connections of the two nations'. His hopes were echoed in newspaper editorials urging all Americans – or at least, all educated Americans – to adopt the rational French measures voluntarily.[14]

For a time success seemed within reach. Fauchet was friendly with President Washington, who was friendly towards France, and the President asked Congress to reconsider the metric system. Washington had stressed the great importance of uniform measures in all three of his earliest State of the Union addresses.

Although this sort of repetition is almost always a bad omen, Fauchet still held out hope. In a coded letter sent back to Paris, he noted that American adherence to the metric system might well prove advantageous to France. 'Would it not make the People here more French if they shared in our knowledge; would it not bind them closer to us with commercial ties if they were subjected to our System of weights and measures?' He did worry, however, that Congress, having learned that the measures were merely 'provisional', would deliberate and delay 'as they so like to do'.[15]

While Congress dithered and America began a diplomatic rapprochement with Britain, Fauchet recklessly supported the Whiskey Rebellion, as a prelude to a great Jacobin revolution in the United States. This infuriated President Washington and prompted Fauchet's recall to Paris. Six months later the House of Representatives voted to adopt national standards based on a modified version of the English foot and pound. These were not the ordinary foot and pound, but standards fixed by scientific experiment, and divisible into sub-units of ten. The Speaker of the House urged passage. So long as each former colony had its own standards of weights and measures, national commerce would remain uncertain. This time, it was the Senate which killed the legislation by inaction.[16]

Would it have helped if Delambre and Méchain had completed their mission in 1794 as planned, and the metre had been declared 'definitive'? Or if Fauchet had been more prudent? It is hard to imagine that anything could have saved America from two hundred years of fruitless debate. Jefferson understood this very well. The United States Congress, he acknowledged, was dominated by a mercantile class hostile to France and fearful of surrendering their customary English units. On a question so immediate to their commercial interests, their views would always predominate.[17]

For its part, the British crown had been trying to reform its weights and measures for as long as the French, with as little to show for its pains. The Magna Carta's lofty promise of uniform measures had been buttressed by stern parliamentary decrees and reaffirmed by the Article of Union between Scotland and England, without curbing a diversity as confounding as the languages of Babel. Travellers needed to learn a new language in every parish or market town, one which 'no Dictionary will enable us to acquire'. The apothecary, the silversmith and the wool merchant all spoke distinct measurement dialects. The county of Hampshire alone had three different acres, plus a different bushel in each market town. This diversity produced 'cabal, delay, fraud, anxiety and indeed everything hostile to the good faith and confidence which ought ever to subsist between buyer and seller, agent and principal'.[18] 'Knaves and cheats' forced the poor to sell wheat in a large bushel and buy back bread in a small one. Indignant administrators condemned these practices as iniquities – which by the rules of transparent exchange they no doubt were – although the local people undoubtedly considered them integral to the just-price economy, which prevailed in much of Britain as well.[19]

British men of science, like their Continental brethren, added their voices to the call for uniform measures. Since the days of John Locke and Christopher Wren, members of the Royal Society had proposed standards based on nature, such as a new 'yard' defined as a pendulum beating for one-second intervals in the Tower of London (equal to 39.2 of the current inches). And in the eighteenth century the new thinkers called 'economists' likewise championed uniform measures as a spur to commerce.[20]

Then, in 1789, an obscure Member of Parliament named Sir John Riggs Miller urged the House of Commons to coordinate its metric reform with the French National Assembly. Miller convinced Talleyrand to allow the pendulum to be measured at a site jointly determined by British and French savants. In Europe's antique universal language, he expressed this antique universal dream.

> *Una fides, pondus, mensura, moneta, fit una,*
> *Et flatus illaesus totius orbis erit.*[21]

> One faith, one weight, one measure, and one coin,
> Would all the world in harmony conjoin.

The hard part about harmony, of course, is getting everyone to sing in tune. Since almost any standard will do equally well, so long as everybody agrees on it, everybody prefers that someone else make the change. Miller faced this obstacle both at home and abroad. Each of Miller's learned allies within Britain had his own idiosyncratic view of which standard would be best. And collectively, Parliament expected the French to follow Britain's lead, especially now that the French embrace of constitutional monarchy would cause them to be 'emancipated from national prejudices'.[22]

So when the French switched to the meridian standard, it killed the enthusiasm of even the most sympathetic British savants. Charles Blagden, Méchain's collaborator in the 1788 Greenwich–Paris survey, saw the Dunkerque–Barcelona project as a transparent bid to exclude all other countries from any say in the new measure. Once the war broke out, the British press began mocking the metric system as another instance of republican rationalism run amok. Miller's proposal died in Parliament.[23]

The same problem kept the Germans away. The patchwork sovereignty of principalities which had multiplied measures in what later became Germany also precluded any centralized solution there. Besides, the German savants likewise preferred a standard based on the pendulum to one based on a meridian whose value, they noted, depended on who conducted the measurements, and where, and with what instruments. It was just this sort of discontent that the International Conference was supposed to pacify.[24]

War kept the British, American and German savants away, but France's victories on the Continent made it impolitic for her neighbours to refuse her invitation. Bonaparte had brought the Italian peninsula under French rule. French occupation had reconstituted the Low Countries as the Batavian Republic. Spain had been forced into a sullen neutrality. Switzerland had been refashioned by the French as the Helvetian Republic. And the left bank of the Rhine had been rechristened the *départements réunis*, 'reunited' to greater France first militarily, then politically and now cartographically through the geodetic labours of engineer Tranchot.[25]

With metre sticks and maps, the French would manage an empire, uniting the tools of commerce and military might in the form of a geodetic metre based on the size of the earth. A transnational metric system would mould the European economy into a Continental bloc, while 'an army of astronomers', outfitted with repeating circles, would assimilate all the nations of Europe to a single grid. As Delambre put it: 'Now that the use of the repeating circle has spread throughout the Continent, one may hope that all Europe will soon be covered with triangles.' Indeed, the French were determined to extend their new metric revolution right around the globe.[26]

Bonaparte was not in Paris for the conference he had helped to convene. The modern Alexander the Great – the world conqueror and world civilizer – had left France on the most exotic metric expedition of all: the invasion of Egypt. At the core of his force of 54,000 soldiers and sailors was an 'academy' of 167 savants, including mathematicians, naturalists, chemists and geodesers. Their goal was both imperial and geo-scientific: to supplant a British Levant with the French civilizing mission and to reclaim antique civilization with the tools of modern science. Among the savants was the twenty-year-old Jerôme-Isaac Méchain, an astronomical assistant to abbé Nouet, the former monk Méchain had trained at the Observatory. Lenoir's son had come along too, to make any needed repairs on the team's repeating circle. While Méchain *père* triangulated his way through the south of France, Méchain *fils* mapped an empire: from Marseille to Malta to Alexandria, then up the Nile to Cairo. While his father sighted wooden pyramids in the Montagnes Noires, the son triangulated the Great Pyramids at Giza. While the father agonized in the monastic retreat at Saint-Pons, the son struck out with an expeditionary force for the fount of all scientific knowledge.[27]

In the summer of 1799, a team of Napoleon's savants headed up the Nile to Syène (Aswan), famous, as the expedition leader noted, for 'its proximity to the Tropic of Cancer and the measurement of the earth conducted by Eratosthenes'.[28] There, on the island of Philae, where the cataracts of the Nile poured from red granite cliffs, the savants carved their global position on the wall of the Temple of Isis:

R. F.

AN 7

LONGIT. DEPUIS PARIS, 30°34'16"

LATITUDE BOREALE, 24°1'34"

Among the sixteen savants who scratched their name below was the young MECHAIN.[29]

Geodesy would not only colonize space, it would colonize time. Built into the embankment of the nearby island of Elephantine, the expedition discovered the 'Nilometre', an ancient standard of length which gauged the great river's level. Comparing this ancient measure with the new metre seemed to suggest that Eratosthenes' estimate for the size of the earth had come within 0.4 per cent of the modern value. And when they looked even further back in time, to the origins of Egyptian civilization a full three millennia earlier still, the French discovered something more remarkable: evidence that the ancient Egyptians had also derived their standard measures from geodesy, building them into the design of the Great Pyramid at Giza. There really is nothing new under the sun. Already some *ancien régime* astronomers had speculated that the Egyptians had derived their standard unit of length from the base of the pyramid, itself said to be ⅟₅₀₀th of 1° of the circumference of the earth. Now, the expeditionary savants had discovered evidence that the perimeter of the Great Pyramid measured 1842 metres, which came to within a minuscule 0.5 per cent of the value of one minute of the earth's meridian. Peering into antiquity, the savants saw their own origins reflected back at them. Whether this was a coincidence or not, no one was prepared to say.[30]

The invasion was an imperial fiasco. Nelson destroyed the French fleet at the Bay of Aboukir, Napoleon slouched back to

Paris and his geodesers were left up the Nile. But the invasion proved a scientific success. The expedition mapped a possible canal through the isthmus of Suez. French archaeologists unearthed the Rosetta stone. And by agreement with the British, the remnants of the expedition, young Méchain among them, sailed back to France in October 1801, with their invaluable logbooks.

Other Frenchmen, meanwhile, were extending their metric rule even further afield. Where the diversity of measures had once hampered colonial trade, the metric system would coordinate a new overseas empire. During the *ancien régime*, the residents of then French New Orleans had complained that ship captains often gave short measures in their deliveries of flour, beef, lard and wine. The captains had always responded that these were not *short* measures, merely *different* measures. Not to be outdone, the colonists ran the deceit in reverse. Merchandise often arrived short-weighted from the Americas 'where they say trickery and bad faith are contagious'.[31] The crown ordered *all* parties to use standard barrels, filled to within one-sixteenth of true weight. But the official measurement bureaux – established to levy taxes and stop smuggling – never monitored more than a tiny portion of the colonial traffic. The metric system now promised to integrate the Caribbean colonies with the metropole, rationalizing transatlantic trade, just as the metre would tame the world.

Already the French state had dispatched circumnavigators armed with Borda's repeating circles to chart the globe. Between 1785 and 1788, the crown sent La Pérouse to explore the Pacific coast from Alaska to California, and then across to Asia and Australia. He sent back reams of precision data before ultimately vanishing in the Indian Ocean. Between 1791 and 1794, the republic sent Entrecasteaux on a mission to learn the fate of La Pérouse. He charted the Indian Ocean and some of the South Pacific archipelagos before succumbing to disease. In the short run, the French metric empire failed, but it would return.[32]

This project of global coordination depended on making the metre 'definitive'. Making the metre definitive meant the International

Commission had to guarantee its precision. So, the Commission focused its attention on the exactitude of Delambre and Méchain's seven-year mission.

But as yet Delambre and Méchain were not ready to present their data – or at least Méchain was not. In the meantime, as a pledge of their exactitude, the French offered to stage a 'theatre of precision' under the Commission's supervision, pitting Delambre against Méchain in a friendly competition to determine the latitude of Paris. Paris was one of those 'superfluous' latitudes chosen to exhibit the curvature of the earth as it arced from Dunkerque to Barcelona. Each astronomer would measure the capital's latitude from his respective site: Delambre from the rooftop of 1, rue de Paradis, and Méchain from the rooftop of the National Observatory, which he now directed. Both men would then adjust their measurements to converge on the Panthéon: offering 'authentic proof' of the excellence of the repeating circle and the skills of the savants who wielded it.[33]

The great dome remained essentially unchanged, a testament to the royal state's engineering prowess and magnificent investment. But the meaning of the building had been altered – again – along with slight changes in the decor. Delambre's old crow's-nest observatory had been torn down. The 52,000-pound statue of Fame had been declared too burdensome for the dome to support. Mirabeau, the first man to be pantheonized – and the first to be depantheonized – was now supposed to be repantheonized – except that no one could find his body. Descartes, France's greatest savant, was also being reconsidered for the honour.[34]

Delambre and Méchain both began observing on 7 December, although they did not embrace their roles with equal enthusiasm. In addition to the faithful Bellet, Delambre was also assisted by Charles de Pommard, the son of Elisabeth-Aglaée Leblanc de Pommard, a longtime intimate of Delambre and a gifted scholar of Latin in her own right, 'learned . . . but not pedantic'.[35] She and her son now resided on the rue de Paradis with Delambre, who doted on the boy. He was already six feet tall, with chestnut hair to enhance his grey eyes, and Delambre thought he combined 'much intelligence with a great love of work'.[36] The young man hoped to become an astronomer – or at least that was Delambre's fond hope. Each night that winter, they climbed to the rooftop above the Marais to observe the stars. Each morning, Delambre handed down their night-time observations to the International Commission.

Méchain did no such thing. He held fast to his data, as he always did – and in return the data tortured him, as they always did. After twenty nights and five hundred sightings, he announced that he would have to start from scratch. His data were inconsistent. This time it was the wintry cold of northern France which was apparently to blame, rather than the heat of Catalonia. His assistant seemed incapable of maintaining the level of the instrument. Or perhaps the refraction correction was askew. In any case, it was all too much. He was despondent, distressed, desperate. To Delambre, he complained that if he did not get acceptable results soon, he would 'renounce it all'.[37] To Borda, he admitted that his results were unacceptable, while 'Delambre obtains results that are as consistent as one could wish'.[38]

The self-doubt which had stalked Méchain through the mountains of southern France had followed him to Paris. Melancholic comparisons sapped his confidence. Every morning he learned that Delambre's results had arrived on the Commission's desk. And every evening he searched for 'the hidden defect' within his data and within himself.[39] He began to avoid his colleagues. He skipped meetings of the Academy of Sciences and the Bureau of Longitudes, over which he nominally presided. He even stopped attending sessions of the International Commission. He announced he was too busy gathering new data to discuss old results.

The Commission began to schedule its meetings at the Observatory so that Méchain could not avoid them. Delambre had to stall to protect his colleague. In an attempt to drag out the process, he revisited observations long after the results had stabilized. Yet Méchain refused to hand his data over for others to sort out. No matter how great his anguish, he insisted on bearing the burden of precision himself. Anything else was an abdication of his responsibility. He was not some lackey sent to gather chestnuts, but a savant entrusted to make delicate judgements. He was not a menial technician, but an emissary of the Academy, whose integrity underwrote his observations. He knew better than anyone which values were valid and which were not. It was his duty to choose . . . But which would he choose: Mont-Jouy or the Fontana de Oro? A full confession or an admission of failure?[40]

As the delays piled up, rumours began to circulate. The foreign delegates were not as docile as Laplace supposed. The Danish royal astronomer, Thomas Bugge, had been the first foreign savant to arrive in Paris. For three months, while he waited for Delambre and Méchain, he had been forbidden to begin his own calculations. The Academy expressly forbade all savants from publicly releasing their own estimates of the metre in advance of the official report. Bugge had begun to feel he was being used. He heard Lalande privately dismiss the whole operation as 'a charlatanism of Borda'.[41] Now, three months after Delambre and Méchain had returned, the Frenchmen had yet to present their geodetic data. Whisperers made slanderous accusations: the data were wretched, the mission had been botched. If the conference was not concluded by January, Bugge declared, he would return to his duties back home.

When rumours circulate through the world of cosmopolitan science, they circulate far and fast. A German astronomer wrote to Lalande with evident *schadenfreude* of 'the scandal of the new measurements'. He had heard from Bugge that the expedition's value for the curvature of the earth was apparently implausible and its geodesic measurements were 'worthless, poorly executed, inconclusive, and untrustworthy – which pains me greatly'.[42] 'These shameful aspects of astronomy are best kept hidden,' he purred with thinly disguised glee.[43] Nearer to home, an obscure amateur astronomer from the French provinces wrote to Delambre to express condolences that his 'zeal and skill had not produced satisfactory results'. Although he did not know the astronomer personally, he offered him this consolation: 'You have been poorly seconded.'[44]

When January ended without Delambre or Méchain presenting their data, Bugge acted on his threat. No sooner had he left for Copenhagen than he was attacked in the Parisian press for 'ridiculing' the metric project.[45] Yet his departure provoked the French into action. Delambre stopped covering for his colleague and formally presented his own data to the International Commission on 2 February 1799. It was an all-day ordeal. The commissioners went through each page of his logbook, querying each station, vetting each observation. In the end, they accepted almost all his data, including some results Delambre himself doubted. But as he himself noted, once data were recorded they became a sacred thing. At this remove from the time and place of observation, it was no mean feat to distinguish a flawed from a valid result. The commissioners had no choice but to trust the logbook and accept its inky assertions, no matter what the author himself now said. By the day's end, Delambre's triangles from Dunkerque to Rodez had been officially endorsed, as had his latitude data for the northern anchor at Dunkerque. Méchain was next.[46]

A few days later, Laplace himself paid a private visit to the Observatory. He had come to deliver an ultimatum. Méchain had ten days to hand over *all* his data. No further delay could be tolerated.

It is easy to picture Méchain's reaction: his eyebrows hitched high in supplication, his eyes searching for some sign of sympathy from a man who could see right down to the whirling nebula of dust that had formed the solar system. There was nowhere left to hide. Every excuse was worn out, every academic courtesy exhausted. Méchain's moment of trial had come. So he agreed to present his data 'without fail' in ten days, under one condition. Rather than supply his original logbooks – which he admitted were in a state of disarray – he would present the summary results for each station as corrected by the usual formulae. To this, Laplace secretly agreed. *That* was how badly the Commission needed Méchain's data.[47]

In the intervening ten days, Méchain discovered that a loose screw on his lower scope had been responsible for his erratic data, or so he informed the Commission. His results had now begun to converge, and he promised to make his presentation ten days hence. Indeed, they were approaching within 0.13 seconds of Delambre's. It was a stunning display of observational prowess. By locating his position on the surface of the earth to within thirteen feet, Méchain had demonstrated that, in the right hands, the precision of the repeating circle was limited only by the observer's patience. Naturally, the Commission agreed to wait.[48]

Ten days later he was still not ready to present his data. And ten days after that, he postponed the meeting again. Then, finally, on 22 March, Méchain presented the results of the southern expedition.

He arrived for the all-day session with his results copied out in a beautiful scribal hand. The commissioners subjected his results to the same ordeal that Delambre had faced. The angles for each station from Mont-Jouy to Rodez were vetted individually before being officially accepted. At times, Méchain considered the review 'a bit severe'.[49] But in the end the Commission was compelled to congratulate Méchain for the remarkable consistency of his triangles. As for his latitude data from Mont-Jouy and the Fontana de Oro, they too were found to be in superb order, and in remarkable conformity with one another. Indeed, their conformity was so great that, at Méchain's request, the International Commission agreed to set the Fontana de Oro data aside as redundant, and use only the data from Mont-Jouy.[50]

Just like that, the nightmare lifted. His anxieties, his fears and his sense of inadequacy all evaporated like phantasms. The International Commission had recognized his work as a masterpiece of astronomical precision. Indeed, one of the foreign commissioners privately approached Delambre to ask him why his results were not as precise at Méchain's. The tables had turned. Méchain had triumphed.[51]

All that remained was to boil down those concatenated results into a single number: the metre. For the next few weeks, each commissioner calculated independently, using his own preferred methods. The mathematician Legendre deployed refined calculations using ellipsoid geometry. The Dutch astronomer

Jan Hendrick Van Swinden made use of traditional geodetic techniques. Delambre employed improved methods that he had recently published.[52]

Borda was not present for these final calculations. The inventor of the repeating circle and the guiding force behind the meridian project did not live to see the metre become definitive. During Méchain's final procrastination the old commander, after a long illness, died. In pounding rain, a cortège of international savants bore his body up a muddy road for burial below Montmartre. His legacy was a conundrum.[53]

As each savant completed his geodetic calculations, it became increasingly clear that the rumours were justified: something was wrong. The meridian results were shocking, unexpected, inexplicable. Against all odds, the meridian expedition had produced something unanticipated: *genuine scientific novelty*.

This had never been their intention. Delambre and Méchain had not been sent out to unearth new knowledge. They had been sent out to refine to a nicer degree of exactitude what was already known. But the world, they had now discovered, was more eccentric than anyone supposed. Was it a scandal, or a discovery?

According to the data gathered fifty years earlier in Peru and Lapland, and confirmed by Cassini III in France, the eccentricity of the earth was approximately $\frac{1}{300}$ – which is to say that the earth's radius at the poles was $\frac{1}{300}$ (or 0.3 per cent) shorter than its radius at the equator. By contrast, Delambre and Méchain's data for the arc from Dunkerque to Barcelona suggested that the eccentricity was $\frac{1}{150}$, or twice as great. Even more startling, when the Commission plotted the curve tracked by the intervening 'superfluous' latitude measures at Dunkerque, Paris, Evaux, Carcassonne and Barcelona, they discovered that the surface of the earth did not even follow a regular arc, but shifted with every segment. It was a stunning discovery. But what did it mean?

The reversal clearly delighted Méchain. It was a vindication of sorts. His colleagues would now regret their refusal to let him triangulate as far as the Balearic Islands or take additional latitude measures. He took the experimentalist's perverse joy in baffling his theoretical colleagues – Laplace most of all. He gloated that, 'The earth has refused to conform to the formulae of my mathematical colleagues, who have insisted until now, with absolute certainty, that it is a perfectly regular spheroid of revolution.'[54] It was perhaps Méchain's one true moment of joy in the entire expedition. It was also a moment of great discovery, as he wrote to his Carcassonne friends.

> Our observations show that the earth's curve is nearly
> circular from Dunkerque to Paris, more elliptical from Paris
> to Evaux, even more elliptical from Evaux to Carcassonne,
> then returns to the prior ellipticity from Carcassonne to
> Barcelona. So why did He who moulded our globe with his
> hands not take more care . . .? *That* is what they cannot
> comprehend. How did it happen that by the laws of motion,
> weight and attraction, which the Creator presumably decreed
> before He set to work, He allowed this ill-formed earth to
> take this irregular shape for which there is no remedy, unless
> He were to begin anew?[55]

Progress, as so often happens, had slipped in sideways where least anticipated. Unexpectedly, the meridian project had produced new and baffling knowledge. Instead of a spherical orange of an earth, or even an oblate tomato of an earth, the geodesers now discovered that they lived on a lumpy pumpkin of an earth.

Everyone knew that the earth was not a perfectly smooth surface, everywhere becalmed at sea level. For a hundred years, savants had known that the figure of the earth was not a perfect sphere, but was flattened at the poles. For the past few decades, they had begun to suspect that its figure was not even an ellipsoid, but rather some more complex form of ovoid. Now they had discovered that its shape was not even that of a curve of rotation, a well-defined figure turned symmetrically on an axial lathe. They had discovered that we live on a fallen planet – a world buckled, bent and warped. And they had discovered this only because they were seeking perfection. Examined from a great enough distance, the earth appeared to be a sphere. Move in closer and it appeared flattened at the poles. Closer still – at the stunning level of precision achieved by Delambre and Méchain – and the earth was not even symmetrical enough to be approximated by a curve rotated through space. Delambre and Méchain had discovered that not all meridians were equal. The meridian that ran through Paris was not the same length as the meridians that ran through Greenwich or Monticello or Rome.

To some extent, even this startling discovery was not entirely unanticipated. Laplace himself, the foremost theoretician of geodesy, had occasionally wondered whether the earth was in fact a perfect spheroid of revolution, as all his models supposed. And Roger Boscovich, the Jesuit geodeser who had surveyed the Papal States in the middle of the eighteenth century, had already suggested that the meridian through Rome did not have the same curvature as the meridian through Paris. Indeed, this doubt had been one of the secret motives for the meridian expedition in the first place, and why the savants had added the 'superfluous' latitude points. Not that the savants had ever admitted as much. 'Sometimes to serve the people', one savant privately acknowledged, 'one must resolve to deceive them.'[56]

There was only one problem. This great discovery invalidated the guiding premise of the entire mission.

Delambre and Méchain had been sent out to measure the world on the assumption that their meridian, standing in for all the earth's meridians, could furnish an invariant and universal measure. Now they discovered that the world was too irregular to serve as its own measure. To be sure, Delambre and Méchain had surveyed only one meridian. That one meridian, however, was sufficiently irregular to suggest that every other meridian would also be irregular, each in its own way. In any case, there would be no simple way to extrapolate from their one small sector of the meridian to the whole, which was the task they now confronted.

In that sense, the results *were* a scandal. But then again, truly new knowledge almost always is.

The members of the International Commission now faced a stark choice. They could extrapolate from the Dunkerque–Mont-Jouy arc to the full quarter-meridian using either the new eccentricity of ¼₅₀, or the older eccentricity of ⅓₃₄. They had every reason to believe that the eccentricity of ¼₅₀ offered the best description of the arc as it passed through France, but they knew that the older data offered a more plausible picture of the overall curve of the earth. They could choose consistency or plausibility. And after some heated discussion, they chose plausibility and the old data. Delambre and Méchain had been sent out to remeasure the world with supreme accuracy, and in the end the single factor that made the greatest difference to the final determination of the metre was based on the very data they had been sent to supersede.[57]

The decision shaved a thin slice off the length of the metre. Where the provisional metre had measured 443.44 *lignes*, the definitive metre measured 443.296 *lignes*. The difference of 0.144 *lignes* (or about 0.325 millimetres, or 0.013 inches) may seem insignificant, about the thickness of three sheets of paper. But it was considerably more than the uncertainty Borda had anticipated. And as paper stacks, so did that difference: it added up to a change of some two miles (3.25 kilometres) in the total quarter-meridian. It was also, we now know, a step in the wrong direction. The definitive metre deviates twice as much from what we now know to be the size of the earth as does the provisional metre. Seven years of labour had only succeeded in making the metre less accurate.

The final step was to embody the metre in a permanent physical standard. Copper had been sufficient for the provisional metre, but for the definitive metre only the ultimate metal would do. Long despised as a contaminant by South American prospectors, platinum was impossible to melt, difficult to purify and nearly indestructible. For just that reason it had acquired a lustrous reputation among savants. It promised to outlast time. Just before the Revolution, an arsenic process had been discovered which made platinum sufficiently malleable to shape into snuff boxes and ornamental vases. After the Revolution, a grander purpose had been found for the new metal: the creation of permanent metrical standards. The Commission of Weights and Measures spent one-fifth of its budget buying and refining some one hundred pounds of pure platinum. Even so, the Commission nearly missed its quota. The final shipment from Spain had been short-weighted by 15 per cent, and the commissioners had to scramble to replace the loss.[58]

It fell to Lenoir to make the final cut. The dwarfish artist was now fifty-five. His repeating circles had made him world famous, a peer of the best instrument-makers of London. In April 1799 he was supplied with the calculated value of the definitive metre and four bars of pure platinum, and told to shape four standards of precisely one metre each. For this purpose he employed a 'comparator' of his own invention, which could gauge objects to within one millionth of a *toise* (0.000072 inches). The task was 'diabolically tricky'. Of the four bars, the one that came closest – within 0.001 per cent of the proper length – was selected as the definitive metre.[59]

In a grand ceremony held on 22 June 1799, this platinum bar was presented to the French legislative assemblies so that the people's elected representatives could add the consecration of man's law to that of nature. It was a solemn ceremony, the occasion for speeches of global import. Laplace reminded his audience that a metre based on the size of the earth made every landowner a 'co-owner of the World'.[60] And the Dutch astronomer Van Swinden expressed gratitude for the iron facsimile that each foreign savant would carry back to his homeland to help 'tie together' the peoples of Europe 'with fraternal bonds'.[61]

Needless to say, no one mentioned the unexpected discovery of the eccentricity of the world, nor its subversion of the seven-year project. And no one mentioned the fact that the platinum metre bar and kilogram weight would have to be whisked back to Lenoir's workshop after the ceremony for further preparation and not returned to their triple-locked box in the National Archives for another nine months. The making of science – like the making of laws and sausages – was best kept out of public view.[62]

But beneath the grandiloquence, it was easy to detect an undertone of plaintive hectoring. Everyone in the chamber that day knew the French people had yet to embrace the new measures. The problem, they all agreed, was not that the people were secretly loyal to the *ancien régime*. The problem was that the people were still attached to their old routines. The President of the Assembly sadly cited this wise saying of Jean-Jacques Rousseau: 'Men will always prefer a worse way of knowing to a better way of learning.'[63]

In the months that followed, the legislature ordered its citizens to start learning. This task of instruction fell mainly to the Agency of Weights and Measures. The Agency's directors included the gifted mathematician Legendre, plus administrators committed to a free-market economy. For several years now they had sought to inspire their fellow citizens with their own passion for the new measures. Even bureaucrats can believe in what they are doing. 'I dream only of weights and measures now,' one said.[64]

Over the previous five years, the Agency had distributed tens of thousands of pamphlets to persuade citizens of the law's simplicity: some a hundred pages long, others broadsheets for shopkeepers' windows. Prieur de la Côte-d'Or designed conversion graphs for those citizens who could read graphs. Commercial publishers had also sold guides to the new measures, including almanacs, paper dial-up 'converters' and educational playing cards. The Agency had also mortised marble-encased metres into the walls of prominent Paris buildings. (The last surviving example can still be seen on the rue de Vaugirard across from the Palais du Luxembourg.) The Agency had even hired a blind man named Duverny to lecture on the metric system under the archways of the Louvre. His message was simple: justice is blind, the scales must balance, and the metric system is easy.[65]

Finally, to help citizens cross into the new metric world, the Agency ordered each *département* in France to draw up a table to translate its old measures into the new. Lengthy as they were, these tables overlooked much of the diversity of the *ancien régime* measures. Local administrators admitted that they had been unable to locate all the old 'master' measures; that they had barely broached the multiplicity of land measures; and that they had

The five of hearts

This Revolutionary playing card – the five of hearts – is named 'Quintidi'
after the fifth day of the week. The card informs us that when the sun is
due south on the meridian the time will no longer read twelve o'clock, but
five o'clock. It explains: 'We reckon by 1, 2, 3, 4, 5 o'clock.' These cards
were made by Jean-Pierre Bézu in 1792 in the town of Château-Thierry,
then known as Egalité-sur-Marne.

(Bibliothèque Nationale de France, Est. Kh383 no. 227.)

necessarily suppressed all mention of the anthropometric practices
that defined most *ancien régime* measurement. From these hundred or so tables, the Agency then compiled an abbreviated
national summary so that 'at last the French will no longer be
strangers in France'.[66] But the real danger was that they would no
longer feel at home in their own parish. Where citizens had once
needed a dictionary to travel from one town to the next, they now
needed one to travel into the future.

The Agency recognized it was not enough to produce pamphlets, marble-encased standards and numerical tables; twenty-five million French men and women also needed to be able to lay their hands on ordinary rulers. Paris alone needed 500,000 metre sticks. Yet one month after the metre became the sole legal standard there, the Agency had only 25,000 sticks in storage. To spur production, it contracted with private manufacturers and transformed churches into factories. They promised to reward citizens who would invent machinery capable of cutting metre sticks 'with precision and promptitude'.[67] If anything was amenable to mass production, surely it was identical standards. But when citizens finally did track down metre sticks for sale, they found that rulers from the same shop differed by a millimetre or more.[68]

So far, the metric system applied only to the city of Paris. Yet even in the capital undercover police reported that merchants still sold cloth by the *aune*, if only because their customers preferred the slightly longer measure. Enforcement was impossible. Every time the police confiscated an *aune* and referred the violator to a criminal court, the criminal courts sent the case back to the police, who could only impose minor fines.[69]

One story making the rounds was the exception that proved the rule. Apparently, a woman from the Pelletier district of Paris had returned home one day from shopping for fabric thinking she had bought an *aune* of cloth, only to discover that she had received a metre. She went to Judge Delorme to complain.

The woman: Monsieur –
The judge (*interrupting*): What do you say? I am no *Monsieur*.
The woman: I beg your pardon, *Citizen*! Last Sunday –
The judge (*impatient*): What do you call Sunday? We have no such thing now.
The woman: Well then, the – the – the Quintidi of the week.
The judge (*angry*): You tire me with your nonsense! I know nothing of *weeks*.
The woman: But, Mons— Citizen, I mean to say – the decade in the month of – of – April.

The judge: Again your nonsense! *April!*

The woman: Of floréal, I should say. I bought two ells.

The judge (*furious*): Enough! You mean a *metre*. Go your
ways. You still have your *Sundays*, your *weeks*, your months
of *April*, your *ells* and your *Monsieurs*! Get out of my court.
You are an aristocrat![70]

After September 1799, when the legislature introduced the
metric system to the region surrounding Paris, complete confusion
reigned. Police inspectors insisted on the new measures; customers preferred the old measures; and shopkeepers stocked both.
This invited the very abuses the new system was supposed to
eradicate, giving shopkeepers yet another means of short-changing
customers. The *Almanach des gourmands*, the city's premier guide to
fine restaurants and upmarket grocers, warned its clientele that
butchers and bakers – especially by the Porte de Saint-Honoré –
were using the new measures to cheat customers, rounding up
prices or dishing out smaller portions.[71]

Yet the savants still could not understand how the common
people could reject the new measures. The new measures were
derived from nature and reason; the entire system formed a logical whole. The members of the Agency warned its critics not to
quibble with the new system:

You cannot attack a part of the system without endangering
the whole. Otherwise many different objections will follow:
some will want a new nomenclature; others will want the
metre to be based on the full circumference of the earth; still
others will prefer the pendulum; and still others will revisit
the idea of a duodecimal system to ease division, etc . . . Now
that the law is promulgated (after long deliberation), it is best
not to attack it, but to give it the respect it is due . . . There
must not be any doubt about the goodness of the law.[72]

However slow their progress, the savants held out hope – the
perennial hope of those already enlightened – that the next

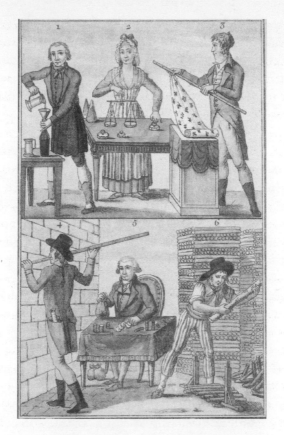

The use of the new measures

These cheerful republican citizens are demonstrating (clockwise from top left) the proper use of the litre, the gram, the metre, the stere (one cubic metre), the franc, and the double-metre.

(Photothèque des Musées de la Ville de Paris. Photo: Svartz.)

generation would see the light. They made instruction in the metric system obligatory in the nation's schools, including the Ecole Normale, where the nation's teachers themselves were trained.[73] They assured their fellow citizens that the metric system would never be imposed by force, nor become an instrument of tyranny. The metric system, they said, 'is simply a police measure to ensure the social order . . . Neither *our good pleasure*, nor *our full power* are part of the lexicon of a reasonable people, whose enduring obedience will only follow if they are enlightened and convinced.'[74]

This liberal creed, however, did allow for the possibility that public opinion was something the nation's leaders could both interpret and direct. That is why the Minister of the Interior saw no paradox in his simultaneous assurance that 'uniformity of measures has always been desired by the people', and his boast that the metric system, as designed by the nation's leading savants, 'would be a splendid instrument for moulding public reason'.[75] The goal remained the same: the metric system would transform the French economy and ease administration by transforming the *thinking* of French citizens, making them into rational calculators who conceived of their interests in a new way. No wonder this transformation was slow in coming. It was also vulnerable to shifting politics, as successive French governments reconsidered the role of the state in the economic life of the nation.

Paradoxically, the only way the state could enforce the metric system was to reregulate the nation's marketplaces. In 1799, a few months after the metre had been declared definitive, the government authorized each major market town to set up its own Bureau of Weights and Measures. Just as the state licensed pharmacists to prevent poisoning, so would the state now license a Bureau of Weights and Measures to prevent the metrical mistrust which poisoned commerce. In return, each privately owned Bureau might charge a small fee for its services. To some, this signalled a return to the hated seigneurial dues of the *ancien régime* and a restriction on the absolute right to trade wherever and however one pleased. The Paris Bureau, run by Brillat and Company, was denounced as 'despotic' and 'tyrannical' after it sent hundreds of government troops into Les Halles to drive out the old-time weighers.[76] Critics warned that the metric system would never take hold if the people were forced to use it at bayonet point. For their part, Brillat and Company claimed that these actions were needed to restore confidence to commerce, make trade fair and prevent the metric system from being held up to ridicule. In this way, the new weights and measures became the wedge by which the government revived the distinction, familiar to the *ancien régime*, between the regulated public *marketplace* (limited in time and location, so that all might have equal access) and the unregulated *free market*.[77]

The metric system was not in itself a guarantor of free commerce (though some kind of uniform measurement system is usually a prerequisite). The metric system could just as easily be used to enforce the state's regulation of trade. But in either case it was designed to break the hold of the old anthropometric measures and the old just-price economy.

The French savants consistently overlooked the rational motives ordinary citizens had for rejecting the metric system: the disruption it caused to community norms, plus the fear of opening local markets to outside competition. In many cases the old units could not even be adequately translated into the new terms, since that would involve thinking of objects apart from the labour and materials that had gone into their making. This was something that many peasant and artisanal producers were understandably reluctant to do, whatever gain they might expect as consumers from transparency in commercial dealings.

Indeed, it was not only conniving merchants and ignorant peasants who rejected the metric system. The nation's most educated citizens clung to their old measures just as tenaciously. The old units had permeated the work routines of *all* French people, including government officials and professionals. For the numerate, it is worth repeating, numbers matter. Many French physicians, having just switched from a 'medical pound' to a 'commercial pound', worried they would have to relearn all their dosages. In 1796 provincial notaries had yet to switch to the new system. In 1797 state surveyors had to be scolded for not using the metre. In 1798 accountants in the Department of the Treasury were still refusing to use the decimal system for sums of money. And in 1799 Paris administrators were still employing the old measures in their official correspondence. Even the national legislators continued to publish new laws in the old measures, in violation of their own laws. The crowning irony came when the central Office of Weights and Measures shipped a set of the new metric standards to a provincial branch office and informed them that the total package weighed sixty *livres*, *poids de marc* (or sixty pounds, old style).[78]

And in those sectors of the economy where the change in measures meant retooling machinery or altering bureaucratic practice, resistance was adamant. The artillery service, the branch of the military most committed to uniformity, precision and modern manufacturing – and Napoleon's old corps – had initially planned to publish a metric edition of their cannon blueprints, but the War Office called the publication too expensive. By 1801, the shoe was on the other foot, and the War Office was pleading with the artillery to adopt the metric units. But metric units, the artillery now complained, would ruin the mathematically precise ratio between the cannonball's weight and its calibre, and would undo the uniformity of matériel they had taken such pains to establish.[79]

As for Napoleon, he refused to learn the metric system. Inspecting a gunpowder factory in Essonnes, he quizzed the plant's manager about the chemical processes in considerable detail. But every time the manager supplied him with the weight in kilograms, Napoleon insisted that he restate the formula in *poids de marc* (old-style pounds). He said he could not think in the new units.[80]

Faced with this obstinacy, the government temporized even as it persisted. One year after the metre was made definitive, the first compromise was made. On 4 November 1800 the metric system was at last declared to be the sole measurement system for the entire nation, and use of the nomenclature of compound words – decimetre, kilometre – was abolished. The metre was still the metre, and its use would be obligatory throughout the nation as of September 1801. But the Greek and Latin prefixes 'which frightened the people' were replaced with 'ordinary names'. After consultation with Laplace and Delambre, the decimetre was renamed the *palme* (the hand's-breadth), the centimetre the *doigt* (the finger's-breadth) and the millimetre the *trait* (the trace) and so on.[81]

The instigator of this compromise was none other than Napoleon Bonaparte, back from Egypt. To honour their colleague's return, the Academy had struck a commemorative medal from the residual platinum left over from the making of the metre. That way, they said, the medal would last 'almost as long as your glory'.[82] Thirteen days after accepting the medal, Napoleon seized power in the coup d'état of 18 brumaire – and held it for the next sixteen years. One of his first acts was to make his old mathematics examiner, Pierre-Simon Laplace, Minister of the Interior with responsibility for enforcing the nation's laws and the metric system. The savants, it appeared, had bet on the right general. Imagine their dismay, then, when they learned of his compromise. Laplace tried to reassure his colleagues: a retreat on the nomenclature did not imply that the entire system would fail. Lalande smirked, 'Monsieur Laplace is not in his place.'[83] After only forty days in office he was turned out in favour of Napoleon's brother. Other retreats were to follow.

The French were not only the first nation to invent the metric system; they were also the first to reject it.

Chapter Ten

THE BROKEN ARC

The fact is that nowhere, these days, is anyone genuinely happy, and that of the countless faces assumed by the Ideal – or, if you dislike the word, the concept of something better – travel is one of the most engaging and most deceitful. All is rotten in public affairs: those who deny this truth feel it even more deeply and bitterly than those who assert it. Nevertheless, divine Hope still pursues her way, assuaging our tormented hearts with the constant whisper 'There is something better – namely, your ideal!'[1]

GEORGE SAND, *Winter in Majorca*

Redeemed from his suicidal melancholy in the remote Montagnes Noires, Méchain had been elevated to the celestial post of the nation's chief astronomer. Honoured by the nation's leaders for his scientific integrity, he had been welcomed into the arms of his loving family and respectful colleagues. No wonder his conscience troubled him.

To expiate his sins, Méchain threw himself into his new administrative role. He complained that the Observatory had been neglected during his absence, and he vowed to transform it into the world's premier astronomical facility. He bought superior telescopes and decided upon their placement in the building. He resumed his own celestial observations, discovering two comets in 1799 and another in 1801 and joining in the exciting hunt for the

newest members of the heavens, the minor planets known as
asteroids. Yet he was miserable.[2]

Friends and colleagues were baffled. He enjoyed every reward
a savant could desire: a capable wife, a warm family, the most eminent position in his field, the respect of his peers and last but not
least – this was Paris, after all – the sumptuous Cassini apartments and full use of the Observatory gardens. Some colleagues
found him reserved, severe, even caustic. Some, behind his back,
took great pleasure in reviling his character. Yet everyone agreed
he was a savant of unimpeachable integrity.[3]

These public accolades only made his secret more unbearable.
The more honours that were showered on him, the more unworthy he felt. He was a hollow impostor, a scientific fraud – and in a
matter of such terrible consequence. He had introduced an error
into the fundamental scientific value, the measure which would for
ever more serve as the foundation for all scientific and commercial
exchange. He avoided his colleagues, retreating to his apartment
whenever one of them came to the Observatory.[4]

Meanwhile, his former partner had been given the honour of
composing the official account of their expedition. The choice
was understandable. Although Delambre was the junior colleague (five years younger in age, with ten years' less tenure in
the Academy), he combined a background in the humanities with
the necessary technical know-how – plus he had actually performed the bulk of the work. Delambre planned to publish
everything: the tale of their adventures, the full roster of their
data, all their formulae and technical apparatus. The work would
be entitled the *Base du système métrique* (*The Foundation of the Metric
System*) and would run to two thousand pages in three thick volumes. It would show the world the exactitude of their labours. To
complete the first volume by the end of the year, he needed
Méchain's data.[5]

No wonder they resented each other. Their mission was
done, yet they remained tethered together. Delambre needed
the complete data of a man who had decided – out of spite, it
must have seemed – to supply the absolute minimum, and in his

own good time too. As for Méchain, everyone knew he had only completed his mission with Delambre's help: reason enough for him to begrudge Delambre's success. On top of all this, Delambre was one of the First Consul's favourites – Napoleon had taken a liking to him on his first day at the Academy – whereas the First Consul hardly knew who Méchain was. Méchain resented Delambre's facile way with words and pretentious classical learning. He despised the ease with which the draper's son moved in the highest circles of the new France.[6]

The fall from camaraderie was hard. For seven years the two savants had trekked the high geodetic plateau, first in opposite directions and then on convergent paths, but always in a spirit of collegial rivalry. But a year in the capital had turned them into petty quarrellers. Paris can do that to people. There is something about the proximity of money and power which makes a person peevish. Late in 1800 Delambre was elected temporary president of the Bureau of Longitudes, making him Méchain's nominal superior. A squabble ensued over who controlled the account books. A question was raised as to whether Méchain could officially claim the title of director of the Observatory. Méchain wrote bilious letters denouncing his 'hyper-pedantic and outrageously ambitious' colleague who had overstayed his term as president of the Bureau of Longitudes.[7] Méchain complained that he had been reduced to begging his colleague for firewood and candlelight. Privately, he referred sneeringly to Delambre as his 'absolute master'. Publicly, he threatened to resign unless he was formally installed as Observatory Director.[8]

Deeper grievances festered. As Méchain now saw it, Delambre had deliberately deprived him of his rightful honours by relegating him to a secondary role on the meridian mission: grabbing two-thirds of the triangles, muscling in on the latitude of Paris and seizing both baseline measurements – including the one at Perpignan, which clearly lay within Méchain's sector. And Méchain had proof of this. Late in 1799 he had become Borda's scientific executor and had thereby gained access to all the commander's papers. Among them he discovered the series of letters between Delambre, Borda and his own wife. One can imagine the reaction of Méchain – a man given readily to paranoia – as he read of the conspiracy unfolding against him: their plan to lure him down from the Montagnes Noires, his wife's secret mission to Rodez, their exchanges of confidences, their vows of silence. They had manipulated him, treated him like an underling and tricked him across the finishing line.[9]

By the time Méchain had reclaimed full recognition as director of the Observatory, Delambre had risen higher still. In 1801 Napoleon added the presidency of the Academy of Sciences to his other titles, making him ruler of both the nation *and* the sum of its knowledge. Napoleon's first act as president was to reorganize the Academy, appointing Delambre to the post of Permanent Secretary. This made the draper's son the most powerful figure in French science: successor to Condorcet, liaison to the highest political authorities, author of its eulogies and hence guardian of his colleagues' reputations.[10]

These circumstances go some way towards explaining why, on 6 September 1801, a member of the Bureau of Longitudes (Méchain, presumably) proposed extending the meridian measure south of Barcelona, as far as the Balearic Islands. The extension had long been one of Méchain's fondest ambitions, and he insisted on his right to report on its feasibility. Such an extension would improve knowledge of the figure of the earth by anchoring the arc's southern latitude on an island, where nearby mountains could not distort the readings. The new arc would straddle the 45th parallel of latitude, making any extrapolation from the partial arc to the quarter-meridian less sensitive to the earth's eccentricity. These were excellent scientific rationales for the expedition, and Méchain agreed to report on them – if, and only if, he was allowed to lead it.[11]

Why would a fifty-seven-year-old man, reunited with his family after seven years of backbreaking travels, wish to undertake such a mission? Delambre and the rest of his colleagues argued that the task be given to someone younger. Méchain was needed in Paris where he had begun to accomplish wonderful things at the

Observatory. Moreover, he had just recovered from a severe illness which had nearly cost him his life, and which he himself blamed on the 'long travails and thousand vexations of my mission, as well as those which followed upon my return'.[12] But the more his colleagues protested the more adamant he became. As director of the Observatory and the nation's senior astronomer, Méchain had the right to send whom he chose. And he chose to send himself.

Méchain had something to prove. He would prove he did not need Tranchot to lay out geodetic triangles. He would prove he could traverse an arc as great as Delambre's. He would prove he did not need his wife's help to complete a mission. Above all else, he would prove he could be trusted. Beneath their accolades Méchain could sense his colleagues' scepticism. (They were professional sceptics.) He had yet to release all his data for the latitude of Barcelona. He still had not handed his logbooks over to Delambre. Only a new expedition could redeem his reputation, the thing he held most dear.

He had a secret motive as well. By extending the arc to the Balearic Islands, he would leapfrog the contradictory latitudes of Barcelona and fix a new secure southern anchor for the meridian. This was not a task he could hand over to someone else. Another savant, triangulating from Barcelona, might discover that its latitude did not match the published results. Already Alexander von Humboldt, on his way to South America, had stopped in Barcelona, checked himself into the Fontana de Oro, and set up his own repeating circle on the hotel terrace. He had taken latitude measurements there, he said, to follow in the footsteps of the illustrious Méchain. Was no place on earth safe from these prying savants? Was no fact of nature secure from their meddling? Mercifully, the young German had devoted only one night to observation, and his results did not contradict Méchain's findings. But he had posted his data privately to Delambre, and Delambre had noted some minor discrepancies.[13]

In the end, however, all these rationales pale before the only motive which ever really justifies scientific labour. To triangulate across the Mediterranean to the Balearic Islands would require measuring a geodetic triangle with sides 120 miles long, whereas ordinary triangles had sides of forty miles at most. It was a stupendous challenge: an expedition across uncharted terrain. Méchain had long held the extension to the Balearic Islands 'close to his heart'.[14] It was a challenge he could not resist.

That is not to say that the mission served no practical purpose. As Méchain himself astutely noted in his proposal to Napoleon, the expedition would cement the 'intimate union' between France and Spain.[15] The Balearic Islands occupied a strategic position in the western Mediterranean. Indeed, the British navy had occupied the island of Menorca in 1798 to intercept Napoleon's invasion of Egypt, and Spain had only recovered the island in March 1802 when Britain and France signed the Treaty of Amiens. That treaty had ended a decade of warfare and opened the world's sea lanes to France. Yet the peace was precarious, with Britain and France already probing for advantage. In September 1802 Napoleon – on Delambre and Laplace's say-so – approved this scientific thrust into the western Mediterranean.

So scientific history repeated itself: the first time as epic, the second time as quixotic farce. The expedition began, appropriately enough, with Méchain on one of his periodic upswings. He spent 1802 assembling a team. This time, he intended to do things right. To maintain the repeating circle, he recruited a young naval engineer named Dezauche. For diplomatic cover he invited a former student of his, Jean-Baptiste Le Chevalier, who had just returned from a year spent in Madrid, ostensibly as an apostle for the metric system, more probably as a spy. And for moral support, Méchain took along his younger son, Augustin, now a strapping lad of eighteen, born in the grounds of the Observatory and home-schooled in astronomy.[16]

Méchain also adapted his equipment for the mission. He refitted his Borda circle with more powerful lenses to triangulate far across the Mediterranean. And he procured outsized parabolic reflectors – some of them specially ordered from London – to take accurate night-time readings. By early 1803 all the human and material elements were ready. He reminded his team to bear in mind that 'never before and never again will anyone undertake so vast and important an operation under the watchful eyes of all the savants of Europe, subject to their criticism and to those of the centuries to come'.[17]

As before, Méchain's final act before leaving for Spain was to hand over a document: not a power of attorney this time – Madame Méchain had already been authorized to administer the mission's 20,000-*franc* budget and run the Observatory in his absence – but rather the data Delambre had waited three years to publish. This consisted of his geodetic results described in indifferent prose, and the same summary data he had already supplied to the International Commission.[18]

Méchain expected to complete his mission in six months, and be back in Paris in ten. He planned to leave in early February, measure the triangles before the summer haze set in, observe the new southernmost latitude on the island of Ibiza during the winter, and resume the directorship of the Observatory by spring. He planned to do everything right this time. But the usual delays – as inevitable as they were unanticipated – prevented him from leaving Paris until 26 April. After passing through Perpignan, he and his team sailed into Barcelona harbour on 5 May 1803 – at which point everything fell to pieces.[19]

Nothing in Spain was ready. Everywhere he turned he found obstruction, incompetence, conspiracy. On his first day in Barcelona the Governor-General informed him that Madrid had yet to supply the passports he would need to travel to the islands. Next he learned that Captain Enrile of the frigate *Prueba* (*Test*), who had agreed to transport him across the straits and assist in the measurements, had been held up in the port of Cartagena, apparently on orders from Madrid.[20]

These obstacles – so his Spanish friends informed him – were no accident. Father Salvador Ximenez Coronado, director of the Royal Observatory of Madrid, hated France, hated the French Revolution and considered the metric system a 'fantastical lie' to pervert Spanish virtue.[21] José Chaix, the Observatory's vice-director, who

had come to assist Méchain, warned the Frenchman that Ximenez
Coronado was 'ignorant, malevolent and a mortal enemy of the sciences and all who cultivate them'.[22] From Madrid he was blocking
any assistance for the expedition. At last Méchain had a real conspiracy to contend with.

Behind these petty intrigues loomed the prospect of renewed
war between France and Britain. Spain hoped to stay neutral,
but seemed likely to be drawn into the conflict over the
Mediterranean sea lanes. Méchain's Paris colleagues asked their
British counterparts to intercede with the British navy and grant
safe conduct to the peaceable scientific mission.[23] In the meantime,
Méchain deferred his sea voyage. Instead, he set off with his team
down the Catalan coast, scouting out new stations in the mountains south of Barcelona.

By attaching a new chain of triangles to his old chain via the
stations of Montserrat and Matas, Méchain intended to bypass
his old measurements at Mont-Jouy and the Fontana de Oro.
During July and August, through the all-consuming Catalan
summer, he sized up stations as far south as Montsia, an isolated
2200-foot peak which marked the border where Catalonia ended
and the Spanish province of Valencia began. There, on the dusty
mountain above the pink flamingo marshes of the Ebro delta, he
was joined by Enrile, and they began triangulating their way
back north towards Barcelona. All that autumn they worked
through torrential rains and gale-force winds. It was just
Méchain's bad luck that the mild Catalan autumn had suddenly
turned ferocious. By late October he found himself back up at the
Montserrat monastery, enjoying its thousand-year-old tradition
of hospitality and climbing once again to the Notre Dame chapel
to take his measurements on top of the organ-pipe stone pinnacle.
(Five years later, the monastery would go up in flames during
Napoleon's invasion.) Then in early November he went back
down to Barcelona to prepare again for his trip across the
straits.[24]

During his six months in Spain he had measured five coastal
triangles, but he had yet to learn whether his mission was feasible.

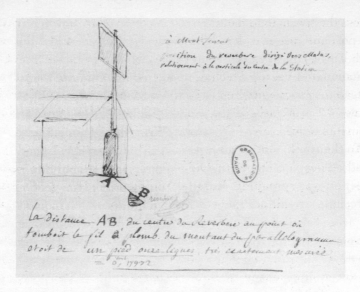

A reflector above Montserrat

This drawing by Méchain shows the position of the reflecting mirror he placed in front of the portico of the small chapel of Notre Dame on the pinnacle of rock high above the monastery of Montserrat in Catalonia.

(Photo: Observatoire de Paris.)

The evidence so far was discouraging. Success depended on measuring a giant triangle across the straits to his arc's final anchor, the southern island of Ibiza. Tightrope-walking his way up and down the Catalan coastline, Méchain had strained to catch sight of the island with all the telescopic power at his disposal. His Spanish hosts had sworn that he would be able to see Ibiza as clear as daylight from the summit of Montsia, but so far neither he nor Captain Enrile had been able to make it out through the autumn mist and rains. The Spaniards, Méchain feared, had lied to him.[25]

The time had come to cross over to the islands, reverse perspective and see for himself whether the coastal range was visible from there. The French ambassador in Madrid had finally procured him a passport. But no sooner had the *Prueba* pulled into Barcelona harbour to ferry her captain and learned passenger across the straits than half the crew died of yellow fever. Terrified, the port authorities ordered the ghost ship to sail for quarantine – with Enrile courageously volunteering to resume command of his contaminated vessel. Méchain begged his friend not to go, and instead to remain with him in Barcelona 'where he could be useful to me'.[26] Enrile insisted that his duty lay with his vessel and the remnants of his crew.

To the south, Andalusia was infested with yellow fever. Three hundred people a day were dying in Málaga alone, and the zone of infection was spreading. Rumour and panic surged up the coast. Barcelona's wealthier citizens fled to the countryside before the city was sealed shut. The French government deployed a cordon of troops along the frontier to prevent any incursion of the disease. Stuck in Barcelona – lodged once again in the Fontana de Oro – his six-month supply of funds exhausted, Méchain's courage began to fail him.

It was like a nightmare. On a December evening a decade earlier, from the tower at Mont-Jouy, he had sighted a signal flare on the island of Mallorca. Then war and injury had snuffed out his ambitions. Now war and disease threatened to cut him off again, while his hand-picked team deserted him. Three days after the team returned to Barcelona, Le Chevalier, Méchain's former student, set off for southern Spain in search of classical antiquities. Then Chaix left too, returning to the safety of Madrid. It was not just fear of disease. Both men complained that Méchain would not let them even look through the repeating circle.[27]

Méchain had never understood the art of leadership: when to share responsibility, as well as when to assume it. For him, the quest for precision was like a voyage through purgatory, with each savant answering for his own sins. And Méchain was too absorbed in his own self-criticism to let others shine. He showed no tolerance for even their most trivial mistakes. When he opened a case of poorly packed reflectors, it was enough to convince him that he had erred, even that once, in trusting others. The ulcerating question, of course, was whether he could trust himself.[28]

To whom could he turn in this moment of self-doubt? To none other than the man he most resented, his erstwhile partner, Jean-Baptiste-Joseph Delambre. A series of plaintive letters, as pathetic as any he had ever written, spilled from his pen. The winter was cold and the Spanish had no notion of indoor heating. What should he do? Which plan should he adopt? Might the Bureau of Longitudes supply him with additional funds? He even offered to return to Paris if the Bureau thought his 'feeble lights' could be better employed in the capital. 'This is truthfully and sincerely my current situation, my dear colleague, and I tell you all this without complaint . . .'[29]

It was as if nothing had changed, nothing had been learned. His son and the loyal Dezauche would stick with him to the end. To replace his renegade assistants he enlisted the aid of a Trinitarian monk named Agustín Canellas, a self-proclaimed astronomer, confident of his worth and eager for a role in a historic expedition. More valuable was the support of a local grandee, the Barón de la Puebla, an amateur astronomer from Valencia. The baron assured Méchain that Ibiza would be visible from the coastal mountain of Desierto de las Palmas, south of Montsia in the province of Valencia; better yet, he offered to establish a signal on its peak while Méchain observed from the islands.[30]

Finally, in early January 1804, Méchain arranged passage to Ibiza for himself and his son on the *Hypomene*, a Spanish frigate named after the youth in Greek mythology who dropped the distracting golden apples as he raced the swift Atalanta. No such swift passage ensued. Méchain's bad luck at sea held. A simple one-day crossing became a three-day torture of calms, contrary winds and high seas. Unable to enter the main port at Ibiza Town, the *Hypomene* swung east around the island to a cove off Punto Grosa. Yet no sooner had they dropped anchor than a troop of armed islanders assembled on the shore to deny them landing. They would not even deliver a letter, for fear it would spread yellow fever. The ship had only two days of provisions remaining on board. Food and water were running low. Every effort to sail out of the cove failed. The *Hypomene*'s captain pleaded with the islanders to convey their situation to the governor, across the island. Two days passed before a response was shouted back: the crew might cut wood and collect water in an isolated spot while the governor looked over the mission's official papers. These were duly transmitted, after first being carefully doused in vinegar. Reassured that the travellers were free of infection, the governor sent word that Méchain and one naval officer might search out a suitable station on the island.[31]

Ibiza today is an 'international resort destination'. Originally settled by Phoenicians, Ibiza Town wraps around its conical hillside like a white Moorish turban, the eyes of its cubical houses turned south towards Africa. The interior of the island is mountainous and was then sparsely populated. Poverty existed side by side with Edenic fertility: figs, almonds, grapes, melons and olives grew in abundance. Palms fringed the coast and pine trees covered the rugged hillsides. But Méchain's unlucky star had followed him on to this island paradise. Ascending a rocky trail to the peak of Los Masons, he fell off his mule, injuring his head and spraining his wrist. He refused to call a halt. 'It was nothing . . .' he wrote to Dezauche, still in Barcelona, 'and you may laugh over it at the table of the Fontana de Oro, or wherever else you wish.'[32] Worse than any physical injury, however, was the disappointment that awaited him at the top of the Los Masons. He had been told the peak afforded the best view of the mainland coast, and indeed he could see a range of mountains to the west, as well as the big island of Mallorca to the north. Unfortunately, he could not make out the peak of Montsia, the southernmost station in his chain of completed triangles. The Spaniards had indeed lied. 'I thumb my nose at them,' he wrote.[33] This left him with two alternatives, each of which presented its own complications. He could return to the coast to extend his mainland triangle chain south into Valencia before triangulating across the straits to Ibiza – although this would mean that his chain would temporarily veer far west of the meridian. Or he could build his chain of triangles by island-hopping: triangulating from Barcelona to Ibiza via Mallorca – although this would mean measuring several giant triangles as

Méchain's expedition to the Balearic Islands

In 1803–4, Méchain made several attempts to extend his chain of triangles south of Barcelona to Ibiza, the southernmost large island in the Balearics. He successfully measured a series of triangles along the Catalan coast as far as Montsia – indicated on the map by solid lines – but he found he was unable to triangulate from the coast across to Ibiza. This prompted him to consider two alternatives. One plan involved 'island-hopping' via Mallorca – indicated by dashed lines – and would have required a number of long-distance triangulations across the straits. The other plan – indicated by dotted lines – required Méchain to extend his coastal triangulations further south into Valencia and only then attempt to triangulate across the straits.

(Map by Chris Robinson.)

well as a baseline on Mallorca. Either way, the season for latitude measurements was ending and his budget was almost spent.

Not only that, but when he viewed the entire circumference of Ibiza from the top of Los Masons, his heart sank still further. He could not locate the *Hypomene*. The ship was no longer in its cove off Punto Grosa, nor had it docked in the main port at Ibiza Town. The boat had vanished, along with his son and his instruments. It was enough to make a savant curse his fate. He prepared himself for the eventuality that 'my last leave-taking from my family and friends was my eternal adieu'.[34]

Again he wrote to Delambre. Might his colleague recommend a course of action? What did the Bureau of Longitudes think of his two alternatives? Would the island-hopping scheme succeed? Would the westward deviation along the coast distort the results? And while he was at it, might he vent his frustration? 'Hell and all the plagues it spews upon the earth – storms, wars, pestilence and dark intrigues – have been unleashed against me. What demon still awaits me? But vain exhortation will solve nothing, nor complete my task.'[35]

The *Hypomene*, he soon learned in Ibiza Town, had sailed to Mallorca for provisions. So while he waited for a response from Paris, he likewise booked himself a passage across to the big island. On 27 January 1804 he sailed into Palma, Mallorca's capital, a busy port of thirty thousand inhabitants, dominated by a hulking ivory-hued cathedral. He spent nearly two months on Mallorca – reunited with his son – an island known since Roman times as 'the fortunate isle'.

Mallorca was more populated than Ibiza, and four times its size. Its northern mountains reached an elevation of five thousand feet, and were covered with snow when Méchain arrived in the winter of 1804. Yet beneath the white peaks its plains were tropical, with groves of oranges, almonds, palms, dates, figs, carob and plantain. Ruined temples lay scattered across the island. In the twelfth century the Balearic Islands had governed a continental kingdom that included Catalonia and much of southern France. The local Catalan dialect embodied millennia of exchange and conquest, with phrases from Syriac, Greek, Latin, Vandal, Arabic and Castilian.

It was indeed an enchanted isle, a refuge from time's march. The town of Palma was ruled by a mechanical 'sun clock', which rang from the Gothic town hall. Legend had it that the sun clock had been brought to the island by Jews from Jerusalem. More probably, the fabled instrument was installed by fourteenth-century Dominicans. The clock divided each day into twelve hours, but hours which lengthened proportionately as the summer days lingered and shrank proportionately as wintry darkness shortened the daylight. Eighteenth-century commentators considered the clock unsuitable for rational administration, but they admitted that the people of Palma found that its bells helped them to regulate the watering of their sumptuous gardens. Standardization lies in the eyes of the practitioner. Sadly, a few decades after Méchain's visit, the sun clock vanished as mysteriously as it had come.[36]

While he waited for the mountain snows to melt, Méchain conducted astronomical observations in Palma with his son, including the viewing of a dramatic solar eclipse. Not until March did he set out across the island for the north-coast town of Sóller in a fertile valley of orange groves. From there, Méchain and his party – his son, Captain Enrile and a band of sailors – rode mules as far as the last manor house, then climbed on foot to the peak of Silla de Torrellas. The path was steep, rising nearly one mile in altitude within two miles of the coast. The lower slopes were planted with olive trees, gnarled and shaped by centuries of wind and pruners' knifes, and ringed by stones to ward off erosion. Further up, the pine forests had been logged to build ships for the Spanish navy. Higher still, in the crags, lay the nests of sea eagles and bearded vultures, which circled apprehensively on the currents. At the summit, they found the traces of the expedition that Méchain had sent across to light signal flares on the mountain a decade earlier, including stakes marking the line of the meridian. Far below, the glassy Mediterranean was streaked with pale blue bands that wandered to the horizon like rivers upon the sea. To the north they could see Barcelona; to the south Ibiza. Which meant that they could indeed triangulate their way through the Balearic Islands – if the Bureau of Longitudes in Paris approved Méchain's island-hopping plan.[37]

Méchain had essentially settled on this solution when Delambre's response finally reached him in mid-March, three months after his enquiry. The Bureau of Longitudes recommended the coastal plan. As Delambre demonstrated mathematically to Méchain, the deviation to the west would not distort the results. This coastal plan, moreover, required only one large sea triangle; whereas the island plan would require at least three. Finally, a baseline could be measured more easily along the mainland shore than on an island. To be sure, admitted Delambre, he was far away, while Méchain was on the spot; hence Méchain alone must decide which plan would ensure the most precise results. He wrote: 'I wait with much curiosity, and look forward with much interest, to the good report of your journey in Mallorca.'[38]

Méchain deferred to his colleague's recommendation. At least the coastal chain was a sure thing, whereas every time he set foot on a boat disaster ensued. This meant, however, that his two months on Mallorca had been a waste of time. He ordered his team to summon their energy for one final foray along the southern coast. Though the prospect was exhausting, he warned his subordinates not to relax their vigilance. Continual self-surveillance is the only protection against error.

> Even I, who can claim some experience and competence [in geodesy], who know a bit about what methods to use and when to take precautions, even I work in constant fear. I mistrust myself. I continually solicit the views and intelligence of my colleagues at the Academy and the Bureau of Longitudes, and nothing pains me more than when they respond that they rely entirely on me, and that no one is better placed than I to judge what must be done, to choose the right methods, and to carry them through. At such times I feel as if they are spitting in my face. Nothing comes easily, nothing is simple, when one seeks precision. All it takes to be convinced of this is to do a little observing of one's own.[39]

In early April 1804 he sailed back across the straits for Valencia, where he hoped to procure passports to scout out stations along the coast. For six weeks, during the finest season for geodetic surveying, Méchain waited for passports in the city of gaudy church spires and pungent yellow dust, an honoured guest in the residence of the Barón de la Puebla, while his Spanish friends did bureaucratic battle with the diabolical director Ximenez Coronado. Méchain was impatient to begin. The sun was stirring vapours from the sea, drawing up miasmas from the coastal plains. The heat accumulated daily. The season of disease was approaching. If Méchain did not measure the coastal stations soon, he would not be able to triangulate across to Ibiza until the following winter.[40]

As soon as his passports arrived in mid-June he set out. In eighteen days he covered some three hundred miles on horseback, accompanied by Enrile's second in command, zigzagging his way through valleys and over ranges. The royal road to Madrid, then under construction, ran straight across the plains for thirty miles until it hit the mountains; there it turned into a narrow track, impracticable for carriages and hazardous for horses.

Along the coast fishermen pulled their triangular-sailed vessels up on to the sand under the palm trees. On the plains irrigation systems, dating back to Moorish rule, fed cotton fields, orange groves and stands of mulberry trees (for silkworms). The mountain slopes to the west were terraced for olives. Lizards swarmed over the rocks, some of them a foot and a half long and fierce enough to intimidate dogs. In all, Méchain located fourteen more stations, including two end-points for a baseline that would run alongside the Albufera lagoon, a shallow brackish body of water surrounded by rice fields and teeming with flamingos, herons and numerous waterfowl. The lagoon, connected to the sea via sluice gates, was notorious for its vapours and pestilent insects. In the morning the wall of the inn would be dark with satiated mosquitoes.[41]

Méchain wrote to his wife that the sun had roasted him alive, scorching his face black as an African's, except where his skin was peeling off. The summer heat was only now reaching maximum

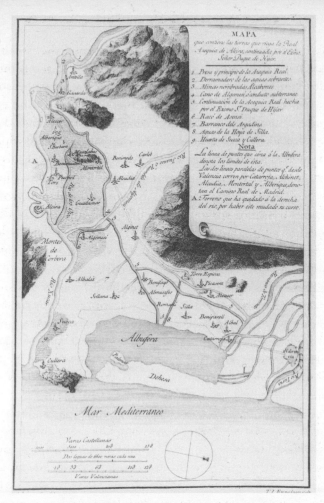

The Albufera marshes, near Valencia

This map shows the region where Méchain caught malaria. The area surrounding the Albufera marshes was cultivated with rice fields, and the lagoon itself teemed with flamingos and other waterfowl. The level of water was controlled by a sluice gate at the entrance to the Mediterranean. Méchain's southernmost station was on the rocky promontory near Cullera, located in the bottom left corner of the map. The town of Valencia is located in the bottom right corner. (Note: This map is oriented so that north is located to the right.)

(Antonio José Cavanilles, *Observaciones sobre la historia natural, geografía, agricultura, población y frutos del reyno de Valencia* (Madrid: Imprenta Real, 1795–7), 1, p. 184. Photo: University of Chicago Library, Special Collections Research Center.)

intensity. During the day the views were obscured by vapours. This meant they would have to use signal flares and so take their triangulations at night. As a precaution, he approached the archbishop of Valencia, a tall Franciscan in tobacco-stained robes, who tended to punch petitioners in the face when they bent to kiss his ring. Would he instruct his priests to warn their parishioners not to harass the strange men with the strange instruments who burned lights at night on the mountain tops? Hostility towards the French was fierce. Despite the presence of Castilian officers – or perhaps because of them – Méchain's party had been threatened on several occasions.[42]

In early July the team dispersed to conduct their night-time triangulations, beginning at the town of Cullera on the southern edge of the Albufera marsh. While Méchain set up his repeating circle on a rocky outcrop seven hundred feet above the rice fields, each of his collaborators – Captain Enrile, the priest Canellas, loyal Dezauche and young Augustin – led their respective groups of sailors to the surrounding mountain-top stations and directed their reflectors back at Méchain.

Two weeks later, after Méchain had tacked inland to La Casueleta to set up a new signal by the Roman aqueduct, his collaborators repositioned their reflectors on the surrounding stations. Then, another two weeks later, when he tacked back towards the coast, setting up his repeating circle on a hillock outside the town of Puig, his collaborators again adjusted their reflectors' positions. By now it was August and the heat was at its most intense. Rather than lodge in the miserable town, a quarter-hour walk down the road, Méchain decided to camp out in tents on the hilltop, two hundred feet above the dense coastal air.

Yellow fever had begun to claim new victims. And mixed in with it was something else: what eighteenth-century physicians called 'tertian fever' (*fièvre tierce*). Already it had claimed the life of one of the sailors requisitioned to transport the instruments. On his way to join Méchain at the station outside the town of Puig the sailor had been hospitalized in Valencia, where he died four days later. Several other members of the expedition had been enfeebled by illness. A naval officer, sleeping in the tent beside Méchain on the hillock outside Puig, was seized by a violent fever in the middle of the night. He had to be transported the next morning to a monastery down the coast, then removed to a more salubrious location. Captain Enrile had also fallen ill, though he had since recovered.

More frustrating still, the monk Canellas had inadvertently cost Méchain two weeks of labour. His error of calculation had resulted in a misplaced signal. Those measurements now had to be redone – further proof, if any were needed, that Méchain could not trust others to do his work for him. The monk had also since fallen ill, of a 'demi-tertiary fever', and had been bled three times. As a result of these delays Méchain was still outside Puig in late August. He had just turned sixty. In the town at the foot of the hill the disease was subsiding, although three or four deaths were still being reported every day from the town of Puzol a mile up the road. They could hear the funeral bells tolling while they laboured. Méchain's son was still atop La Casueleta; the death toll in the nearby town of Chiva had reached five a day.[43]

In his last letter to Delambre – another ten pages in his dense crabbed hand – Méchain admitted that he was exhausted. 'For until this moment I have not proved successful, and my unlucky star – or, rather, fate – which, as you say yourself, my dear colleague, seems to preside over this mission, hardly gives me leave to hope that I will bring it to a happy conclusion.' Yet he did not fear hard work or the scorching heat; he did not fear anything – except failure. He would continue until he succumbed, he said – which admittedly seemed likely, given that one by one everyone around him had already fallen ill and 'I am not made of harder stuff than they, nor am I younger, nor more robust, nor better acclimatized'. He was even ready to come home if the Bureau of Longitudes could find a savant to replace him, one 'more able, less maladroit, and luckier than me'.[44] He did not think that finding such a person would prove difficult. He had but one consolation, paltry though it might be: he had nothing to reproach himself with. He made every effort to fulfil his mission. As he wrote to a friend:

As for the rest, I tell you that though I do not seek death, I am far from fearing it. I would watch its approach without the least regret and in my current state would even consider it a gift from heaven . . . Never, no never, though I have spent much of my life in suffering and shed many tears over my loved ones and myself, never, I say, have I found myself in a situation so hopeless, so terrifying and so wrenching. This dreadful commission, whose success appears so far off and so improbable, will more than likely be the end of me, and worse yet, that of my family, and become my tomb and that of my honour.[45]

It was as if, having failed to die on his first trip to Spain, he was now mounting a more determined effort. In frustration, he handed over the final measurements at Puig to Dezauche – the first time he had trusted someone else with the repeating circle – so that he might prepare the next station. He ordered his son to shift his position sixty miles to the north and set up his reflector on the high mountain of Arès. He would himself move inland to the Sierra de Espadán, a three-thousand-foot mountain peak covered with pine trees.

Three days after he set up camp at Espadán, Méchain felt the first thrill of fever. At night, while he waited for the signal lights, icy chills crawled under his clothes. His body shivered in tune with the stars. His appetite had vanished. All that week he took no food, only hot tea. At night, the high dry air was filled with the scent of wild herbs: rosemary, thyme, lavender and mesquite. One evening, overcome with exhaustion, he fell asleep before the reflectors were lit, and the night watchman did not dare wake him when the lights finally came on. In the morning, Méchain bitterly condemned this failure.[46]

On 12 September, although the observations were not quite complete, his companions convinced him that he had to leave the Sierra de Espadán. He was emaciated and his fevers, though intermittent, were growing more intense. He agreed to be taken to the provincial capital of Castellón de la Plana, an eight-gated city of eleven thousand inhabitants, a mile from the coast and the hometown of his

A view of the Valencia coast

This panorama over the Spanish coast of Valencia captures Méchain's last view as he descended from the Sierra de Espadán to the town of Castellón de la Plana (labelled 'b' in the illustration).

(Antonio José Cavanilles, *Observaciones sobre la historia natural, geografía, agricultura, población y frutos del reyno de Valencia* (Madrid: Imprenta Real, 1795–7), 1, p. 110. Photo: University of Chicago Library, Special Collections Research Center.)

new friend the Barón de la Puebla. He could see its solitary octagonal belltower as he rode down through the fertile fields of sugar-cane and hemp. Once in town, he took a room in an inn. At first, his illness did not appear serious, but he passed a horrible night. Summoned by an urgent letter, young Dezauche rushed down from his station to join his expedition leader the next morning, just as the baron arrived from Valencia. Together, they transferred him to the baron's local residence. There he passed his final days, recorded with vivid sympathy in Dezauche's private journal.[47]

Friday 14 September 1804: I go to see Monsieur Méchain. I find him well enough, but very weak because he refuses to take anything, not even chicken broth, and has not eaten in eight days.

Saturday 15 September 1804: I go promptly in to see M. Méchain and find him quite well, even gay, but still very weak.

Sunday 16 September 1804: At nine in the morning, a servant begs me to come quickly to the baron's. There, I learn that M. Méchain passed a horrific night, and that since morning his mind has been distracted. I go in to see him. He opens his eyes very wide but does not recognize me.

Monday 17 September 1804: At seven in the morning, I go to see M. Méchain. I find him in the grip of a violent fever, in delirium, his mind vacant, not knowing what he says or does. The people looking after him find it difficult to confine him to bed. At nine, two doctors arrive, neither of whom inspires much confidence. They prescribe cinchona. I write to his son to come immediately. The doctors return at noon, and again in the afternoon. The patient has been enervated by his morning outburst. The doctors agree that he has a nervous tertian fever, which does not mean much to me. They summon the baron and me into the antechamber to tell us that the fever has turned deadly and that they cannot answer for their patient surviving another attack that night. Then, to ease their conscience, they ask that M. Méchain be confessed. The news stuns me. When M. Méchain himself is sounded on this topic, he indicates that he does not feel as bad as all that. I spend all day with my patient, turning him to urinate, and carrying him to the toilet when he needs to use it. Having heard the likely course of his illness from the doctors, I resolve to spend all night with him, not wishing to leave him to the servants, who are country people and cannot understand him. Around four-thirty in the afternoon I find him much better. His mind is again clear, and his speech orderly. At nine at night I give him cinchona, at ten-thirty a broth, and so on in alternation until three in the morning. My patient is getting better and better. He has no fever but is still excessively tired. Finally, at six-thirty in the morning, I leave him in a good state and hand him over to the care of the baron.

Tuesday 18 September 1804: At seven-thirty I return to M.
Méchain; he is doing very poorly and has lost consciousness.
Four doctors, whom I summon, all say he has malignant
tertian fever. At noon they say he has an ardent fever. That
night, I have him given extreme unction out of fear that he
will die at any minute. Then the doctors apply the Spanish
fly – a blistering agent – to the back of his head. They put a
compress on each of his feet. I send for M. Lanusse, the
French commercial attaché in Valencia, and ask him to bring
a surgeon and some cinchona, since the supplies here are no
good.

Wednesday 19 September 1804: I spend all night awake
watching over my patient. He will not drink. He rejects any
medicine as soon as it is given him. He is unconscious; his
eyes, half-closed, are yellow, as is his complexion. At six in the
morning, the doctors return to examine the patient. His arms
are trembling and they decide he is apoplectic. He is still
unconscious, his eyes open and vacant. He still refuses to
swallow. At ten in the morning, they remove the blistering
agent, which has taken well; the wounds are dressed with a
poultice of pear leaves and honey. At noon, the doctor
prescribes a febrifuge to reduce the nervous heat in his chest.
He revives somewhat, enough so that when I say to him, 'My
dear friend, drink, it is for your own good', he looks at me
through half-opened eyes and parts his lips for me to feed him
some cinchona on a spoon. At one o'clock, I am summoned
for my own lunch, and with hope for his improvement, I go.

As soon as the meal is done I return to my patient. It is two
o'clock. I am astonished. I find him in agony, with a very
strong bronchial rattle, his eyes almost entirely shut, his
mouth wide open, his tongue very dry, and in the grip of a
fever worse than any yet. I immediately call the four doctors.
They arrive and agree there is no further remedy; he is all but
dead.

At ten o'clock at night, the baron, M. Lanusse and I are
in the antechamber, discussing what arrangements to make

for the calamity which awaits us, when suddenly we see Augustin Méchain enter. He embraces us all in turn, me last, then asks where his father is. I tell him that he cannot see his father at present because he has been in a fever all day, and is exhausted now and resting. He repeats his request more emphatically, and I tell him again that it is impossible for him to see his father. At this, the unhappy young man throws himself on to a bed and cries: 'Where is my father? My poor father! I want to see my father! Oh, I know what you're up to: this is his bed, he lay here and now he's dead. My poor father, I will never see you again!' We assure him that his father is not dead, only very sick. The young man wants to sleep in the house, but I insist that he goes to M. Bigne's instead, and have the gentlemen escort him there.

I then return to the father's sickroom. The doctors agree that there is not much time left, but to drag things out they place a compress of hot wine on his stomach and occasionally moisten his lips and tongue with wine and water. That is our procedure for the rest of the night. At midnight, the death rattle ceases, his pulse fails, and at five in the morning, on Thursday 20 September 1804, on the third complementary day of the year XII, he dies in my arms: I receive his final breath.

And so ends the life of one of Europe's greatest talents, a man who was very good to me and who will be mourned by all Europe.

Augustin Méchain fell sick that night. Unable to sleep, he collapsed in the morning. His fever was slight, although he suffered from an attack of nerves which caused his legs to twitch uncontrollably. He sobbed in his bed. Later that day he had a still more violent attack, and it took five men to hold him down while he cried out for his father and mother. He subsided only after being bled from the arm. Dezauche stayed all that night in a cot beside the young man.

The funeral was the next morning. Dezauche put on his naval uniform. The cortège was led by the provincial governor, the Barón de la Puebla, and the members of the expedition, followed by Spanish nobles and military officers, French expatriates and three hundred monks. The cortège was met under the carved portico of the cathedral by the wives of the nobles and expatriates, all dressed in mourning attire. After the mass, Méchain was buried in the cathedral cemetery – in a lead casing, in case the French government or his family ever wished to retrieve the body.[48]

What eighteenth-century physicians called 'tertian fever', because the fever returned every third day, we today call malaria. The disease was endemic in Valencia, especially around the Albufera marshes, which Méchain had recently traversed. The ailment had been diagnosed since ancient times, and by the end of the seventeenth century physicians had discovered a palliative: an extract of the bark of the cinchona tree of South America. This is what they were giving Méchain. But the bark's active ingredient, quinine, comprised less than one-sixtieth of cinchona, and hence the potion was not always as effective as it would be today.[49]

In his final delirium, Méchain had obsessed over the fate of his mission – and his papers. During all his years of geodesic travels, he had always carried his manuscripts with him in a trunk. These calculations, logbooks and notes were the summation of a life of scientific labour. They were his continual point of reference, consulted and adjusted with each new insight. As Méchain was famously reluctant to publish his findings, these papers were all the more precious. No wonder, reasoned his subordinates, that he spoke of them so often during his final collapse.[50]

Now that Méchain was dead, they had no choice but to abandon the mission and return home to Paris. However, they took special care to bring the manuscripts with them. They deposited some of the larger surveying instruments in Valencia for the use of his replacement (should the Bureau of Longitudes appoint one), then packed one portion of his papers for shipment back to Paris. Augustin took the rest with him on his mournful journey home.

News of Méchain's death reached Paris on 8 October. A few weeks later, Augustin himself arrived. Without hesitation he personally delivered the papers to Delambre, his father's former partner. The remainder, sent by post, were handed over to Delambre by Madame Méchain some four months later. All told, they included several thousand pages of formulae, observations and calculations, scribbled, revised and rewritten. As Méchain's scientific executor, Delambre had the task of going through those papers and salvaging what knowledge he could.[51]

The rest of Méchain's small collection of scientific books and instruments was quickly auctioned off. Neither son had any intention of pursuing a scientific career. And after what they had seen of their father's fate, who could blame them? With her husband's death, Madame Méchain was obliged to vacate her apartments in the Observatory. She took up residence in Paris' ninth *arrondissement* on a modest pension. The Méchain scientific dynasty had lasted five short years.[52]

Augustin Méchain composed a brief obituary to his father. He wrote of a man who had died 'far from his country, his wife and his old friends', but who had found in his last moments 'the consolation one expects from untainted love . . . in the arms of those who accompanied him'. 'They bathed him in their tears,' he wrote, 'they were his friends and did not blush to call him *maître*.' His father had possessed the qualities which mattered most. He was 'virtuous, frank, affable, modest, a good husband, a good father, a good friend. He loved his country, his fellow man, and the arts. His friends and the sciences will deplore his loss and record his memory to the most remote period of posterity.'[53]

In a still briefer obituary, Lalande spoke of the young man he had brought into astronomy, and who had died a martyr to that science.[54]

These touching obituaries were followed by a grand eulogy from Delambre, delivered before the assembled Academy of Sciences, with the bereaved family in attendance. Delambre paid this tribute not just as Méchain's partner, but in his capacity as Permanent Secretary. In a tradition dating back to the masterful orations of the seventeenth century, a scientific eulogy was more than a recital of the technical achievements of the deceased; it was a secular sermon on the moral qualities of the natural philosopher, whose life, like his work, was permeated with the virtues of self-sacrifice, disinterestedness and stoic candour. These were the virtues which enabled the savant to contribute to the accumulation of true knowledge, thereby serving both the nation and humanity, and making him worthy, like the statesman and the general, of immortal remembrance. A eulogy consoled a family, assured colleagues of the sanctity of their calling and inspired the young to join their ranks. Making sense of death is the survivor's privilege – and his burden.

Delambre recounted Méchain's life as a tale of arduous labour and ultimate sacrifice, driven not by overblown ambition but by the obstinacy of service. Méchain came from a humble family, he reminded his audience, but had risen by dint of hard study. Patient observation and fastidious calculation had led him to discover eleven comets. These same qualities had earned him a role in the grand mission to measure the world. Delambre did not make Méchain's labours appear glamorous; on the contrary, he emphasized their tediousness.

It was those same virtues which had enabled Méchain to complete his grand mission. Delambre walked his audience through the stations Méchain had travelled on his route to martyrdom: his arrest on his first day's travel out of Paris, his stamina through the mountains of Catalonia, his fateful accident at the pumping station, his detention in blockaded Spain, his long struggle to return to France, his battles with the ignorant peasants who tore down his signals, his triumphant return to Paris – and then, when his life promised ease at long last, his self-sacrificing return to the field of his scientific labour. Delambre did not assert that Méchain's achievement was due to his genius or intellectual creativity. No such claim was possible. Rather, he ascribed his success to a kind of obstinacy. It was Méchain's obsession which had produced the most precise measurements in the history of astronomy. For proof, one need look no further than his repeated efforts to confirm his latitude measurements at Barcelona. 'And never has a verification been more thorough, more satisfactory, and for just that reason, all the more superfluous.'[55]

Delambre acknowledged that Méchain had occasionally seemed to tarry on his mission. At times he had been tempted to set aside his burden, overcome by a melancholy brought on in part by his injuries, in part by the grievous trauma which had befallen his nation. In his darkest moments Méchain had even contemplated emigration, so painful was the thought of returning to Paris where several of his colleagues had met a terrible fate. Yet the same obstinacy which had propelled him to complete his mission had also driven him to return home to perfect his study. He was a martyr to the endless quest for precision, Delambre concluded, not because he sought personal glory, but out of modesty and fierce self-doubt. Méchain had always been dissatisfied with his own work, piling on new observations, adjusting his formulae, refining his calculations. As a result, he had avoided the finality of the printed page, even when it came to their joint labour on the meridian. 'Never did he consider these observations, the most exact ever achieved in this domain and conducted with unsurpassed certainty and precision, never did he consider them sufficiently perfected – and so he worked continually to refine them.' This scrupulousness had long delayed the publication of the *Base du système métrique*. But now that all Méchain's papers were in his hands, Delambre promised to prove a faithful guardian.

From this day forth, my most cherished occupation will be to extract from this archive everything which may contribute to the glory of a colleague with whom I was honourably bound in a long common labour. And if I have not succeeded today in painting a picture of the departed astronomer worthy of his merits and the feelings I have for him, I am at least certain that whatever I publish of his work will do far more for his memory than even the most eloquent oration.[56]

It was a sincere and moving eulogy. While it glossed over certain embarrassing details, it was true to the dead man's spirit. It read Méchain's character as the source of both his triumphs and his limitations. The family expressed their gratitude, and, at their request, Delambre had his eulogy published so that they could distribute it to their friends. No one, however, reclaimed Méchain's body and it remained in the cemetery of Castellón de la Plana.[57]

In January 1806, the same month the eulogy appeared in print, the first volume of the *Base du système métrique* was also published. There Delambre paid even greater homage to his deceased partner by listing Méchain first as the expedition leader. The volume offered a lengthy preface laying out the history of the meridian expedition, followed by the record of all the triangulation data from Dunkerque to Mont-Jouy. It deferred the latitude measurements to the second volume.

But between his delivery of the eulogy and its appearance in print, in the gap between the writing of the first volume of the *Base* and its publication, Delambre made a discovery – a scandalous discovery. The publisher had been pressing Delambre to deliver the first volume of his book manuscript, and so he had postponed his examination of Méchain's papers. Now, as he worked his way through them, he discovered the discrepancy between the latitude results for Barcelona and those for Mont-Jouy, and worse – far worse – a systematic effort to cover up that discrepancy, suppress observations and rewrite scientific results. It was a discovery and it was a scandal, for while it clarified many of the mysteries which Delambre had delicately glossed over in his eulogy and in his preface to the *Base*, it also presented him with an acute dilemma. The platinum metre had been constructed; the metric system had been published and made law. The metal bar sat contentedly in its triple-locked box in the National Archives. The bar did not equal the metre; it *was* the metre. What did it now matter that the data that had gone into its making had been erroneous? What should Delambre reveal?[58]

Chapter Eleven

MÉCHAIN'S MISTAKE,
DELAMBRE'S PEACE

The fault, dear Brutus, is not in our stars,
But in ourselves, that we are underlings.
WILLIAM SHAKESPEARE,
Julius Caesar (Act I, scene ii)[1]

The historian owes the dead nothing but the truth.[2]
JEAN-BAPTISTE-JOSEPH DELAMBRE,
History of Modern Astronomy

What is error? And who decides when it is too great to bear?

Delambre finally grasped what in retrospect appeared obvious. Méchain had deceived him, had deceived them all, and had confessed as much in countless letters, if only they had read between the lines. Throughout their expeditionary years, Delambre had continually reassured his colleague: your measurements are excellent, your measurements are as good as my own. Meanwhile in his own mind, he had dismissed Méchain's worries as melancholic self-deprecation, tinged perhaps with jealousy. Then, when Méchain presented his final results to the International Commission, Delambre considered himself vindicated: Méchain's values for the latitude of the Fontana de Oro beautifully matched his latitude for Mont-Jouy. Everyone had always known that Méchain was a worrier, a pessimist, an obsessive – the very traits which made him a man of unimpeachable integrity. His false alarms had only confirmed their judgement.

Except Méchain had fudged the data.

Delambre had vowed to record every last detail of their expedition 'without the least omission, without the least reticence'.[3] When he presented a copy of the *Base* to Napoleon, the Emperor was magnanimous in his verdict. 'Conquests will come and go,' the conqueror said, 'but this work will endure.'[4] Delambre was now composing the second volume, to be devoted to latitude measurements. This time, rather than simply transcribe his partner's summaries, Delambre had decided to pull the data directly from Méchain's original logbooks.

Except there were no logbooks, only loose scraps of paper.

Méchain's manuscripts, carted back from Spain, testified to his agony. Time and again he had reworked the data, trying to make them conform to expectations, or what he thought others expected of him. It went far beyond the Barcelona data. He had recorded all of his observations on loose sheets of paper rather than in a bound notebook with numbered pages. He had also recorded them all in pencil. As Delambre wryly remarked, 'Loose pieces of paper can be lost; pencil marks fade.'[5] More to the point, loose pieces of paper can be torn up; pencil marks can be erased. In some cases Méchain had recopied observations on to pages dressed up to look like originals, whereas the true originals had vanished. In other instances he had erased values, or rewritten his pencil marks to alter the numbers beneath.[6]

Delambre's task was to fashion this mess into a permanent record. He retraced each pencil mark in ink, pasted the sheets into a bound volume in chronological order and appended marginal notes to explain their provenance. He reconstructed Méchain's journey like a historian, fashioning a logbook where none had existed before. The result was as revelatory as the meridian journey itself.

Méchain had suppressed and altered data. Sometimes, to disguise an anomalous geodetic reading, he had folded a discordant series into a longer series, as if it had been observed on the same

Méchain's 'logbook' assembled with notes by Delambre

In the years 1806–10 Delambre reconstructed Méchain's logbook by pasting into a bound register the loose sheets on which Méchain had recorded his data. Delambre organized the sheets in chronological order, retraced Méchain's pencilled data in ink and indicated the provenance of each document. On this particular page Delambre has pasted Méchain's celestial observations at Mont-Jouy for 15 December 1793. In the margin day, making the result appear more consistent than was warranted. More often, he had simply discarded those series which did not accord with his prior results, or which prevented his triangles from converging on 180°. In one instance Méchain had dropped a series which appeared to him anomalous, whereas, Delambre discovered, Méchain had simply miscalculated and the data were sound.

As he reconstructed the original values, Delambre also recorded these values in the margins of his personal copy of the *Base* (now located in the Karpeles Museum in Santa Barbara, California). Page after page of the Karpeles edition documents the data Méchain suppressed or doctored. If anything, the fudging intensified as Méchain approached his encounter with the International Commission. Yet amid the chicanery, Delambre noted a paradoxical integrity at work. In no instance had Méchain's alterations distorted the final result by more than two seconds, meaning that his adjustments were minor compared to the uncertainties caused by the observer's inability to correct entirely for the refraction of light in the earth's atmosphere. He had doctored his results, not to alter the outcome but to make himself look good – that is to say, to look better than his rival colleague. Beneath the printed text on page 510 of the Karpeles edition of the *Base*, Delambre wrote in ink:

> All the variants I have provided, based on the manuscripts of Méchain, are values for which no observer can answer. Undoubtedly Méchain was wrong not to publish these observations as he found them, and to modify them in such a way as to make them appear more precise and consistent than

---

Delambre notes: 'Here are some changes that Méchain has made to the angle measurements for which it is difficult to imagine a legitimate rationale.' He goes on to explain that Méchain's calculations on this page leave no doubt whatsoever that the corrections are not legitimate but serve only to make the data appear more precise than they actually were.
(Observatoire de Paris.)

they were. But he always chose his final values in such a way as to ensure that the average was not altered, so there was no real harm in his action, except for the fact that another observer who published unadulterated numbers would be judged less capable and careful.[7]

In situations where he had nothing else to rely on, Méchain had clung to those who appeared more self-confident. It now appeared that Méchain had also doctored his data for the latitude of the Panthéon so as to approach Delambre's value. The stunning convergence had been a sham, the moral theatre all done with mirrors. The irony – which neither Méchain nor Delambre could know – is that Méchain's suppressed data more closely approximate today's accepted latitude for the Observatory.[8]

The Barcelona latitude data were something worse. There Méchain had largely been obliged to keep faith with the Mont-Jouy data, having mailed his original results to Paris. (Though even here he had adjusted the observed values after the fact.) But he had left no paper trail for the Fontana de Oro data and was therefore free to rework those results endlessly. The earliest version, which Delambre considered otherwise irreproachable, indicated a residual discrepancy of 3.2 seconds between the adjusted latitudes of the Fontana de Oro and Mont-Jouy – the unreported gap which had tortured Méchain for a decade. Yet in later versions Méchain had systematically jacked up his Fontana de Oro observations by three seconds to account, he claimed, for the width of the sight-line in his scope. This tweaked the two latitudes back into alignment. But as Delambre noted in the margins of the reconstituted logbook, Méchain had applied this *post hoc* adjustment inconsistently: adjusting the values for some stars and not others, and neglecting to 'correct' his Mont-Jouy data at all. The inference was clear. Méchain had adopted this 'specious' adjustment to convince the Commission to expunge his Fontana de Oro results, not because they were wrong but because they *appeared* to be right – and were hence redundant.[9]

Amid the flimflammery, Delambre again saw a paradoxical integrity at work. If Méchain had been disingenuous with the Commission, he had done so in order to keep his doctored data out of the final determination of the metre. He had used sub-terfuge to spare the Commission an agonizing choice. He had lied (if only by omission) to keep his results honest.

Méchain's goal had been Utopian in its simple-mindedness: he had tried to undo the mess he had made of his mission. He had tried to return to a time before his discrepancy, a time before his accident, a time before the war. No doubt, if he could have found a way to do so, he would have reversed the whole Revolution. It was, in a sense, a perversion of the redemptive promise which science had made since the days of Francis Bacon. Having eaten of the fruit of the Tree of Knowledge (the original error, you might say), human beings were now permitted to use that knowledge to work their way back to Eden. Méchain had sinned to reclaim his innocence. He had tried to erase the past.

Delambre refused to go along with this erasure. Having assured the world's savants that he would publish all the data of the metric expedition, he (mostly) kept his word. He set out to filter the past through his own scrupulous hands, removing the spurious adjustments, recalculating Méchain's data and creating a new set of tables sufficiently trustworthy to publish. In November 1807, in Volume 2 of the *Base*, he presented the data from Mont-Jouy alongside the data from the Fontana de Oro. As for the remaining discrepancy of 3.2 seconds, Delambre wrote, 'it is a fact worthy of astronomers' full attention'.[10] He even said so in the foreign press. Delambre had decided to treat Méchain's 'error' as a discovery, not a scandal.

Yet there were parts of this story that Delambre did not want the public to know. The lay public did not need to know that Méchain had fudged his data or lied to his colleagues. Too many savants already doubted the metre's precision. The metric system had enemies enough. Delambre's solution was to deposit the orig-inal manuscripts of the meridian expedition in the archives of the Observatory, and to announce their disposition in the *Base*. On 12

August 1807, in the octagonal meeting room of the Observatory, where the portraits of Delambre and Méchain now hang, a legal protocol, witnessed by three members of the Observatory, detailed the inventory.[11] In a note appended to one of Méchain's reconstructed logbooks Delambre explained his rationale for deciding which material to publish:

> I have carefully silenced anything which might alter in the least the good reputation M. Méchain rightly enjoyed for the care he put into all his observations and calculations. If he dissimulated a few anomalous results which he feared would be blamed on his lack of care or skill, if he succumbed to the temptation to alter several series of observations . . . at least he did so in such a way that the altered data never entered into the calculation of the meridian.[12]

Then, three years later, after publishing the third and final volume of the *Base*, Delambre went one step further: he deposited all the private correspondence between himself and Méchain in the Observatory's archives as well. These letters, however, he thought it 'prudent' to place under seal, so that they could not be read unless some serious doubt arose as to the validity of the entire enterprise.[13]

Having called the Barcelona discrepancy a discovery and not a scandal, however, put Delambre under some obligation to explain the discrepancy. There were several possibilities. One might blame the stars or the earth; one might blame the instrument or the methods; or one might blame the observer.

Méchain had blamed the stars – at least at first. His original motive for revisiting Barcelona's latitude at the Fontana de Oro was an inconsistency *within* his Mont-Jouy data caused by the star Mizar. He feared that the refraction tables were invalid for towns at lower latitudes, especially for stars which dipped close to the horizon, as Mizar did. Delambre certainly shared this concern and never made use of Mizar. However, even with Méchain's Mizar data removed, the latitudes of Mont-Jouy and the Fontana de Oro did not agree. As for the suggestion made by later astronomers that Méchain had erred because Mizar is in fact a double star, it turns out that Méchain was well aware of this fact, and always focused on the larger body 'in case someone supposes I did not take care'. The fault lay not in the stars.[14]

For his part, Delambre preferred to blame the earth. As he pointed out, the meridian project had confirmed that the figure of the earth was irregular, and that not all meridians were equal. Moreover, by the time he published Volume 2 of the *Base*, he could cite a new meridian survey in England which confirmed these irregularities. Delambre hypothesized that Méchain's readings at the two nearby sites had been distorted by local inequalities in the earth's crust or by nearby mountains. These inequalities, he suspected, had deflected the plumb line the savants had dropped to define the vertical of the star's transit across the celestial meridian. The plumb line definition of vertical, however, was doubly ambiguous. First, because the plumb line pointed to the gravitational centre of the earth, on a non-spherical earth that is not quite the opposite of the perpendicular to the immediate surface of the planet. In other words, astronomical straight up is not the exact opposite of the direction in which gravity tugs down. Indeed, the gap between the two at any point offers a measure of the local difference between the astronomic latitude and the geodetic latitude, reflecting the earth's eccentricity at that point. Second, the plumb line might be deflected by local gravitational effects (due to mountains and the like) beyond those caused by irregularities in the figure of the earth. This concern was not new. Newton himself had tried to estimate the gravitational pull of mountains. And the French savants had deliberately selected a meridian arc which extended from Dunkerque all the way to Barcelona to avoid any distortion caused by the Pyrénées. Delambre now speculated that Mont-Jouy itself might be to blame for distorting its own measurement.[15]

Today, the science of geodesy consists principally of mapping these gravitational effects. Ballistics engineers estimate the pull of mountains on their rockets. Some of the maps which chart the contours of the geoid are classified as military secrets. Beneath the earth's surface deep processes have roiled the planet. However, these gravitational differences seem unlikely to explain so wide a divergence in measurements at two sites a mile apart. The earth's irregularities are not so finely grained as that.

It is of course possible that Borda's marvellous instrument was the culprit. Yet Méchain was famous for the excruciating care he took when setting up the apparatus. Nor is there any evidence to support the charming suggestion, made by one Catalan historian, that a patriotic saboteur, posing as an astronomical assistant, queered Méchain's instrument to prevent him from acquiring data on Barcelona's defences. Méchain had always monopolized all the observational work. And while Méchain did use a cumbersome method of calculation that involved much tedious labour, Delambre found that even upon recalculating Méchain's data, the discrepancy remained.[16]

In the end, Méchain came to blame himself – to his eternal shame and torment. This conclusion was not shared by Delambre. After extensive review, he declared the Fontana de Oro data as credible as the Mont-Jouy data.[17]

There is, however, one other possibility. What if nothing and no one was to blame? Indeed, what if there was no meaningful discrepancy at all? That is: what if the error lay neither in nature nor in Méchain's manner of observation, but in the way he understood error? Twenty-five years after Méchain's death, a young astronomer named Jean-Nicolas Nicollet showed how this might be the case.

Nicollet reanalysed Méchain's data in a series of steps. First, he threw out Méchain's data for the inferior transit of the star Mizar, whose passage near the horizon was indeed overly distorted by refraction. Second, he recalculated Méchain's other stellar heights using accurate tables of stellar declination, tables that had been completed at the very end of Méchain's life and which circumvented the iterative method by which both Méchain and Delambre had assessed their latitudes. Both these changes were relatively minor, however, compared with the way Nicollet reconceptualized Méchain's treatment of the data.[18]

Méchain and his contemporaries did not make a principled distinction between *precision* (the internal consistency of results) and *accuracy* (the degree to which those results approached the 'right answer'). The two are not the same: precise results may appear 'reliable' in the sense that they give very nearly the same answer when measured again; yet they may lack 'validity' in that they deviate consistently from the 'right answer'. Of course, in practice, distinguishing between the two can be extremely difficult because the 'right answer' is unknown.

Repetition using the Borda circle was designed to improve *precision* by reducing those errors that stemmed from the imperfect senses of the observer or the imperfect construction of the instrument's gauge – the sort of errors we would today characterize as falling into a random distribution. The Borda circle, however, was still subject to errors caused by the basic set-up of the instrument as a whole; the sort of errors we would characterize today as those constant (or systematic) errors which made results *inaccurate*, whatever their level of precision. Constant errors generally go undetected, of course, as long as they stay constant. And in an intuitive way Méchain and Delambre, like all *ancien régime* astronomers, understood this. That is why they were so vigilant about maintaining a consistent set-up for their apparatus from one series of observations to the next. What they failed to appreciate was that the same repetition which enhanced precision might reduce accuracy. For instance, constant manipulation of the circle might wear down the instrument's central axis and, over time, cause the circle to tilt ever so slightly from the perpendicular. It was this unanticipated drift in the constant error, Nicollet suggested, which was the source of Méchain's discrepancy. This had kept his results for each winter location more or less internally consistent (i.e. precise), while making the results from the two successive winter locations discordant (i.e. inaccurate). Without a concept of error to help him identify the source of this contradiction, Méchain was in torment.

Oddly enough, Nicollet noted, it was Méchain's own obsessiveness which made it possible to confirm the cause of the discrepancy – and to correct for it. The trick was to compensate for any change in the instrument's verticality by balancing the data for stars which passed north of the zenith (the highest point of the midnight sky) against those which passed south of it. Because Méchain had measured so many extra stars, such an operation was possible.

To calculate the latitude, Méchain had first calculated the average latitude implied by each star he measured, and then averaged all the averages, giving equal weight to each. Nothing could be simpler – or more naïve. Nicollet, by contrast, first analysed the data for the stars Méchain had measured which passed north of the zenith (his many observations of Polaris and Kochab, plus those of Thuban and Capricornus), taking the average of the average latitude implied by each. Then Nicollet separately did the same thing for the stars Méchain had measured to the south of the zenith (his sparser observations for Pollux and Elnath). Clustered in this manner, the results seemed to lack *precision*: at Mont-Jouy the average latitude implied by the north-going stars differed from the average latitude implied by the south-going stars by 1.5 seconds. At the Fontana de Oro they differed by a dismaying 4.2 seconds. But when the northern average and the southern average were themselves combined at each location, they suggested a remarkable *accuracy*: the combined latitude for the Fontana de Oro agreed with the combined latitude from Mont-Jouy to within a stunning 0.25 seconds – twelve times as accurate as Méchain's 3.2-second discrepancy! In sum, Nicollet proved that there was no discrepancy and that Méchain's reported value for Mont-Jouy was within 0.4 seconds (or forty feet) of the answer indicated by his data, when properly analysed.

Nicollet was a typical French astronomer of the early nineteenth century: a student of Laplace's, well versed in error theory. In 1828, when he reviewed Méchain's results, he was forty-two years old and working part-time at the Paris Observatory. Unfortunately, his statistical skills did not stand him in good stead outside astronomy, and he lost his fortune a few years later playing the stock market. He emigrated to America, where his astronomical and mathematical skills earned him the leadership of the first geodetic survey of the upper Missouri and Mississippi valleys. A generation after Lewis and Clark passed through the territory of the Louisiana Purchase, Nicollet compiled the first accurate maps of the Upper Midwest.[19]

The irony was that the very stars Méchain cursed himself for measuring had vindicated his exactitude. By contrast, Delambre had measured only Polaris and Kochab, both of which passed north of the zenith, so his results could not be corrected retrospectively. Between them, Méchain and Delambre had identified most of the sources of error which had produced the discrepancy: the refraction correction, the verticality of the circle, the set-up of the instrument. What they lacked was a way to disentangle these errors. Méchain had not so much erred as misunderstood what error meant. Yet by his very misunderstanding he had inadvertently contributed to our own understanding of error, for ever altering what it means to practise science.[20]

What *is* error? And who decides when it is too great to bear?

Modern science accepts error as its lot. It does not demand truth from its practitioners, only honesty. It assumes that the truth will emerge eventually from a collective effort – so long as everyone is honest. Certainly scientists care passionately about getting the right answer. But when theory and experiment align too tightly, suspicion is warranted. Thus, the statistician R. A. Fisher concluded that Gregor Johann Mendel's pea-breeding data could hardly have come as close to the one-to-three genetic ratio as he claimed. The same holds for Robert Millikan, who won a Nobel Prize for an electron experiment in which he suppressed anomalous data. In Mendel and Millikan's day – from the nineteenth through the early twentieth century – such fudges were common, though frowned on. Today they continue, despite official censure. In Méchain's day they were not only common, they were considered a savant's prerogative. It was error which was seen as a moral failing.

Méchain took his science personally. His observations were his to publish or suppress, keep or destroy. He felt no obligation to parade his accomplishments before an anonymous public. Rather, he sought to impress his fellow savants with his ability to approach perfection. Recording his observations in pencil and on loose scraps of paper only facilitated the task of refining the data. Even after he had handed over his scribal report to the Commission, he considered the raw expedition data to be his private property, part of the accumulated experience he carried with him in his trunk wherever he travelled. The data were his legacy and his tombstone; they were all he had. The truth belongs to everyone, but error is ours alone.

Delambre's scrupulousness was cut of a different cloth. For him, investigators answered to both their colleagues and their sponsors. The Revolutionary government had sponsored their lavish meridian mission, and it deserved a full accounting. That was why Delambre recorded his results like a public official: in ink, in a bound notebook with numbered pages, in order of observation and with each page signed and dated. He believed that just as the republic operated with transparency – with roll-call votes, public laws and open trials – so science should open itself to scrutiny. Delambre considered his data public property. All he asked was that he get credit for his work. Thus, he was forthright about the discrepancies he discovered and the approximations he used. He never pretended that his results were definitive. What mattered was that he had tried to be as exacting as he could, and that the results were sufficiently precise to solve the problem at hand.

Once, on the eve of presenting some astronomical tables to the Academy, Delambre discovered a trivial error in his calculations and spent the next three weeks working day and night to eliminate an error which was 'to all intents and purposes imperceptible'.[21] It was the most tedious task in a career of monumental labour, but when it was done he could hand over the tables with a clear conscience. Then, when a fellow savant detected yet another error – as one colleague soon did – Delambre could freely and publicly acknowledge his mistake and correct it yet again.

As to whether perfection existed 'out there', curled up in nature's womb awaiting delivery, that was a theological question, and Delambre was a pagan. He had been raised in a devout home, he was schooled by the Jesuits, and he had considered taking holy orders. But he was neither a believer nor an atheist in the manner of his *maître* Lalande. He was a sceptical Stoic, for whom perfect knowledge lay beyond man's grasp. Why then should anyone expect him to produce a perfect metre?[22]

This was something he had known all along. He had known it when he stood before the fiery volunteers of Saint-Denis to explain his absurd mission. Even then he had secretly agreed with the volunteers – in part, anyway. His mission *was* absurd. Why set out to measure the world while the old order was going up in flames, while millions of soldiers were rushing to die in battle? Why measure the world to create a unit of length when a standard 'metre' could be created by legal fiat or simple agreement? It was absurd to travel so far to find what lay so near. Yet someone had to do it. Someone had to dedicate himself to the task of reconstructing the world. Otherwise there would be nothing left standing after the soldiers had finished their slaughter and the vandals had levelled the towers to the heights of the 'humble cottages' of the sans-culottes. Someone had to construct a new order, a horizontal grid to enable people to keep track of where they were, and what they made, and how much they bought and sold.

Delambre knew the International Commission's boast of perfection was a sham. The Commission had claimed to know the length of the metre to six significant digits, or within 0.0001 per cent. Delambre now acknowledged that this was 'a precision to which we ought not to presume'.[23] Now that the sanctified platinum bar was safely stored in the National Archives – smug and untouchable in its triple-locked box – he thought it only honest to admit as much. Thus, in 1810, in the third and final volume of the *Base*, he described a range of plausible values for the earth's eccentricity and a corresponding range of plausible values for one ten-millionth of the quarter-meridian. He suggested that the best value for the earth's eccentricity was probably nearer to ⅟₃₀₉ than the Commission's ⅟₃₃₄. He took into account *both* of Méchain's values for the latitude of Barcelona, thereby shifting the length of the arc by another 0.01 per cent. He concluded that a better length for the metre would be 443.325 *lignes* (rather than the official 443.296 *lignes*).[24]

It was a trivial adjustment, less than the thickness of a piece of paper. But it was an act of remarkable integrity. Only one decade after the metre had been declared 'definitive' its chief creator, writing in the official account of its creation, acknowledged that scientific progress had undermined its validity. Today we recognize this as a step in the right direction – not because Delambre's new value closed about a third of the shortfall of the definitive metre (after all, his new value was *still* less accurate than the provisional metre of 1793), but because it was a calculated homage to the transience of human knowledge.

And Delambre did not stop there. He suggested rounding off the official length of the metre to 443.3 *lignes*. This may seem another trivial adjustment. But by lopping off two decimal places, Delambre implied that the metre was only accurate to within 0.01 per cent. And he noted that the revision had yet another reason to recommend it: the value of 443.3 was easy to remember because it was composed of two 4s followed by two 3s. As for those who still insisted on working with six significant figures, he suggested that they consider the metre to be 443.322 *lignes* long, because that value too was easy to remember: two 4s, two 3s, two 2s.[25] Nothing illustrates more clearly Delambre's acceptance of the arbitrariness of conventional standards. As he noted in a private letter to a foreign savant: 'Say what you like of the degree of precision we achieved, all I can assure you is that I conveyed every detail of our mission with the greatest possible sincerity and without the least reticence.'[26]

In the end, Delambre took longer to write the history of the metre than to measure France, and in a sense he travelled further in the process. He began writing in 1799 in the wake of the disturbing results of the International Commission. The 1806 volume began as an adventure story; the 1807 volume plunged him into a tale of scandal and discovery; and the 1810 volume concluded with a demonstration of the open-endedness of knowledge.

Delambre had come to accept, as Méchain could not, the evanescence of earthly knowledge. So to the extent that he conspired in his colleague's cover-up he did so for the opposite reason: because he realized that getting the perfect answer did *not* matter. Which is to say that Delambre understood that Méchain had agonized – and died – for nothing. We live on a fallen planet, and there is no way back to Eden. Delambre had decided to live on the surface of the earth, buckled, bent and warped though it was.

In coming to terms with the imperfection of earthly knowledge, Delambre had a powerful new intellectual tool at his disposal, one which he and Méchain had inadvertently inspired, but which he alone had lived to take advantage of. For the past century savants had sought to fit imperfect data to a perfect planetary curve. Geodesers had agreed that the earth was an oblate ellipsoid, but they had been unable to agree on its degree of eccentricity, which now seemed, moreover, to vary from place to place. Or were the data faulty? Assume the data then. Assume that the data had been gathered by fallible (but exacting) investigators using fallible (but ingenious) instruments on a (possibly) lumpy irregular earth, and then ask yourself: what was the best curve through the data, and how much did the data deviate from that curve? Or that was the question Adrien-Marie Legendre asked.

Legendre's answer, the method of least squares, has since become the workhorse of modern statistical analysis. It was also among the most important breakthroughs in modern science – not because it produced new knowledge of nature, but because it produced new knowledge of error.[27]

Legendre cloaked his personal life in obscurity and invested his clarity in his mathematics. A contemporary of Laplace and Delambre, he was elected to the Academy at the age of thirty for his work on number theory and analysis. In 1788 he showed the geodesers on the Paris–Greenwich expedition how to correct for the curvature of their triangles. Appointed with Cassini and Méchain to the Revolutionary meridian project, he withdrew in favour of Delambre. During the Terror he went briefly into hiding, only to emerge with a bride half his age. Later he co-directed the Agency for Weights and Measures and was one of the savants to calculate the length of the metre for the International Commission. He was as baffled by the outcome – an unexpected eccentricity of $\frac{1}{150}$ – as the rest of them. Five years later – one year after the death of Méchain – progress slipped in sideways yet again.

For centuries savants had felt entitled to use their intuition and experience to publish their single 'best' observation as the measure of a phenomenon. During the course of the eighteenth century, they had increasingly come to believe that the arithmetic mean of their measurements offered the most 'balanced' view of their results. Yet many savants continued to feel, like Méchain, that any measurement which strayed too far from the mean ought to count for less than those near to it, and hence could be suppressed without apology. And even the most rigorous savants were flummoxed when they confronted multi-variable phenomena for which they had diverse observations – such as the curvature of a non-spherical earth based on observations at various latitudes, or the elliptical orbit of a planet, especially if that orbit were perturbed. Some mathematicians had tried to find rules to compensate for results that deviated too much. The Jesuit geodeser Boscovich had proposed one method, and several other astronomers had tried their hand. And Laplace had introduced a cumbersome method of minimizing the maximum deviation. But these methods remained awkward and unjustified.[28]

Legendre suggested a practical solution. He suggested that the best curve would be the one which minimized the square of the value of the departure of each data point from the curve. This was a general rule, and it was a feasible calculation. It was a practical dictum, and it prompted a radical reconceptualization. Legendre's least-squares method played off the intuition that the best result should strike a balance among divergent data, much as the centre of gravity defines the balance point of an object. As he noted, the least-squares method also justified choosing the arithmetic mean in the simplest cases.

In 1805, just as Delambre was completing the first volume of the *Base*, Legendre tried out his method on what was now the world's most famous data set, the one he had puzzled over ever since Delambre and Méchain had handed it to the International Commission. Legendre assumed that the earth's meridian traced out an ellipse; he then used the least-square rule to find the eccentricity which would minimize the square of each latitude's deviation from that curve as it arced – in a kind of high-wire balancing act – along the data Delambre and Méchain had gathered at Dunkerque, Paris, Evaux, Carcassonne and Barcelona. And when he did, he found that the deviations of the various latitudes from that optimal high-wire curve remained sufficiently large to be ascribed to the figure of the earth and not to the data. And in Volume 3 of the *Base*, Delambre echoed his analysis: it was the earth which was warped not the data.[29]

As Legendre presented it, the great advantage of his least-squares rule was that it could be easily and systematically applied. It gave savants a workable method for weighting data. And in a few years, it became something more. It became a method with meaning.

Four years after Legendre's paper, the mathematical genius Karl Friedrich Gauss claimed that he had been using the least-squares rule – which he called 'my method' – for nearly a decade. As often happens, this simultaneous discovery was no coincidence. Both men were working on the same geodetic problem. Indeed, Gauss was working on the same data set, the meridian data gathered by Delambre and Méchain, which had been published in Germany in 1799. They were also reading the same mathematicians, especially Laplace. And, as often happens, this simultaneous discovery prompted a bitter dispute over priority: first, because both parties wanted credit for the discovery; finishing in the rear meant running the risk of being accused of plagiarism; second, because the two parties differed on the meaning of the discovery. In this instance, there seems little doubt that the two men arrived at the method independently, although it was Legendre who published first. Nor is there any doubt that it was Gauss who suggested the method's deeper meaning.[30]

Legendre presented his method of least squares as workable and plausible. Gauss *justified* it by showing that it gave the most probable value in those situations where the errors were distributed along a 'bell curve' (known today as a Gaussian curve). This probability-based approach prompted Laplace to show in 1810–11 that the least-squares method had the following advantages: it best reduced the error as the number of observations went up; it indicated how to distinguish between random errors (precision) and constant errors (accuracy), and it suggested *how likely* it was that the chosen curve was best. This was new. In their search for an illusory perfection, the savants had learned not only how to distinguish between different kinds of error but also that error could be approached with quantitative confidence. The years between 1805 and 1811 saw the rise of a new scientific theory – not a theory of nature, but a theory of error. It was this theory which would allow Nicollet to redeem Méchain's honour by distinguishing between those errors that were random from those that were systematic.

Some experiments were inherently erratic; others could be refined. Some investigators took pains with their observations; others were sloppy. The appeal of the new approach was that colleagues could now begin to distinguish between these two forms of uncertainty, and judge each other with the same impersonal techniques by which they judged nature. It was at this time that Delambre, Laplace and the rest of the French savants started to come to terms with something their British colleagues had just discovered: that even the most fastidious astronomers were subject to idiosyncrasies in their observations (depending on their reaction time and the like), and that these idiosyncrasies introduced a constant bias into their observations, a 'personal equation' as it came to be called. This recognition that they themselves were fallible instruments was followed by a programme to tame error. Astronomers began to calibrate themselves against one another and divide their labour to average out personal influences. Over the course of the next few decades astronomy became something of a bureaucratic science in which a staff of junior observers (career-minded young men) and an office full of calculators (underpaid young women) toiled for a senior astronomer who directed their efforts, analysed their data and then published the results under his name.[31]

Approach the world instead through the veil of uncertainty and science would never be the same. And nor would savants. During the course of the next century science learned to manage uncertainty. The field of statistics which would one day emerge from the insights of Legendre, Laplace and Gauss would transform the physical sciences, inspire the biological sciences and give birth to the social sciences.[32] In the process, 'savants' became 'scientists'.

Méchain lived and died a savant. Measurement mattered to him as much as it did to the *ancien régime* peasants, bakers and families who grew wheat, baked bread and bought loaves in the marketplace. Whether it was the height of a star or the weight of a loaf, measurement expressed *value*. It was a moral act, an exercise in justice. For the savant, the pattern of the heavens revealed a comprehensive plan. To measure the shape of the earth or the height of a star was to glimpse its place in that pattern, just as the weight of a loaf sustained the just price for bread.

Men like Delambre, Laplace, Legendre and their generation had a foot in each world. I have called them savants, but the term no longer fits. They were henceforth engaged in a struggle to quantify their uncertainty. They would ask: how confident are we that we know what we think we know? They sought to rid themselves of value-judgements about nature and to cordon off meaning from their measurement of the world. They had launched themselves on a very different kind of career. In 1792 Jean-Paul Marat had been the first person to tag savants with the name of *scientifiques* (scientists) when he referred sneeringly to the academicians' self-serving project to measure the earth in order to create uniform weights and measures. For better or worse, the savants were now on their way towards becoming scientists.[33]

As for Lalande, he chafed under the new régime even more than he had under the old. He was the last of the *philosophes*, now in his seventies: a free-thinker who preferred monarchy, an atheist who admired the Jesuits, a feminist who propositioned young women, ugly as ever and still just as vain. The new régime had little patience for these old-time contradictions. Lalande had initially welcomed Napoleon's rise to power, proud that the general called him 'Grandpapa'. Napoleon had studied astronomy under one of Lalande's students, and had written to Lalande in a manner calculated to flatter the old man. 'To divide one's night between a beautiful woman and a clear sky, and then spend the day matching theory and observation, that is my idea of heaven on earth.'[34]

Lalande's ego was pure as platinum. When his star chart hit fifty thousand – thanks to the labours of his daughter and his nephew – he published a massive compendium under his own name. Surveying the fifty-four auditors who attended his lectures at the Collège de France, he admitted in his journal that 'Lalande is still the one who interests me most'.[35] In the summer of 1798, he handed the lovely Citoyenne Henry, the world's first woman aeronaut, into a balloon for her maiden voyage, in spite of a ban on female ascension. The next spring, he tried to visit Germany by balloon while observing the stars above the atmosphere's veil. A wit composed the following verse to accompany him:

> Observe the dwarf of academicians
> Whose pride could fill a room.
> He wanted to hear it straight from the winds
> If they talked of him
> On the moon, the moon, the moon.[36]

The balloon never made it past the Bois de Boulogne, and he cancelled the trip. He was always good copy. When a traveller brought back news of an African people who, like him, ate spiders, the newspapers advised Lalande that he would now have to switch to 'insects of distinction'.[37] Yet he was living proof that vanity could serve noble causes. Lalande did not care what people thought. As the editor of the *Dictionary of Atheists*, he maintained its honour roll of eight hundred adherents: from Socrates to Lalande. In 1799 he signed up several of his colleagues, including 'Buonaparte of the Academy of Sciences'.[38]

This was risky. After the vandalism of the Revolution, piety was making a comeback. Wags claimed that Lalande had turned to atheism out of revenge, because God had made him so ugly. 'Look at his knock knees and rickety legs, his hunched back and little monkey's head, his pale wizened features and narrow creased forehead, and under those red eyebrows, his empty glassy eyes.'[39] Lalande answered insults with epigrams:

> That men are witless, wicked fools
> Proves there's evil in this house of rot.
> A scoundrel's word can make heads roll;
> If God existed, Man would not.[40]

Not all his colleagues appreciated this frankness on their behalf. Lalande's antics put Delambre in an awkward spot. At first the dilemma was amusing. Delambre had agreed to serve as godfather to Lalande's granddaughter, Uranie, but her baptism had been postponed until Delambre returned from his mission. By then the child was seven, and able to respond on her own behalf. Asked by the priest if she renounced Satan and all his works, she said, 'I do renounce'. Asked if she renounced all worldly vanities, she said, 'I do renounce'. Asked if she swore to live and die in the Catholic Church, she said in a loud clear voice, 'I do renounce'. Everyone in the church laughed, including the priest.[41]

Then Napoleon decided to buy peace among the French by reconciling with the Catholic Church. These delicate negotiations resulted in the Concordat of 1802, and were to culminate in the Pope's arrival in Paris for the coronation of Napoleon as Emperor. In the midst of these delicate accommodations, Lalande had the temerity to reissue his *Dictionary*. The Emperor was furious. But was it atheism that enraged him, or something worse? The *Dictionary* dared to preach peace: 'It is up to the *philosophes* to spread the light of science, so that one day perhaps they may curb those monstrous rulers who bloody the earth; that is to say, the warmongers. As religion has produced so many of them, we may hope to see an end to that as well.'[42]

From the battlefield of Austerlitz – after the finest victory in his career – the Emperor wrote a searing rebuke: Lalande had fallen into dotage, atheism destroyed the moral order and Delambre, as Permanent Secretary, must convoke the Academy to silence their senior colleague. Delambre tried to make Lalande's compliance appear voluntary, preserving a semblance of intellectual freedom while bowing to authority, but Lalande refused to be silenced. In 1806 he published another edition of the *Dictionary*, albeit without Napoleon's name.[43]

That year Lalande fell ill with a chest ailment. Right to the end, he was insufferable and self-mocking in alternate breaths. In his final moral testament he wrote: 'I have sometimes amused myself by saying that I thought I possessed all the human virtues. This phrase of mine has been bitterly cast up against me as if I had claimed "to have all the human virtues". In fact, what I said was that "I *thought* I had them", which is quite a different matter. Nevertheless, I was perhaps wrong to have said as much; but my conscience required it of me.'[44]

In the evening of 3 April 1807, after his daughter had read him the evening papers, he sent her to bed, saying, 'I don't need anything else'.[45] At two in the morning he died. Even beyond the grave, he shocked the public. Two days after his death, his family had to deny rumours that Lalande had asked for his dissected body to be put on display in the Museum of Natural History. The eighteenth century was finally over.

This earthly transience – of knowledge, of men, of régimes – was not necessarily a cause for melancholy. On the contrary, in a world of confounding factors, in an age of Revolutionary terror, Delambre felt joy, only joy. Joy had accompanied him through all his travels, through all his labours, through all his life. Not the ecstatic joy of transcendence, but the modest joy of immersion. He had long ago passed the age of illusion. Delambre could accept that we live on a fallen irregular planet, in a world of imperfection and error, because by the collective labour of honest scientists this imperfection could be contained, and error tamed. In 1806 he wrote to a friend who had suffered much during the Revolution:

> I have all my life experienced a happiness so gentle, so peaceful and so untroubled, that if I were truly persuaded that it is man's lot to discharge a debt of suffering and pain, I would fear for the future. But I like to think that there are exceptions, and I dare to hope that I will be one of them. My good fortune, I think, is due to my character and my temperament. The only passion I have ever known is the one which has never yet caused misfortune; and that is work. My passion for work is not diminished. I continue to give myself to my labours with all my strength.[46]

Delambre's face had thickened, but his eyes had strengthened with the years. His legs were chancy, but his hand was firm. The traveller who had once traversed much of France could no longer cross a Paris street; in 1803 he had been hobbled by a rheumatic fever. He had seen much in the way of suffering, but his optimism was unabated.[47]

In 1804, after a liaison of several years, he married Elisabeth de Pommard, the mother of his young assistant. He was fifty-five; she was in her forties, a spirited widow with some well-situated property to the west of Paris. They made a good match. She read Virgil's epics in Latin, Addison's essays in English and Metastasio's librettos in Italian. For several years before they exchanged vows, she and Delambre had traded low-interest loans. While the d'Assy family remained in the country, Delambre and his wife could live on the sumptuous rue de Paradis. There they read the classics, followed the Amazon voyages of their young friend Humboldt, and planned a brilliant career for her son, whom Delambre loved as his own.[48]

Young Pommard, who had once planned to be an astronomer like his stepfather, enrolled instead in the Ecole Polytechnique to study the earth: mining and mineralogy. Two years later, he left to join Napoleon's finance bureaucracy. Pommard was serving in

Naples when he died in 1807, at the age of twenty-six, an incon-
solable loss to his mother and her new husband. Delambre
transcribed this English translation of an Athenian poem for his
wife:

> Ah! Love how soft and tender
> Begins thy happy reign,
> But when our heart surrenders
> Thou'rt bitterness and pain.
> If from the Daylight flying
> The shaded woods we rove
> Or thro' the slow night sighing
> Of breath but for my love.
> Tempt not the soft illusion
> Ye wand'rers free of air
> 'Tis nothing but delusion
> The voice that calls you there
> Or tells the heart to languish
> The youthful bloom to glow
> Then damps that heart with anguish
> And fills that heart with woe.[49]

Yet sorrow too is transient. And Delambre would find
renewed satisfaction here on earth. He spent the final decades of
his life as a power-broker of imperial science: dispensing favours,
deciding careers, disciplining colleagues. He was simultaneously
Permanent Secretary of the Academy, Lalande's successor at the
Collège de France, a member of the Bureau of Longitudes and
Treasurer of the University of Paris. He also became the nation's
premier historian of science. On the title pages of his books his
honorific titles took up half the page. He became a star of the
cumul system, the deplorable French practice of gathering multiple offices into one thick fist. Although the salaries came to a
hefty sum, Delambre denied any worldly ambition. 'Official positions have come to me unbidden, and I received what I did not
covet.'[50]

Delambre as Permanent Secretary of the Academy

Delambre in his fifties at the height of his academic influence.

(Archives de l'Académie des Sciences. Photo: Charmet.)

Generosity during war, honesty under empire, science in an age of puffery: these are the trials of integrity. He conducted fewer astronomical sightings now, and concentrated instead on the inner science of synthesis. For one thing, he had lost easy access to his private observatory. In 1808 he and his wife moved out of the rue de Paradis and into the official residence of the Treasurer of the university. After her son's death, wife and husband relied on one another more than ever. She learned enough mathematics to help him with calculations. Delambre had an endless appetite for work. In 1806 he published a revised set of solar tables, the most exact to date. In 1813 he published an *Abridged Astronomy*, and a year later a three-volume *Treatise on Astronomy*. According to Gauss, these latter works were dull and craftsman-like, mathematically simplistic and lacking conceptual elegance. In other words, they were textbooks.[51]

Delambre made himself useful to the Napoleonic régime, even as he kept his distance from palace intrigue. In 1803, as the Treaty of Amiens had faltered, Napoleon ordered detailed maps of landing sites along the south coast of Britain, as well as information on which French coastal tower was best suited to supervise the invasion. No one knew more about the geodesy of the north coast of France than Jean-Baptiste-Joseph Delambre. Within a week, he had supplied a set of tables of optimal viewing sites, based on his own research and that of the 1788 Greenwich–Paris survey.[52] As a servant of the state, he was duty-bound to put peaceful science to bellicose ends. But then, throughout the war, he worked with his opposite number in Britain, Sir Joseph Banks, President of the Royal Society, to save their respective colleagues caught up in the conflict. Banks helped the French geodesers sail home from Egypt; Delambre helped release those trapped on the Continent. Their nations might be at war, but scientists could maintain civility. Delambre shipped multiple copies of the *Base du système métrique* to Britain – accompanied by hopes for more peaceful times to come.[53]

In 1809 Napoleon directed the Academy to conduct a prize competition for the best scientific publications of the decade. In the category of applied science, the Academy unanimously nominated 'the work of Delambre on the meridian'.[54] The work of *Delambre*? Méchain's sons wrote to defend their father's honour and to have his name included. A committee of the Academy, asked to adjudicate, noted that while Delambre and Méchain had split the latitude measurements equally, Delambre had measured 89 out of 115 triangles and both baselines. More to the point, Delambre had refined all the geodesic methods, recalculated all Méchain's latitudes and written virtually the entire text of the *Base du système métrique*. Despite the fact that Méchain's name came first on the title page, Delambre alone deserved the prize. Deserved perhaps, but would not accept. Delambre withdrew the *Base* from consideration on the grounds of conflict of interest.[55]

Two years later, the metric system itself was withdrawn. The Revolutionary calendar was the first to go, its own creators delivering the *coup de grâce*. Rather than knit the world together, the calendar had only isolated France. Even in France, it was universally ignored. Parisians still celebrated 1 January, and the ten-day working week had proved curiously unpopular. Napoleon Bonaparte had another objection; he wanted Catholic legitimacy for his new régime, and the Church wanted its Sundays and saints' days back. Shortly after he was crowned Emperor, he asked his Senate to reconsider the reform. Pierre-Simon Laplace, kicked upstairs to become senator-for-life, agreed that the calendar should be abolished – for its scientific flaws, he said. At midnight on 10 nivôse of the year XIV, the date in France reverted to 1 January 1806.[56]

The rest of the metric revolution did not last much longer. Since 1801, the metric system (shorn of its Classical prefixes) had served as the nation's official system of measurement – without changing the shopping habits of French men and women. The imperial government exhorted its subjects to do better; it supervised the annual production of three hundred thousand metric rulers; it commanded its police to punish the recalcitrant; and it printed detailed instructions on the proper way to dole out wheat, firewood, wine, olive oil and the countless other commodities that were still being shipped in *ancien régime* containers. Yet Napoleon's administrators watched helplessly as trading continued in the old units. Again and again they found themselves denying rumours that the government was on the verge of revoking the metric system.[57]

The rumours were true. In 1805 the French savants, led by Senator Laplace and Secretary Delambre, lobbied against further dilution of the metric system. The Minister of the Interior, a chemist, helped his colleagues hold off the day of reckoning. Five years later, when the system came under renewed attack, the academicians took a different tack; now they praised Napoleon's imperial conquests, arguing that they offered a unique chance to disseminate a universal metrical language. Senator Laplace took this appeal directly to the Emperor himself, pleading with his former examination pupil to retain the decimal division, even bowing so low as to suggest that the measures be renamed the 'Napoleonic Measures' if it would help. It did not.[58]

With preparations underway for his invasion of Russia, Napoleon decided to minimize economic turmoil at home. On 12 February 1812, France adopted the so-called 'ordinary measures'. The empire's legal standard would still be defined in relation to the platinum Archive Metre, but the workaday measures would approximate to those of *ancien régime* Paris. Length, for instance, would be measured in a *toise* (fathom) two metres long and divided as before into 6 *pieds* (feet) of 12 *pouces* (inches) each. In principle the system of decimal weights and measures would still be taught in the nation's schools and used for public works and wholesale transactions. In practice, however, Napoleon had revoked yet another revolutionary achievement.[59]

The goal was now imperial uniformity, pure and simple. Napoleon had no patience for the Revolutionary fantasy that a new language for the objects of the material world would create an autonomous and egalitarian citizenry able to calculate its own best interest. Instead, he grasped at central rule. All the other elements of the metric system were discarded to achieve that single goal. Among the few French intellectuals brave enough to decry this act was Benjamin Constant.

> The conquerors of our times, peoples or princes, want their
> empire to possess a unified surface over which the arrogant
> eye of power can wander without encountering any inequality
> which hurts or limits its view. The same code of law, the same
> measures, the same rules, and if we could gradually get there,
> the same language; that is what is proclaimed as the
> perfection of the social organization . . . [T]he great slogan of
> the day is uniformity.[60]

This was the tragic lesson of the past twenty years. This was the prospective measure of the world. Where absolutist régimes had once been satisfied with the outward show of homogeneity, modern dictators aspired to *inner* uniformity, levelling any difference which interfered with allegiance to the whole. Condorcet, the dead optimist of liberation, had naively imagined a world in which universal law, derived from nature's truth, could produce equality and freedom without contradiction. Constant, the living pessimist of liberation, had witnessed how uniformity, enforced by mass mobilization, could suppress difference of thought and custom. Both men, of course, were right. Their aspirations and fears remain the two poles of the axis around which the modern world still revolves. And both underestimated just how difficult it would be, for good or ill, to realize that uniformity.

Such a formidable goal lay beyond even Napoleon's grasp. Across the empire his 'ordinary measures' were rejected, just as the metric system had been. To the people of the annexed territories, the measures were just another attempt to coax them into a unified Continental economic block. In Rotterdam, where a prodigious commerce flowed down the Rhine, citizens ignored broadsheets converting Dutch to French measures. The Imperial Prefect there despaired of 'the character of the inhabitants, and the notions they have about the new measures'. But back in Paris, the Minister of the Interior was not surprised; he knew first hand the tenacity with which the common people – French or Dutch – defended their particular ways of doing things.[61]

Failure stirs resentment. Defeat embitters allies. From his exile in remote Saint Helena, Napoleon slandered his former colleagues for foisting the metric system upon him – and upon the French people. 'It was not enough for them to make forty million people happy,' he sneered, 'they wanted to sign up the whole universe.' The savants had wanted to overturn every custom, rewrite every rule, remake every French citizen into an image of themselves, and all for the sake of a miserable abstraction. They had behaved like foreign conquerors, he said, 'demanding, with raised rod, obedience in all things, without regard to the interests of the vanquished'.[62]

Delambre accepted the collapse of Napoleon's empire – and the demise of the metric system – with equanimity. He had faith in the long run of history. As the coalition armies entered Paris in the spring of 1814, he was at his desk working.

On the day of the siege, in spite of the cannonade audible in my study, I worked peacefully from eight in the morning until midnight. I was confident that the army would not be so foolhardy as to defend the town long, and would open their gates to the allies, who, piqued with pride, would comport themselves with generosity. Some days afterwards I saw foreign troops crowd on to the quays of Paris, pass under my window, and fill the streets and boulevards . . . The future does not offer a bright prospect for savants, but they should know how to content themselves with little. My savings will assure my own independence, and my wife's small fortune offers a still more reliable resource. You know my needs are simple. Work occupies all my time and all my faculties. My happiness does not depend on having a little more comfort; and I do not expect that I will have to change my personal habits.[63]

The fall of the empire cost Delambre several of his positions and three-quarters of his salary. He did not regret their loss, though he had to change residence again, this time to 10, rue du Dragon, convenient to the Academy's new home in the Collège des Quatre-Nations. Besides, Louis XVIII had reappointed him to the position that mattered most: Perpetual Secretary of the renamed *Royal* Academy of Sciences. And he retained his chair at the Collège de France and his post at the Bureau of Longitudes. As he explained to the new royalist administration, his fellow astronomers had kept their noses out of politics; they ought not to be shunted aside just because the régime had changed. Their political neutrality (some might say their political submissiveness) entitled scientists to keep their posts.[64]

The obligations of the present had already made Delambre an accomplished historian. In a sense, he had been preparing for this labour all his life. He had spent his youth indoors, hiding his eyes from the sun, studying ancient and modern languages. He had spent his scientific career poring over old texts, combing the work of dead astronomers for data to compare with his own. (In that respect, every astronomer is something of a historian.) Since becoming Permanent Secretary, he had composed eulogies, reports on his colleagues' accomplishments, plus a *Report to the Emperor* on the progress made by science during the past two decades. Even his preparation of the *Base du système métrique* had involved historical reconstruction.[65]

He now dedicated his final years to a comprehensive history of astronomy 'from Hipparchus and Ptolemy to us'.[66] One by one, in chronological order, he would pass each astronomer, ancient and modern, through the filter of current knowledge, extricating their genuine contributions from the ephemeral speculations of their age so that, as Delambre put it, his own *Treatise on Astronomy* would cap the whole. It was a six-volume, four-thousand-page scientific-extraction machine, and it was the first great history of science.[67]

His theme was the rise of precision: the relentless drive for exactitude. His method was empirical: a close reading of original works. Delambre sharply criticized those historians who had conjured up an antique people whose comprehensive astronomy had since been lost. There was no evidence of such a people. Nor did he agree with his colleagues who believed that the ancient Egyptians had derived their system of weights and measures from the size of the earth. Delambre had taken great interest in the French geodesers' trip up the Nile, but he rejected their conclusions. Their pyramid studies had unjustifiably projected current fantasies on to the past.[68]

The historian's great duty was impartiality, and impartiality began at home. Thus, in his article on Descartes – France's greatest savant – Delambre adopted a sharply critical tone. 'The historian owes the dead nothing but the truth,' he wrote. 'It is not our fault if *in astronomy* Descartes produced only chimeras.' And he proceeded to document just how often in his physics Descartes had violated his own norms for clear and consistent evidence. By ignoring this evidence Descartes' many admirers had 'cast a kind of ridicule upon the French nation by reviving the memory of the very errors they sought to hide behind the veil of official secrecy'.[69] In the history of science, as in science proper, progress was only possible with the forthright acknowledgement of error.

How fitting, then, that one of Delambre's last acts as Permanent Secretary was to superintend Descartes' reburial – and determine whether they had buried the right man. One hundred and fifty years earlier the great philosopher had died in self-imposed exile in Sweden, but his body had been exhumed and returned to France soon after. Since that time his remains had been buried in a church, exhumed again and then transferred to an Egyptian-style sarcophagus for storage in a national museum during the Revolution (alongside the royal statuary from the basilica of Saint-Denis) while politicians debated whether he was worthy of being pantheonized. By 1819 the Panthéon had been reconsecrated as a Catholic church, Voltaire and Rousseau's tombs were shunted aside, and the motto 'AUX GRANDS HOMMES, LA PATRIE RECONNAISSANTE' was covered over by scaffolding. Descartes, it was decided, would be better off re-reburied in the church of Saint-Germain-des-Prés. Delambre watched as they opened the sarcophagus and removed the small interior case engraved 'René Descartes, 1596–1650'. Inside, there was not much to see. 'Only the femur was recognizable; the rest had been reduced more or less to dust.'[70]

Imagine everyone's horror then, when, two years later, Sweden sent France a precious gift: the skull of its greatest genius, Descartes. Had a horrible error been committed? Had the wrong man been buried? It was another discrepancy to resolve. The Permanent Secretary was seventy-two years old by then, and in failing health. Yet he marshalled the documentary and forensic evidence, and adjudicated the matter with the same rigour he had brought to science and history. The gift skull, he concluded, was fake; the authentic skull of Descartes was presumably dust. After the physical evidence was gone, only inference remained. That would have to be enough.[71]

That year Delambre made careful preparations for his own death. He knew all too well what historians are capable of. He destroyed the bulk of his personal papers. He also set aside his correspondents' letters so that his wife might enquire after his death whether they wished to have them returned, lest their confidences fall into indelicate hands. He also composed a short manuscript autobiography, which later became the basis for the biography published by his student and scientific executor Claude-Louis Mathieu, and thus (as he knew it would) the basis for all subsequent biographies.[72]

All this was part of a conscious strategy. He was acutely aware that he would one day become the subject of historical investigation. And he did what he could – within the bounds of honesty – to shape that story. Thus, he planted the clues to the true story of the metric system in plain sight, publicly announcing that the logbooks of the meridian expedition were on file in the archives of the Observatory. He did not destroy Méchain's letters, but placed them under seal in the archives. He made it possible for historians to tell a story that he could not tell in his own lifetime.

Jean-Baptiste-Joseph Delambre died at home, at 10, rue du Dragon, at ten o'clock in the evening on 19 August 1822. The Permanent Secretary was buried in the Père Lachaise cemetery so that a new Permanent Secretary might live. Jean-Baptiste-Joseph Fourier's first eulogy paid homage to his predecessor. This was not a eulogy by an *ancien régime* savant, but a speech by a modern scientist intent on glorifying science. Fourier puffed up the meridian project with exclamation marks: he called it the greatest application of science in living memory – yet he gave all his data in the old units of measurement, the law of the land in Restoration France.[73]

History is made by the dead as well as the living. Their obsessions perch like fetishes upon our mantels and upon our consciences. Science likes to think of itself as the one human endeavour free of idolatry. It imagines that it erases the past each time new knowledge sweeps the mantel clean. But the errors of the past can shape the direction of science as much as its truths.

Delambre left the sixth and final volume of his comprehensive history of astronomy unpublished at the time of his death. He warned his friends that *The History of Astronomy in the Eighteenth Century* would tell 'the whole truth'. If they found some of its judgements severe, they should remember that history was no eulogy. He had written the book, he said, to 'discharge my conscience'.[74] Perhaps for that reason he profiled only dead astronomers and delayed publication until his own death. Claude-Louis Mathieu, Delambre's student and his scientific executor, shepherded those pages into print five years later.

Among the final astronomers to be discussed was Pierre-François-André Méchain. This was no eulogy. Delambre had learned a great deal about his former colleague since his funeral oration seventeen years before. So he began at the beginning: there was no evidence for the story that Méchain had made his start in astronomy by selling his telescope to Lalande to pay off his father's debts. He reassessed his colleague's career: Méchain had never been a scientific innovator and had borrowed all Delambre's formulae for his calculation of the meridian. He hedged on the exact date when Méchain had departed for Barcelona: delays in the making of the instruments, Delambre now said, meant that the operation 'could not commence' until 25 June 1792 – which is not the same thing as saying that Méchain actually left Paris on 25 June. He reapportioned credit for the grand mission: Tranchot deserved full recognition for his work. (The engineer had died in 1815 while triangulating at Montlhéry, the station just south of Paris where Delambre had begun his survey thirty years earlier.) He revisited the discrepancy at Barcelona: Méchain's 'fatal decision' to conceal the 3.24-second gap in the latitudes had made a mystery of what any other astronomer would have frankly acknowledged. And he supplied further revelations: Madame Méchain had been obliged to coax the astronomer to finish the mission; Méchain had refused to return to Paris until promised the directorship of the Observatory; Méchain had hidden his data from the Commission and obstinately clung to the notion of returning to Spain; Méchain had kept his secret intact until his papers were brought back to Paris.[75]

Yet, after all this, Delambre still considered Méchain a man 'admirable in every way', and assured his readers that they could have confidence in the meridian the two men had jointly measured.

No one can claim to have known Méchain in his capacity as an astronomer better than I. For ten years we maintained an intensive correspondence. I long had his papers in my hands, and made a careful study of them, going so far as to revisit every calculation which bore upon our mission. In this way, I assured myself that Méchain, enamoured above all by exactitude, but also very jealous of his reputation, had the misfortune to believe that the repeating circle could produce a degree of agreement and precision which was, in truth, impossible. When his observations presented unexpected anomalies, instead of reconsidering this view, he began to doubt his own abilities. Indeed, he feared that his own (unjust) opinion of himself would come to be shared by others, and would eventually overshadow his reputation. But this was not to be, and he remains an astronomer for ever worthy of our admiration.[76]

Delambre had left one more manuscript unpublished. In *The Size and Shape of the Earth*, he carried the history of geodesy up to his own day. In it, he rectified Méchain's suppressed observations so as to 'release ourselves from the obligation to disclose the false-hoods to which we had, in some sense, been made complicit'. Yet he also recorded them to console those geodesers who would follow in the path that he and his partner had marked out, and thereby 'disabuse them of the chimera of perfection, which mankind has yet to achieve and will probably never achieve'.[77] These revelations were considered impolitic by Mathieu, then campaigning to revive the metric system. Not until 1912 did an edition of this work find its way into print. It was at that time that the sealed letters between Delambre and Méchain were opened in the archives of the Observatory, where they lay unread for the rest of the twentieth century.[78]

Chapter Twelve

THE METRED GLOBE

I know no harm of Bonaparte, and plenty of the Squire,
And for to fight the Frenchman I did not much desire;
But I did bash their baggonets because they came arrayed
To straighten out the crooked road an English drunkard made . . .[1]

G. K. CHESTERTON, 'The Rolling English Road'

The origins of measures, we may presume, go back to the dawn of
human history. Well, not quite the dawn. According to Josephus,
the Jewish historian of antiquity, the origins of measurement go
back to Cain. This degenerate son of Adam not only killed his own
brother, he was the first land surveyor and city planner. Then to
round out his sins 'he put an end to that simplicity in which men
lived before, by the invention of weights and measures'.[2]

Measures are a consequence of man's fall, a human invention
for a world outside Eden, where scarcity and mistrust rule and
labour and exchange are our lot. Measures are more than a creation of society, they *create* society. As the outcome of years of
negotiations over the proper way to conduct exchanges, their
ongoing use reaffirms our social bonds and defines our sense of
fair dealing.

Inaugurated during the French Revolution and rescinded during
the French empire, the metric system has gone on in the past two
centuries to be readopted by France and embraced by every other
nation on earth – except the United States, Myanmar (formerly

Burma) and Liberia. In 1821 John Quincy Adams (the son of a different Adam) was asked to report on whether the United States should adopt the metric system. Adams had made a close study of Delambre's *Base*, and greatly admired the meridian expedition. He declared that the International Commission of 1799 had marked a new epoch in human history, pointing towards a future in which 'the metre will surround the globe in use as well as in multiplied extension; and one language of weights and measures will be spoken from the equator to the poles'.[3] Adams' prediction has been borne out. A system once spurned in its homeland has become the world's measure – though not in Adams' homeland. How did this happen?

Its advocates have called the triumph of the metric system inevitable, and this aura of inevitability has always been their most compelling argument. If everyone else is going metric, there is a huge incentive to join the crowd. This, however, begs the question of how its advocates managed to convince the world that the metric system was inevitable. As late as the 1950s, visitors to a science museum in Paris were warned that the Anglo-Saxon measures were about to 'implant themselves' in France.[4] How was the world convinced that the metric system would triumph?

To outward appearances, the spread of the metric system has tended to follow upon political upheaval, at least as a matter of law. The metric system was first legally adopted in France during the Revolution, imposed on Western Europe during the French empire, adopted by the newly unified nations of nineteenth-century Europe as a sign of their sovereignty, and then pressed upon their colonies by metropolitan administrators. At the same time, the actual on-the-ground implementation of the metric system has taken a much more gradual course, tracking lumbering social developments in education, manufacturing, trade, transportation, state bureaucracy and professional interests. From the beginning, Adams anticipated it would be thus. A change in metrical standards, he warned, was 'one of the most arduous exercises of legislative authority'. Writing the legislation was easy, 'but the difficulties of carrying it into execution are always great, and have often proved insuperable'.[5] Yet even this process of gradual implementation depended essentially on political will. Only sovereign states had the authority to coordinate so far-reaching a transformation in the lives of their citizens. And unless the change were coordinated there was little point in converting. When Adams wrote to Thomas Jefferson to ask him for his views, the former President, who had long given up hope of metrical reform, put his finger on the essential dilemma: 'On the subject of weights and measures, you will have, at its threshold, to encounter the question on which Solon and Lycurgus acted differently. Shall we mould our citizens to the law, or the law to our citizens?'[6]

But if standards are a matter of political will as much as of economic or technical readiness, then reaching an agreement on standards depends as much on myths as on science, especially on myths *about* science. It was an open secret among nineteenth-century astronomers that Méchain had obtained contradictory results for the latitude of Barcelona. And any scientist who looked in a table of physical constants could see that the Archive Metre fell a hair short of one ten-millionth of the distance from the North Pole to the equator. These two flaws were not, in fact, connected. The metre was flawed because the expedition's governing premise was flawed – the premise that the French sector of the meridian measured by Delambre and Méchain in 1792–9 could be considered representative of the world's shape as a whole. Scientific progress had falsified the metre, as Lalande had hoped it would. Yet in spite of this, Delambre and Méchain's epic mission succeeded – not because it had produced accurate results, but because it was epic.

Ultimately, the restoration of the metric system in nineteenth-century France depended as much on filial piety as on the promised reign of reason, as much on the grandeur of the past as on the allure of the future. But past and future could not meet in the present until the French Revolution (and its metric revolution) had reclaimed an honoured place in French history. The Revolution of 1830, which deposed the Bourbons and inaugurated the 'bourgeois monarchy' of Louis Philippe, made such a present possible. In 1837 the government revived the metric system, both as a promise to modernize France and a public assertion that the new régime was a worthy successor to the first great Revolution. The two men who did the most to promote the legislation had similarly mixed motives. One was Charles-Emile Laplace, the physicist's son, who had inherited his father's title and sat in the House of Peers. The other was Claude-Louis Mathieu, Delambre's scientific executor, now a representative in the House of Deputies. Their argument was simple: the metric system would make France a modern, prosperous nation in the years to come, and it could be implemented immediately, thanks to the glorious achievements of France's past.

The story of Delambre and Méchain's mission played a prominent role in this political campaign. Their exactitude in the face of social chaos exemplified what was noble and salvageable from the first great Revolution. Their comical troubles with the common people – those benighted folk who had accused them of espionage and sorcery – implied that the people's rejection of the metric system had been based on a similar misunderstanding. Above all, their meridian expedition had been a monumental undertaking, a celebrated piece of France's Revolutionary legacy that must be preserved. In this sense, the meridian expedition succeeded as a matter of politics, even if it had failed as a matter of science. The great virtue of the meridian expedition, it now turned out, was that it could *not* easily be repeated – as a simple pendulum experiment might have been. The meridian expedition, by its very grandeur, difficulty and expense, had fixed the metre – permanently. The same meridian project which had scuttled international cooperation in the 1790s by alienating Thomas Jefferson and the British savants now made the metre impervious to change. The expedition had removed the metre from the flux of scientific progress and locked it away in the National Archives as a platinum fact.[7]

The legislation, which was passed with overwhelming support in 1837, made the metric system obligatory throughout France and its colonies as of 1 January 1840. France had elected to mould its citizens to the law. When one representative – a prominent physicist, as it turned out – asked that the law permit units divisible by eight as well as ten, so as to help those who sorted goods by halves and quarters, an anonymous deputy shouted back from the floor, 'On the contrary we must break their bad habits'.[8] For some, this metric victory signalled the final repudiation of the *ancien régime*, both in the workplace and the halls of power.

> Challenging routine and hatred,
> Taking its stand on useful things,
> The measure of the Republic
> Has overthrown the foot of kings.[9]

But what one person dismisses as routine or habit, another calls a livelihood. While the legislature deliberated in Paris, a riot broke out in Clamecy, a small town in Burgundy on the banks of the new canal connecting the Loire to the Seine. Dock-workers there smashed decimal measures, and the government had to call in the cavalry. The dissension had been sparked less by the new measures *per se* than by the suspicion that the transition would come at the dock-workers' expense and open the town to ruinous competition.[10] A plaintive song began making the rounds in 1840:

> What's it good for, this new law?
> From this day forth can we no more
> Order a pound of yellow tallow,
> Nor butter by the quart?
>   Will every corner grocery
>   Hire a staff of sorcerers?
>   Or will the Paris Academy
>   Supply us with our stock boys?
> *Chorus:*
> I'm no fan of our legislators'
>   Decimal
>   Systemical.
> Long live the measures of yesteryear!
> And damn the new weights and measures![11]

Fifty years later, a priest in Corrèze, a region along Delambre's sector of the meridian, could still complain that the metric system was unknown there. In 1900, in the area surrounding Amiens, Delambre's hometown, many citizens still used the old measures to measure cloth. In the 1920s land in the Midi was still parcelled out in units which varied from district to district depending on the quality of the soil.[12]

By then, however, the world of the old measures was dying. Across the decades of the nineteenth century, knowledge of the metric system had radiated out from schools, cities and railway lines. As provincial and foreign immigrants poured into the cities, their children acquired a public education sponsored by the central state. As towns became important markets for the countryside, farmers packaged their produce accordingly. Rural France found itself being lured out of the village marketplace and into the world of the market principle. World War I was the turning point, in metrical matters as in so much else. The younger generation stopped speaking the various local patois; now they spoke only French. In the decades that followed, electrification reached the farms, along with government subsidies. Full conversion took nearly two centuries, but today the metric system feels as natural throughout France as the old measures once did. In the process, the thinking of the French people has also changed.

Everyone in France is now 'enlightened'. They accept the metric system as the only possible system of weights and measures, and are barely aware that there has ever been any other. In market towns, grocers will still sell you a *livre* (a pound) of beans. This is no longer a local variant, however, but simply the popular name for five hundred grams (though tourists are still advised to watch for a surreptitious thumb on the scale). Today's French citizens are much wealthier than their ancestors. More educated. More numerate. More calculating. The young people are all leaving for the cities. Local distinctiveness is receding into the distance. The metric system is all they will ever know.

Yet France was not the first country to convert to the new measures. By the time France restored the metric system in 1840, it had already been obligatory for two decades in Holland, Belgium and

Luxembourg. This was a consequence of the French empire – and of its defeat. The diversity of measures in the Low Countries had long frustrated administrators there. After France annexed those territories they shared her metric régime – and her populist revolt. The collapse of the Napoleonic empire threatened complete metrical chaos. The people of the Low Countries may have resented French rule, but the restored monarchy saw the advantages of its centralized form of administration, especially for a fractious territory that thrived on commerce. King William I of Orange ordered the decimal metric system obligatory throughout the Low Countries by 1820. And when Belgium separated from Holland in 1830 it not only retained the metric system but reverted to the original nomenclature.[13]

Thus the metric system simultaneously became a tool of political unification at a national level and facilitated the sort of international commerce that would – in the long run anyway – dilute national sovereignty. Italy is a good example of how this pattern played out. The French armies had forced the peninsula into larger political groupings, ruled metrically by the iron metre rulers that the Italian savants brought back from the International Conference. The French retreat disrupted a reform that had met with little popular success. But once the French resurrected the metric system in the nineteenth century, Piedmont and Sardinia quickly declared the metric system obligatory as of 1850. Over the next decade other Italian city-states joined the bandwagon. This embrace of a common measure pointed towards the creation of an Italian nation-state – which took place in incremental stages between 1861 and 1870, and which declared the metric system the sole national standard in 1863.[14]

The Spanish case shows how the metric system united not only nations but those nations' colonies and their successor states around the world. Spain had been among the first nations invited to join the metric system. After all, the meridian arc had one foot in Catalonia. That invitation was declined. A Spanish law of 1849, which set the metric deadline for 1852, was extended half a dozen times. In 1852 Portugal likewise called for a ten-year transition, a deadline likewise extended. In the meantime, however, the metric system was legally adopted throughout the newly independent states of Latin America. Decrees in favour of adopting the metric system were passed in Chile (1848), Colombia (1853), Ecuador (1856), Mexico (1857), Brazil (1862), Peru (1862) and Argentina (1863). Each of these laws had to be reiterated on many occasions, and the local populations retained their old measures for many years, but these laws gave the metric system the aura of inevitability, which was always its greatest asset.

So far, legal enactment of the metric system had followed in the wake of revolution and war. In each case, the impetus came from an upstart régime seeking to legitimize its rule. Yet the popular adoption of the metric system followed a quite different pattern; it accompanied the expansion of networks of education, transportation and trade, together with the spread of a money economy. By the middle of the nineteenth century, there were those who wanted to press for a coordinated conversion on a global scale.

It was an era of international commerce and great-power rivalry. Bilateral agreements regulated trade between states, even as alliances divided them. Professional groups reached across national borders, even as nationalism grew more shrill. A worldwide postal treaty signed in Paris in 1863 defined weights of international parcels in metric grams. The globe was striped by time zones and stitched together with undersea telegraph cables. Statisticians convened international meetings – at Brussels (1853), Paris (1855), London (1860), Berlin (1863), Florence (1867) and The Hague (1869) – to insist that their respective governments adopt the French metric system.[15]

The virtues of international standards of weights and measures were first showcased to the general public at the spectacular Crystal Palace Exhibition of 1851. The judges there complained that they could not pick the prize winners fairly because the entries were presented in thousands of incommensurable weights and measures. Some concluded that the best solution lay in one of the exhibits, a set of metric standards submitted by the Paris Conservatoire des

Arts et Métiers. At the Paris World's Fair of 1867 visitors could walk through a glass-and-iron pavilion and gape at the diversity of the world's measures, culminating in the metric standards. A guidebook pointed visitors towards the obvious conclusion.[16]

Suddenly, the Utopian dream seemed within reach. In the 1860s Britain, the United States and the German states all appeared to be on the verge of joining the metric bandwagon. In 1863 the House of Commons passed, by a vote of 110 to 75, a law mandating the metric system throughout the British empire. The parliamentary session ended before the House of Lords could act, but a new vote was slated for the next year. In 1866 the United States Congress voted to make the metric system legal – though not obligatory. America's metric advocates expected to win full conversion in the next session of Congress. And in 1868 the German Zollverein – the Prussian-led customs union that laid the groundwork for German unification – agreed to require the metric system as of 1 January 1872.

For France, this was a momentous opportunity – with commensurate risks. Eager as they were to welcome the world's great economic powers to their metric network, the French feared that these nations would dictate the terms of their entry in such a way as to invalidate the original standards. Having argued so passionately that the fundamental unit must be based on nature, the French now feared being hoist by their own rhetoric.

The pivotal test came from Germany, France's alarming new rival. The metric system appealed to the various German states for the same reasons it appealed to the Italians. It was just as the French savants had foretold: the metric system was acceptable to everyone because it favoured no one. Prussia wanted to unify all the German states under its rule. The Prussian state may have been militarily and administratively potent, but it wanted the rich, industrialized states of western Germany to agree to unification willingly. In 1861, when Austria (Prussia's rival) conferred with those western states on common weights and measures, the Prussians refused to join the discussions. But by 1867, when Prussia had won the upper hand against Austria, it could afford to behave more magnanimously. Prussia agreed not to impose its own measures and instead to adopt the metric system as a natural, neutral standard sanctioned by science.[17]

But was the metric system neutral or was it French? Was it natural or historical? Was it sanctioned by science or by law? Was it derived from the size of the earth or was it just a corrupt platinum bar housed in the Paris Archives?

These questions all came to the fore at the first international geodetic conference, held that very year in Berlin. The geodesers in attendance knew better than anyone the shortcomings of the original determination of the metre. Since Delambre had published the final volume of the *Base* in 1810, his scientific successors had further refined their knowledge of the earth's shape. Each passing decade had widened the gap between the Archive Metre and the known size of the earth.[18]

During the past half-century, moreover, each European nation had triangulated its own territory, mapping its terrain with reference to the regular ellipsoid which best represented the earth's curvature through its own particular lands. It was as if each European nation inhabited its own eccentric planet. Some of these nations, Prussia first and foremost, were now eager to knit their maps together with those of their immediate neighbours. For this they needed a common standard and uniform procedures. The Germans suggested a technique for doing this: Gauss's method of least squares, by which the triangles of every nation could be brought into optimal alignment. In 1861 General Johann Jakob Baeyer of the Prussian army and longtime director of its cartography department had secured permission to establish a Central European Geodetic Association in Berlin. As he noted, 'By its very nature, such an enterprise cannot be the work of a single state; but what one [nation] cannot realize alone, many may achieve together. And if in the process central Europe should unite for this purpose, devoting all its might and resources, a great and important work will be called into being.'[19]

It was an alarming and exhilarating call, reminiscent of nothing so much as the calls for European unity which had come out of

Revolutionary France seven decades earlier. The geodesers were determined to bring their numbers into alignment, to live, as it were, on the same planet. The association became the nucleus of the world's first international scientific association.[20]

But when the Germans sent out invitations to expand the association from Central Europe to the entire Continent, the French response was divided. Some scientists saw it as a chance to revitalize French geodesy, which was still using the seventy-year-old techniques of Delambre and Méchain. Others saw it as an attempt to subordinate France's triangles to a pan-European grid and to subvert the one true Archive Metre as determined by Delambre and Méchain, whose results ought only to be revised 'with caution and intelligence'.[21] The French government refused to send delegates to the Berlin convention. Relations between the two Continental powers were deteriorating rapidly and the French did not want their metre publicly impugned. When one French scientist rashly suggested that the Academy launch its own preemptive expedition to remeasure the earth – and get it right this time – his colleagues quickly shut him up. A standard was fixed, or it was not a standard.[22]

The international geodesers understood this. At the meeting in Berlin, General Baeyer and the rest of Europe's geodesers agreed that the metre should remain the standard of length – not because it was based on nature but because it was widely accepted. If the natural origin of the metre was a fiction, it was a useful fiction. 'In truth, the metre draws a good deal of its prestige from the notion, flattering to human pride, that our daily measures are drawn from the dimensions of the globe we inhabit.'[23] They insisted, however, that the Archive Metre was defective, and that a new metre bar be made to replace it.

For the previous seventy years, nations adopting the metric system had been obliged to beg France to calibrate their weights and measures for them. This gave France an unseemly custodial power, and the bar itself had been worn down by the continual comparisons. In 1837 a Bavarian scientist found that the ends of the bar were scratched. In 1864 a microscopic inspection revealed that the surface there was pitted. Moreover, chemists had discovered that the platinum, once considered 'pure', was actually adulterated with allied metals (such as iridium), complicating the bar's rate of expansion with temperature. In sum, there was reason to fear that the bar's length had altered since 1799, could not be accurately described now, and would change in the future. Nothing, it turned out, was more ephemeral than yesterday's cutting-edge science. The geodesers of 1867 agreed that the new bar should differ 'as little as possible' from the Archive Metre. But they also wanted a permanent international agency to take charge of this new standard so that no one nation could claim the standard for itself.[24]

This sent the French into a paroxysm of self-doubt. Would Germanic precision supplant French precision, as Germany was supplanting France? It was as if the entire nation had taken its cue from Méchain, making his error collective and his paranoia general. Some French scientists, notably at the Bureau of Longitudes, welcomed the chance to put the metric standard on a secure footing, but the Minister of Commerce rejected any move to replace the metre. And scientists at the Observatory and Academy concurred. The Archive Metre was in excellent shape, they reported, and remained the only possible standard. Indeed, they went so far as to deny that their Revolutionary forebears had ever claimed that the metre should be based on nature at all, or that all meridians were the same length, or that the length of the Paris meridian might be measured definitively. Thus they thrice denied the founding premises of the metric system. Yet they concluded that they would rather invite their foreign colleagues to Paris than see them establish a rival system.[25]

So seventy years after Napoleon I promoted the first international scientific conference, Emperor Napoleon III, his nephew, sent out invitations for a second metric conference to be held in Paris. This time, scientists from all the world's nations – including the Americans, the British and the Germans – were invited. 'Today, as in the distant days of the great International Commission of Weights and Measures, it is by inviting French and foreign savants to work together in complete equality that we may best preserve the metric system's universality and obtain truly international models, perfectly identical with those in the French Archives and capable of serving the scientific needs of each nation, while preparing the world for the general adoption of the metric system.'[26]

In July 1870, two weeks before the conference was to start, Prussia and France went to war. The German delegates stayed at home, but scientists from fifteen other nations, including the United States and Britain, held their first assembly in Paris on 8 August. At the time, the French army was falling back on Metz. Under the circumstances everyone agreed that any final determinations would have to wait until *all* their colleagues were present. Then the French brought into the open the question which haunted their scientific nightmares: did their guests really expect to base the new metre on the size of the earth? It took several days for the German-born Swiss delegate Adolph Hirsch (a co-organizer of the Berlin geodesy conference) to reassure his French colleagues that 'no serious scientist in our day and age' would contemplate a metre deduced from the size of the earth. The new metre bar would be built to match the old one.[27]

The Prussian army won the war; the French emperor abdicated; the Prussian king became emperor of Germany; and France (after a ghastly blood-letting) became a republic. The French lost Alsace and Lorraine, but they regained democracy. In 1872 the new French republic reissued invitations for an international metric conference. The German empire sent several delegates. Wilhelm Foerster, their chief representative, was an affable enthusiast for world metrical harmony. For nearly a month, scientists from thirty European and American nations discussed the form, content and distribution of the replacement measures. It was all very collegial. They agreed that the new bar should be made as similar to the old one as possible, right down to its impurities: a mix of 90 per cent platinum and 10 per cent iridium. They also resolved to make as many standard metres as there were nations, and only then to select one to serve as the definitive standard: a first among equals. Finally, they proposed a permanent International Bureau to superintend these activities.[28]

Forging the new metre

In the 1870s scientists at the workshop of the Conservatoire Nationale des Arts et Métiers in Paris experimented with the construction of a new standard metre. The new definitive metre was not completed until the late 1880s.

(*Illustration* (16 May 1874), 316. Engraver: H. Dutheil. Photo: Roman Stansberry.)

The 'Convention of the Metre' of 1875 remains the framework for all international metric standards, including those for electricity, temperature and other phenomena. Although the French delegates were not keen on a permanent International Bureau of Weights and Measures, they offered to house the institution in Paris rather than see it wrested away by Berlin. The Pavillon de Breteuil, which they donated for its headquarters, had been almost completely destroyed during the recent Prussian siege of Paris and was rebuilt at international expense.[29]

It would take fifteen years of scientific controversy – including a spat over the platinum-iridium alloy which nearly caused another Franco-German split – to construct new metres to the

Bureau's specifications. By then virtually every European nation had mandated gradual introduction of the metric system. When the bars were shipped out in 1889, the old Archive Metre, built to match the meridian data collected by Delambre and Méchain, lost its universal status and fell, like the Châtelet *toise* before it, into the pit of history. It became just another stick of precious metal, worth the market price for platinum, plus whatever value human memory ascribed to it. But needless to say, the French government did not melt it down. They preserved it, as before, in the National Archives, a historical artefact like any other, to be read as evidence of the past.[30]

Delambre and Méchain, the myth now went, had heroically measured the earth so that the metre could be set at one ten-millionth of the quarter-meridian. Their error was forgotten, even as the embodiment of their error was preserved. It was preserved in 1889 when the new platinum-iridium bar replaced the old Archive Metre. It was preserved again in 1960 when the International Bureau redefined the metre in terms of the wavelength of light emitted by a specific energy transition in the krypton-86 atom. And it was preserved again in 1983 when the Bureau redefined the metre as the distance travelled by light in a vacuum in $1/299{,}792{,}458$ seconds (with time, the fundamental unit, now defined by an atomic clock). Thus the new quantum mechanics, famous for its principle of measurement uncertainty, has again provided the Bureau with a standard based on nature that can be specified with exceeding (but never final) precision. Yet each redefinition, including the most recent, has been concocted so as to preserve the length of Delambre and Méchain's original metre of 1799.

The truth belongs to everyone and no one. It is public property and ephemeral, or else it is not the kind of truth we call science. But error is forever because, having happened once, it exists, like an unhappy family, in its own particular way. Delambre and Méchain built their lives into the metre: they travelled the meridian, they selected its stations (including their own private sites: the country château at Bruyères, the observatory on the rue de Paradis), they peered through the scopes, and their inky fingers calculated the angles of the earth and the stars. The metre is their epitaph because only a person's mistakes are truly his own. Yet in accepting their metre, we have made their error our own, which is to say public, singular and true. Theirs was truly an error for all people, for all time.

By the middle of the twentieth century, the vast majority of the world's nations – with the major exceptions of the British Commonwealth and the United States – had joined the metric system. Each time, the precipitating event was political upheaval. Shortly after it became a republic in 1912, China announced that it would switch to the metric system during the next decade; the law was enforced after the revolution of 1949. Tsarist Russia recognized the metric system in the late nineteenth century, but it was the Soviet Union that in 1922 made metric measures mandatory. Despite earlier legislation, Japan and Korea did not seriously convert to the metric system until after World War II. The system spread through Asia and Africa in the wake of colonization, and later, in the wake of decolonization; either way, metric uniformity appealed to those who wished to legitimize their territorial rule and create a national administration, even as they opened up their territory to extranational market forces. Thus Jawaharlal Nehru took India metric soon after the British left in 1947. And the more the metric system spread, the more irresistible became the logic of joining the world's pre-eminent international network.[31]

Britain was the first economic power to adopt the metric system without passing through radical political upheaval first. No doubt, this explains why it was also the last. Though advocates of the metric system had been trying to hustle Britain in the metric age for over a century, it did not sign on to the metric system until it was on the verge of joining the Common Market in the 1970s.

A few British savants had courageously embraced the metric system in the midst of the Napoleonic wars. Upon reviewing the first volume of Delambre's *Base du système métrique* in 1807 the eminent Scottish geologist John Playfair urged his countrymen to heed the innovation across the Channel. Playfair quibbled about the details, of course, as was a savant's prerogative. He would have preferred a duodecimal scale to decimal division. A meridian standard was uncertain because not all meridians were equal or regular. On the other hand, he admired Delambre's integrity. By identifying the meridian's irregularities as a real feature of the globe, not observational error, the French had provided geologists with further evidence of the earth's extreme age and the accidents of history which had distorted its shape. And on the laudable goal of universal measures, he closed ranks with his fellow savant. Delambre and Méchain had created a system worthy of emulation by all European nations. Playfair quoted Ovid: *Fas est et ab hoste doceri*, 'It is proper to learn, even from an enemy.'[32]

Delambre repaid the compliment. In the final volume of his *Base* he offered an appreciation of Playfair's review, and expressed the regret that the meridian arc had not been extended past Dunkerque to Greenwich. A longer arc would have been more accurate and, by including more territory, still less the property of any one nation.[33]

This burst of cosmopolitanism would prove to be the last cross-Channel metric cooperation for fifty years. The Victorians concentrated instead on unifying their imperial measures, itself a daunting task. British measures, though not as diverse as those of *ancien régime* France, still snarled central administration and hampered long-distance trade (while greasing the wheels of local life). The nineteenth-century Parliaments began making good on vows dating back to the Magna Carta – 'one measure of wine shall be throughout our realm, and one measure of ale' – eliminating local variants, conducting spot inspections, imposing fines. Even after the 1834 Parliament fire incinerated the 'Exchequer yard', the British affirmed their commitment to a physical standard rather than a natural standard subject to continual modification. By mid-century, the British had largely eradicated local variations and anthropometric measures, while still allowing each trade to enjoy their own idiosyncratic units: from the jeweller's carat to the doctor's scruple.[34]

At the time, of course, some Britons saw the metric system as the next logical step. A society formed in 1857 – the British Branch of the International Association for Obtaining a Uniform Decimal System of Measures, Weights and Coins – lobbied to remove 'one more of those barriers which at present divide the nations of the earth'.[35] The society spoke on behalf of most scientists, some engineers and a few forward-thinking manufacturers. The British Association for the Advancement of Science endorsed the metric system. 'Science suffers by the want of uniformity, [and] international commerce is impeded by the same cause.' It was time to put national pride aside. As only the French system had a chance of being adopted by all nations, Britons 'have no choice but to conform'.[36] Engineers like Joseph Whitworth (famous for his screw-thread standard) argued for decimalizing British measures. And everyone denounced the British Babel of pounds, shillings and pence. Trade with France was being liberalized. Businessmen wanted to spur exports. The Council on Education, the Associated Chambers of Commerce, even the Prince Consort, all favoured metric reform. Witnesses before a parliamentary select committee asserted that the metric system had been adopted abroad without a hitch. In 1863 a bill making the metric system obligatory throughout the empire passed the House of Commons by a vote of 110 to 75 – only to lapse when the session ended before the House of Lords could act. The bill was slated for reintroduction in the next Parliament.[37]

Instead, reform would take another century. The economic logic was clear enough. Having at last achieved internal imperial uniformity, most Britons saw little advantage in switching units. Moreover, because Britain and America shared common measures, any gain in commerce with outsiders would be matched by a compensating loss with a major trading partner, unless the two nations coordinated their reform, an unlikely prospect. But all this economic logic might not have halted the metric tide had not an anti-metric lobby quickly emerged to articulate these concerns. The next year they were ready. Manufacturers testified to the cost of converting machinery. Politicians worried about sowing confusion and resentment among shopkeepers and customers. Protectionists made a virtue of complexity: the morass of British units advantaged British traders who could understand foreign measures, whereas foreigners could not understand the British ones. Above all, the anti-metric camp touted the imperial measures as an authentic expression of British history. They scorned the metric reformers as elitists, and presented themselves as practical and patriotic men – always good politics, especially when they could claim to be anti-French to boot.[38]

Having assumed that victory was within their grasp in 1863, metric advocates were obliged to settle in 1864 for a bill merely 'tolerating' the metric system. A principal reason for their defeat was betrayal from within the scientific ranks. Britain's two most eminent astronomers, Royal Astronomer George Airy and Sir John Herschel, joined forces to undermine the scientific basis of the metre. They dredged up the discrepancy in the Barcelona latitude and charged Méchain with 'disingenuous concealment' in the affair.[39] They pointed out that the Archive Metre fell at least 0.1 millimetre short of its natural length. If Britain really wanted a measure based upon nature, Herschel argued, it ought instead to adopt the length of the earth's axis from pole to pole. This distance was real, unlike an imaginary meridian upon the surface of the geoid. It was truly international, unlike a meridian arc which traversed only one nation. And finally, as it equalled 500,500,000 inches in length, it would, with a tiny adjustment, provide a natural basis for the imperial inch.[40] Thus, with British ingenuity, everything could be brought thoroughly up to date – without changing anything at all.

The metric reformers did not give up without a fight. The Decimal Association (pro-metric) squared off against the British Weights and Measures Association (anti-metric). Lord Kelvin (pro-metric) squared off against Herbert Spencer (anti-metric). Exporters (largely pro-metric) squared off against manufacturers of scales (anti-metric). In the ensuing pamphlet war, doggerel did as much damage as calculation. The engineer William Rankine attacked the metric reformers with this ditty:

> Some talk of millimetres, and some of kilogrammes,
> And some of decilitres, to measure beer and drams;
> But I'm a British Workman, too old to go to school,
> So by pounds I'll eat, and by quarts I'll drink, and
>   I'll work by my three-foot rule.
>
> A party of astronomers went measuring the earth,
> And forty million metres they took to be its girth;
> Five hundred million inches, though, go through
>   from pole to pole;
> So let's stick to inches, feet and yards, and the good
>   old three-foot rule.[41]

The Decimal Association battled back (most unsuccessfully) with the horrendous 'Chains of Habit':

> But not until France shook us out of our trance
> Did anyone dare to look even askance
> At the horrible mix of the furlong and pole
> That poisons our youth and our life as a whole.[42]

With verse like this, perhaps the metric advocates deserved to lose. Not until 1965 did the British government announce a ten-year transition. And not until 1971, when the British agreed to join the Common Market, did serious 'harmonizing' begin.

More than thirty years have elapsed and the process drags on, as do the protests. On 1 January 2000 a new era dawned in Britain: shopkeepers were obliged to sell in metric units. A few months later, Steve Thoburn, a Sunderland grocer who sold bananas by the pound, had his scales confiscated. The British Weights and Measures Association took up the cause. The tabloids worked themselves into a frenzy of indignation. This is the sort of local resistance that has everywhere accompanied the introduction of the metre. The conversion of Britain involved only a simple switch from one set of impersonal measures to another, yet even this degree of disorientation can *feel* like a loss of sovereignty. And in a sense,

Britain's long-delayed adoption of the metric system has been precipitated by a political rupture: the decline of the Commonwealth and Britain's entry into the European Community.

Once Britain committed itself to conversion, the rest of the Commonwealth nations followed suit. In 1970 Canada announced that it would not wait upon its huge trading partner to the south. A voluntary transformation was envisaged, coaxed along by an educational campaign, with animated films such as *Ten: The Magic Number*. When consumer goods were packaged in metric units (toothpaste first), some Canadians protested. But generally Canadians have been puzzled by their own 'sheepish' acceptance of the metre – to the point where many consider it a point of national pride that their country has gone metric, while America has not.[43]

Americans have been arguing about the metric system since shortly after the signing of the US Constitution. Article 1, section 8, granted Congress authority 'to fix the Standard of Weights and Measures'. Who would have thought so quantitative and banal a subject would elicit such passion? Industrialists and scientists, mystics and nativists, curmudgeons and enthusiasts, schoolteachers and politicians have all battled over the world's measure. To date, America's metric advocates have always failed.

Paradoxically, it is America's modernity – its freedom from 'feudal' institutions, its origin in colonial rule, and hence the *relative* uniformity of its measures – which explains the country's failure to switch to the ultra-modern metric system. As a large and homogeneous economy, the United States already enjoys most of the coordination advantages that come from common standards, reducing its incentive to join the rest of the metric world. And even though doing so would undoubtedly bring long-term economic gains, the American government is notoriously beholden to business groups and populists, who insist on short-term pay-back. America is the only country in the world that still thinks it can afford to stay outside the metric system even as it participates in the world economy.[44]

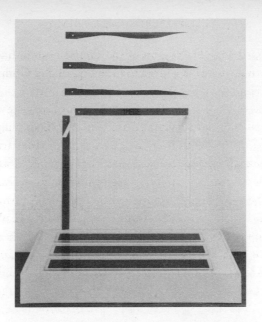

Marcel Duchamp, '*Trois Stoppages Etalon*'

Later in life the Franco-American artist Marcel Duchamp referred to this piece, usually translated as 'Three Standard Stoppages,' as 'a joke about the metre'. It was also his investigation into the relationship between universal standards and individual creativity, as well as a seminal moment in the inauguration of 'found' art. The piece was begun in 1913–14, when Duchamp 'dropped' – or said he dropped – a piece of thread one metre long from a height of one metre on to a wooden slat and preserved the resulting curvy shape by painting the string over with varnish. For each of the three repetitions of this 'experiment' he then cut a wooden template to match the curvy shape. These he later placed one above the other (as pictured), eventually adding vertical and horizontal straight-edged rulers labeled '1 METRE' to document his procedure. The assembled piece was not completed until 1953 when it was displayed in the Museum of Modern Art in New York. In the intervening decades Duchamp deployed the curvy wooden templates as his own personal standard to design a series of other artworks. According to the artist, this was his first attempt to use 'chance as a medium'; yet closer examination shows that he did not drop the threads at all, but in fact carefully arranged them on the wooden slats. There are many ways to read this fascinating piece. It subverts the ideal of exact measurement even as it demonstrates the role of universal standards in the creation of the most personal and idiosyncratic art. It may also suggest the 'stoppage' of the guillotine.

(Museum of Modern Art, New York. Photo: Art Resource.)

Not until the 1830s did the United States even define a national standard. A Swiss geodeser named Ferdinand Rudolph Hassler, who had arrived in the US in 1805 bearing one of the definitive iron metres of the International Commission, became the first director of the National Bureau of Standards. He affirmed the decision to stick with the English weights and measures. In 1863, within a month of its founding, the National Academy of Sciences began pressing for the metric system. Yet Congress was content merely to legalize the metric system and allow Americans to adopt the new measures voluntarily – an approach that remains US government policy to this day. America, in Jefferson's terms, has preferred to mould the law to its citizens, rather than its citizens to the law – at least where commercial interests are at stake.[45]

In America, as elsewhere, even the prospect of the metric system has provoked a nativist response. The humorist Josh Billings mocked the International Metric Convention as a harbinger of universal standardization. 'Never,' he wrote, 'did so many Kaisers, Kzars, Kings, kum kling knit together in so Klean a Kawse to work so Kommendable a kure.'[46] Others considered the metric system an abomination. Charles Latimer was a devout Christian, a successful railway engineer and an avid pyramidologist who believed that the 'sacred inch' had been built into the Great Pyramid at Giza and then transmitted across the millennia to the United States. He also had a visceral contempt for atheism, the French and the metric system. He would even have preferred a Statue of Liberty 'measured in good earth-commensurable Anglo-Saxon inches, not in French milli-meters.'[47] No doubt, the anti-metric arguments of industrialists and engineers did more to dissuade Congress in the long run. But Latimer could plausibly boast that he had stopped Congress from passing metric legislation in the 1870s and 1880s.

The most recent US campaign to adopt the metre began in the 1970s when it became clear that the United States would be the last major metric holdout. In 1971 the National Bureau of Standards issued a report entitled *A Metric America: A Decision Whose Time Has Come* without even the courtesy of a question mark. Efficiency gains and international trade, it argued, made metric conversion well worth the short-term costs to consumers, manufacturers and government agencies. Multi-national corporations wanted to assemble goods from parts made in all corners of the globe. Certain industries – alcohol and automobiles – had already adopted the metric system. But when the 1975 Metric Conversion Act emerged from Congress it lacked enforcement powers, financial wherewithal or a timetable for conversion. As President Gerald Ford memorably announced at the signing ceremony: when it comes to the metric system, US industry 'is *miles* ahead of official policy'.[48]

And so it has remained. Official attempts to convert freeway signs to kilometres have only riled the citizenry. Newspaper editorials mock metric advocates as petty dictators, or worse, boring. *Chicago Tribune* columnist Bob Greene founded WAM! (We Ain't Metric!): 'WAM!'s guidelines are eloquent and simple: We are against the metric system because we don't like it. We won't learn it because we don't want to.'[49] President Reagan disbanded the Metric Board. A 1992 follow-up pamphlet from the National Bureau of Standards, *A Metric America: A Decision Whose Time Has Come – For Real*, had the ring of desperation. Gallup polls showed that as awareness of the metric system doubled between 1971 and 1991 (from 38 to 80 per cent), the number of those who wanted the US to adopt the system dropped by half (from 50 to 26 per cent). It was a trend that would have made Condorcet weep.[50]

The kilometre signs have come down and gas is again being pumped in gallons, now that the service stations have added a digit to the left of the decimal point to keep track of the dollars. But America keeps rolling silently towards the metric system. Its car parts are sized in metric units. So are its bicycles. It is no longer enough for American exporters simply to label their products in both American and metric units (soft metric); trade groups abroad are demanding that goods be delivered in even metric units (hard metric).

Oddly, as more Americans are lured into using the metric system, it may be that the nation will lose the very uniformity of weights and measures which has long made the metric system seem unnecessary in the United States. The most spectacular fatality of this new mixture was the crash of the Mars Climate Orbiter in 1999.[51] Has the time come for America to join the metric globe at last?

No doubt the world economy would operate with greater efficiency if we all spoke the same language; yet the world would be impoverished by the loss of the diversity this represents. Foreigners often complain – and the French complain vociferously – that America is currently spearheading a capitalist globalization that is levelling all the differences that make life worth living. Well, in this instance it is America which is different. This book has demonstrated that measures are social conventions, the outcome of a political process. Many Americans already use the metric system in domains where the economy operates on a global scale, as do many (but not all) engineers, physicians, scientists and other technical professionals. These people are already metrically 'bilingual', which is, in its own way, a good thing. But Americans have shown little willingness to give up their traditional measures in their daily lives. Sooner or later it will seem time for Americans to give up their old units, not because the rest of the world uses the metric system, but because America does.

Epilogue

THE SHAPE OF
OUR WORLD

'You're such a fool! Of course I don't need to see you, if that's what you mean. You're not exactly a sight for sore eyes, you know. I need you to exist and not change. You're like that platinum metre bar they keep in Paris or thereabouts. I can't imagine that anyone actually ever wants to see it.'

'That's where you're wrong.'

'Well I don't want to see it, not me. I'm just glad to know it exists, that it measures exactly one ten-millionth of the quarter meridian. I think about it every time they measure a room or sell me cloth by the metre.'

'Really?' I answer coldly.

'You know, I could remember you as a merely abstract virtue instead, a sort of limit. You ought to be thankful that I remember your face each time we meet . . .'

Anny suddenly smiles at me so tenderly that tears fill my eyes.

'I've thought about you more often than that platinum metre bar. Not a day goes by that I haven't thought of you. And I remembered exactly what you looked like right down to the last detail.'[1]

JEAN-PAUL SARTRE, *Nausea*

The baseline of Melun, which Delambre measured in the early spring of 1798, is now known as the N6, a stretch of the national route which runs north-east out of Melun towards Paris some thirty miles away. In 1882 a party of French geodesers returned to Melun. They checked the masonry pyramids that Delambre had erected at each terminus and rebuilt them as a memorial to his labour. But they did not remeasure his baseline, for fear, it would seem, of impugning the accuracy of the original Archive Metre. Instead, they calculated the length of the baseline indirectly, by triangulation, and found it to be within one centimetre of Delambre's value. A discrepancy of one centimetre over the course of ten kilometres (six miles) amounts to an error of 0.0001 per cent. Whatever progress scientists have made since the eighteenth century, it is still worth pausing to marvel at the level of precision achieved by Delambre and Méchain. And to appreciate what they taught us – both by their efforts and inadvertently – about error.[2]

It is often said that modern science has 'disenchanted' nature, stripping it of demons and divinities, making it mechanistic, devoid of moral lessons, and, above all, amenable to explanation. But having disenchanted nature – rendering it fully comprehensible and empty of meaning – science itself could hardly avoid disenchantment as well. In pushing measurement towards the final approach to precision, the savants discovered that error was inevitable, and that treating it would mean turning against themselves the same apparatus of disinterested analysis that they had long deployed against nature. In the process, a calling hedged with personal virtue became a career, and a new ethic of error management came to guarantee the accuracy of results. Of course human beings – both lay folk and professionals – will always read moral lessons into nature. And character still counts in science: a scientist's reputation remains his or her most prized possession. But colleagues now assess one another's results with dispassionate tools. Error has become a problem that is addressed by a social process.

Today the memorials at Melun are gone, destroyed in a road accident. The French countryside has been transformed by the world that the metric system helped make possible. The N6 still

runs as straight and true as the king's engineers could make it, a
headlong journey through French history. Once it leaves the
medieval centre of Melun, the N6 – named rue General Patton –
passes a nineteenth-century brick Catholic school and charity
hospital, before it runs parallel to a strip-mall of shabby bistros
and petrol stations. It then crosses the Cercle d'Europe, a grassy
roundabout decorated with the flags of the European
Community's member nations, where, on the morning I cycled out
of town, a caravan from the Circus Zavetta was encamped, with
three Indian elephants grazing on the verge. Beyond the roundabout, the highway traverses a stretch of early third millennium
consumer paradise: a sleek BMW dealership, a gigantic
Conforama *hypermarché*, followed by a series of megastores selling
furniture, bathroom tiles and the like. The scene can be found on
the outskirts of a hundred French provincial towns. The Paris
media complain about 'globalization', and the tourists marvel at
the vegetable stalls on market days, but the French are Europe's
most avid superstore shoppers.

Then the scene changes. Beyond the megastores lie rolling
fields of rape, yellow in the sunshine. Soon the fields lick at the
roadside, shaded by twin rows of plane trees, quite possibly the
same six hundred trees Delambre trimmed to get a clear line of
sight down the highway. In the middle of an empty field stands an
old inn, A l'Attaque du Courrier de Lyon, named after a notorious
postal-express robbery which took place near by in 1796, two
years before Delambre conducted his measurements.

Then the scene changes again and the N6 swerves briefly (its
only deviation) to pass over the A5 expressway and skirt the main
Paris–Lyon TGV high-speed train line. An industrial park, located
at this triple junction, ships liquid air and other high-tech products
to markets around the world. Once it arrives on the other side of
the expressway, the N6 reverts to cobblestones and the highway
narrows to pass through the ancient village of Lieusaint, where
Delambre's baseline came to a halt. The town is a remote suburb
of Paris, an hour's commute by rail. The village centre consists of
a small church, a pizza parlour and an Arab grocery. Many who

live there today are North African immigrants. Some kids were
kicking a football in a field behind a brick wall when I passed
through. They asked me about my bike; they told me about life in
Lieusaint (which translates as 'holy place'). It's miserable here,
they said with a smile.

France has changed in two hundred years. French products go
out into the world, and the world has come to France. The world
is wrapped in a single metrical language, yet France is still divided
by languages and cultures – as is the world.

The creators of the metric system believed that human beings
were shaped, first and foremost, by their experience of the world.
They wanted citizens to be able to assess their own best economic
interest, without which they could never be free. Give people the
tools to treat the material world in a rational and consistent
manner, they believed, and in time the people themselves would
become rational and consistent. They wanted the metric system to
create a new kind of citizen, much as we expect the internet to
teach new political virtues to the citizens of the Information Age.
Their goal was to make productivity the visible measure of economic progress and price the paramount variable in exchange. In
many ways, their vision has triumphed. The euro, the common
currency of much of Europe as of 2002, is a direct heir of the
metric system. Nowadays it seems as if price has at last become
the measure of all things.

But even the global markets which set prices are social cre-
ations governed by human institutions and human desires. And as
the labours of Delambre and Méchain amply attest, even our
modern impersonal measures are the product of human ingenuity,
human passion and the choices of particular people in particular
times and places. So in the end there is no escaping Protagoras'
25-century-old motto: 'Man is the measure of all things.'

NOTE ON MEASURES

All measures in this book are given in the Anglo-American units, unless otherwise specified. I provide below some rough equivalents in *ancien régime* units. Needless to say, they should not be construed as exact equalities since *ancien régime* measures with the same name varied by as much as 50 per cent within France. The best modern table of correspondences is Ronald Zupko, *French Weights and Measures Before the Revolution: A Dictionary of Provincial and Local Units* (Bloomington: Indiana University Press, 1978). Zupko's 224-page dictionary is necessarily incomplete and should be supplemented by the hundred or so tables of correspondences drawn up by the various French *départements* between 1793 and 1812. These tables are likewise incomplete. An entire sub-field of history – historical metrology – is devoted to extracting old measures from archaeological and documentary evidence. At times, this evidence about the measures of the past is then mined to reconstruct the daily lives of the people of the past.

Linear measures

toise ≈ fathom ≈ 6 feet
aune ≈ ell ≈ 3 feet
pied ≈ foot
pouce ≈ inch
ligne ≈ ¹⁄₁₂ of an inch

Other measures

livre ≈ pound
boisseau ≈ bushel
pinte ≈ quart

On degrees

During the years 1793–8 a circle of 360° was occasionally defined as having 400°. All angle measurements in the text, as well as all latitudes and longitudes, are given in the 360° system. Note that in this standard system each degree (symbol: °) is divided into 60 minutes (symbol: ′) and each minute is divided into 60 seconds (symbol: ″). Thus a latitude of 36°44′61.26″ should read as lying 36 degrees, 44 minutes and 61.26 seconds north of the equator (since all latitudes in this book are north of the equator).

On dates

The text gives the Gregorian calendar date for events rather than the Revolutionary calendar date, although both are given in the end-notes where appropriate. Years in the Revolutionary calendar are given in Roman numerals.

On money

The *ancien régime* unit of currency was the *livre* (pound) divided into 12 *sous* (shillings) each worth 20 *deniers* (pence). The republican government resurrected the old name *franc* (to be divided decimally into 100 *centimes*), and after some discussion settled on a valuation making one *franc* nearly equal to one *livre*. In fact, the decimal *franc* was initially worth one-eightieth more than the old *livre*, so that it would come out to a round weight of 4.5 grams of silver instead of the old lower tolerance of 4.419 grams of silver. In the Revolutionary period, however, the names *franc* and *livre* were used with some degree of interchangeability as the legislation changed. With the Napoleonic banking reform of 1803 the *franc* became the fixed currency and its silver content was raised to 5 grams. *Assignats* were a paper money created early in the Revolution with a face value of 1 *franc*; however, they soon lost value and the government began to publish tables of discounted values for an inflation rate which totalled 20,000 per cent over the course of four years. The *mandat* was another paper currency temporarily introduced to replace the *assignat*. The best discussion of finance and the economy during the Revolution is François Crouzet, *La grande inflation: La monnaie en France de Louis XVI à Napoléon* (Paris: Fayard, 1993).

ACKNOWLEDGMENTS

The literary convention that the author's spouse is the last to be thanked is surely meant to imply 'last but not least'. So in a book that makes much of priority, let me state for the record that I owe thanks 'first and most' to Bronwyn Rae. My daughter, Madeleine, is my other star of incalculable worth – although her full height cannot yet be taken. I thank my parents for first introducing me to matters scientific and French, though they cannot be held entirely responsible for the resulting concatenation. My sister, a scientist of integrity, and my brother, a reasoning moralist, have taught me that such concatenations are still possible, and that the age of the savant is not quite past.

For the attention they have devoted to this book, I would also like to thank my editors, Richard Beswick and Viv Redman, as well as the rest of the staff at Time Warner in London. I also thank the staff at The Free Press in New York. I greatly appreciate the optimistic and energetic guidance of Christy Fletcher and her colleagues at Carlisle and Co. For help with photography, I would like to thank Roman Stansberry, Gilbert Coudon, Jim Lane and Aron Vinegar. For the fine original maps, thanks are due to Chris Robinson. For help in gathering research materials, I would like to thank Dario Gaggio, Sander Gliboff, Arne Hessenbruch, Stanislav Rosenberg and Dana Simmons.

Colleagues and friends at Northwestern University and in the invisible college of scholars have taught me that the writing of history, like the making of science, is a social activity. The manuscript has benefited enormously from critical readings by Guy Boistel, Peter Gaffney, John Heilbron, Susan Herbst, Sarah Maza, Joel Mokyr, Ted Porter, Jessica Riskin, Dana Simmons, Mary Terrall and Mike Tobin. Finally, in a book about error – and its transformation from a moral failing into a social problem – let me hasten to add that any remaining mistakes are my own and I take personal responsibility for them.

NOTE ON SOURCES

This book is based on original sources, the vast bulk of them unedited manuscripts written in French. Unless indicated otherwise, all translations are my own. The main cache of papers regarding the metric expedition of 1792–9 is located in the E2 series of the archives of the Observatoire de Paris. Important additional Delambre papers are located in the Bibliothèque de l'Institut National, the library of Brigham Young University, Provo, Utah and the Karpeles Museum in Santa Barbara, California, as well as in holdings in Amiens, New York, London and Utrecht. Important additional Méchain papers are located in the Biblioteca Universitaria di Pisa and in the Kongelige Bibliotek, Copenhagen, as well as in holdings in Laon, Milan and Madrid. Other valuable papers on these two savants, on their contemporaries and on the metric system in general can be found in the archives of the Académie des Sciences in Paris, as well as in major and minor archives and libraries in Paris, the French provinces and throughout the world. Below I list those institutions where I found the documents cited in the endnotes and the acronyms and other abbreviations I use when citing them. Following each citation of an archival source, I provide the carton reference number for scholars who wish to locate that document. I want to thank the archivists and librarians who assisted me with my research, and especially the staff of the interlibrary loan office at Northwestern University.

As the notes which follow stick closely to the material presented in the text, I would like to acknowledge several key intellectual debts here. The foremost historian of metrology is the great Polish economic historian Witold Kula, whose book *Measures and Men*, translated by R. Szreter (Princeton: Princeton University Press, 1986) helped me to formulate the argument of this book, although I differ with him in my conclusions. My thinking on political economy has been shaped by the work of Karl Polyani, whose *The Great Transformation* (New York: Farrar and Rinehart, 1944), a neglected and sometimes baffling classic. In the past decade a new approach to the history of science has transcended the simplistic division between technical histories and contextual studies. The most helpful works on the intersection of the exact sciences and social-cultural values have been the writings of Lorraine Daston, Simon Schaffer, Steven Shapin, M. Norton Wise and especially Theodore Porter, whose *Trust in Numbers: Objectivity in Science and Public Life* (Princeton: Princeton University Press, 1995) has inspired me time and again. For general histories of the metric system, no one has yet surpassed two rather dated works from the beginning of the twentieth century: Guillaume Bigourdan, *Le système métrique des poids et mesures* (Paris: Gauthier-Villars, 1901) and Adrien Favre, *Les origines du système métrique* (Paris: Presses Universitaires de France, 1931). The one strong recent treatment is the wry article by John L. Heilbron, 'The Measure of Enlightenment' in *The Quantifying Spirit in the Eighteenth Century*, edited by Tore Frängsmyr, John L. Heilbron and Robin E. Rider (Berkeley: University of California Press, 1990), pp. 207–42. My own analysis of the political and economic significance of the metric system can be found in Ken Alder, 'A Revolution to Measure: The Political Economy of the Metric System', in *The Values of Precision*, edited by M. Norton Wise (Princeton: Princeton University Press, 1995), pp. 39–71.

Abbreviations of Archives and Libraries

AAS	Archives de l'Académie des Sciences, Paris.
ADAu	Archives Départementales de l'Aude, Carcassonne.
ADC	Archives Départementales du Cher, Bourges.
ADPO	Archives Départementales des Pyrénées-Orientales, Perpignan.
ADSe	Archives Départementales de la Seine, Paris.
ADSM	Archives Départementales de Seine-et-Marne, Melun.
ADSo	Archives Départementales de la Somme, Amiens.
ADT	Archives Départementales du Tarn, Albi.
AHAP	Archives Historiques de l'Archevêché de Paris, Paris.
AMAE	Archives du Ministère des Affaires Etrangères, Paris.
AMSD	Archives Municipales de Saint-Denis, Saint-Denis.
AML	Archives Municipales, Lagny.

AMNM	Archivos del Museo Naval de Madrid, Madrid.
AN	Archives Nationales, Paris.
AOAB	Archivio dell'Osservatorio Astronomico di Brera, Milan.
AOP	Archives de l'Observatoire de Paris, Paris.
APS	American Philosophical Society Library, Philadelphia.
BA	Bibliothèque de l'Arsenal, Paris.
BEP	Bibliothèque de l'Ecole Polytechnique, Palaiseau.
BI	Bibliothèque de l'Institut de France, Paris.
BL	Bureau des Longitudes, Paris.
BLL	British Library, London.
BLUC	Bancroft Library, University of California, Berkeley.
BMA	Bibliothèque Municipale d'Amiens, Amiens.
BMC	Bibliothèque Municipale de Carcassonne, Carcassonne.
BMCF	Bibliothèque Municipale de Clermont-Ferrand, Clermont-Ferrand.
BML	Bibliothèque Municipale de Laon, Laon.
BMR	Bibliothèque Municipale de Reims, Reims.
BMSD	Bibliothèque Municipale de Saint-Denis, Saint-Denis.
BN	Bibliothèque Nationale, Tolbiac, Paris.
BNR	Bibliothèque Nationale, Richelieu, Manuscripts, Paris.
BNRC	Bibliothèque Nationale, Richelieu, Cartes, Paris.
BUP	Biblioteca Universitaria di Pisa, Pisa.
BVCS	Bibliothèque Victor-Cousin, Sorbonne, Paris.
BYU	Brigham Young University, Provo, Utah.
CNAM	Conservatoire Nationale des Arts et Métiers, Paris.
CUL	Lavoisier Collection, Cornell University, Ithaca, New York.
CUS	David Eugene Smith Collection, Columbia University, New York.
DLSI	Dibner Library, Smithsonian Institution, Washington, DC.
ENPC	Ecole Nationale des Ponts et Chaussées, Paris.
KBD	Kongelige Bibliotek of Denmark, Copenhagen.
KM	Karpeles Museum, Santa Barbara, California.
NL	Newberry Library, Chicago.
NYPL	New York Public Library, Rare Books and Manuscripts, New York.
SBB	Staatsbibliotek Berlin, Berlin.
SHAT	Service Historique des Armées de Terre, Vincennes.
UBL	Universiteitsbibliotheek Utrecht, Utrecht.
WL	Wellcome Library, London.

Abbreviations of Serial Works

AP	*Archives parlementaires de 1787 à 1860; recueil complet des débats législatifs et politiques des chambres françaises.* First series, 1789 to 1799. Paris: Dupont, 1875–.
AP2	*Archives parlementaires de 1787 à 1860; recueil complet des débats législatifs et politiques des chambres françaises.* Second series, 1800 to 1860. Paris: Dupont, 1862–1913.
ASPV	Académie des Sciences, *Procès-verbaux des séances de l'Académie tenues depuis la fondation de l'Institut jusqu'au mois d'août 1835.* Hendaye, Basse-Pyrénées: Imprimerie de l'Observatoire d'Abbadia, 1910–1922.
CT	*Connaissance des temps ou des mouvements célestes, pour le méridien de Paris, à l'usage des astronomes et des navigateurs.* Paris, 1766–. Exact title and publisher varies. After 1795, edited by the Bureau des Longitudes.
CR	*Comptes rendus hebdomadaires des séances de l'Académie des sciences.* Sessions since 1835. Paris: Gauthier-Villars, 1835–1965.
HAS	Académie des Sciences, *Histoire de l'Académie des Sciences.* Paris: 1666–1792.
MAS	Académie des Sciences, *Mémoires de l'Académie des Sciences.* Paris: 1666–1792.
MC	*Monatliche Correspondenz zur Beförderung der Erd- und Himmels-Kunde.* Ed. F.-X. von Zach. Gotha: Beckersche, 1800–1813.
MI	Académie des Sciences, *Mémoires de l'Institut.* Paris: 1795–1815.
Moniteur	*Moniteur universelle* [also known as *Gazette nationale*]. Paris: Agasse, 1789–1810.
PVCIP	James Guillaume, ed., *Procès-verbaux du Comité d'Instruction Publique de la Convention Nationale.* Paris: Imprimerie Nationale, 1891–1907.
RACSP	François-Alphonse Aulard, ed., *Recueil des actes du Comité de Salut Public.* Paris: Imprimerie Nationale, 1889–1951.

Abbreviation of Institutions

ATPM	Agence Temporaire des Poids et Mesures.
CIP	Comité d'Instruction Publique de la Convention Nationale.
CPM	Commission des Poids et Mesures.
Dépt.	Département.
Min. Aff. Etr.	Ministère des Affaires Etrangères.
Min. Int.	Ministère de l'Intérieur.

NOTES

Prologue

1. Leviticus, 19, verses 35–6, in *The Holy Scriptures* (Philadelphia: Jewish Publication Society, 1955).
2. Arthur Young, *Travels During the Years 1787, 1788 and 1789* (Dublin: Gross 1793), 2, pp. 43–4. For the number of measurement names, see ATPM, *Aux citoyens rédacteurs de la Feuille du Cultivateur* (Paris: Imprimerie de la République, III [1795]), p. 11. For the number of measurement units, see Ronald Zupko, *French Weights and Measures Before the Revolution: A Dictionary of Provincial and Local Units* (Bloomington: Indiana University Press, 1978), p. 113.
3. KM, Delambre, *Base du système métrique*, 1, title page.
4. For the comparison with the printing press and the steam engine, see John Quincy Adams (Secretary of State), 'Weights and Measures', US Senate, 22 February 1821; 16th Congress, 2nd Session, in Walter Lowrie and Walter S. Franklin, eds, *American State Papers: Documents* (Washington, DC: Gales and Seaton, 1834), no. 503, Class 10, vol. 2, pp. 656–750; see p. 688.
5. Napoleon, *Mémoires pour servir à l'histoire de France sous Napoléon, écrits à Sainte-Hélène*, ed. Gaspard Gourgaud and Charles-Tristan Montholon (London: Bossagne, 1823–4), 4: 211.
6. Delambre, *Rapport historique sur les progrès des sciences mathématiques depuis 1789* (Paris: Imprimerie Impériale, 1810), p. 68. Delambre, ed., *Base du système métrique décimal, ou mesure de l'arc du méridien compris entre les parallèles de Dunkerque et Barcelone, exécutée en 1792 et années suivantes, par MM. Méchain et Delambre* (Paris: Baudouin, 1806, 1807, 1810); cited hereafter as Delambre, *Base*. For prior histories of the metric system, see the Notes on Sources.
7. AOP E2-9, Delambre's final comments in Méchain's notebook, c. 1810.
8. Delambre (c. 1810), marginal note on AOP E2-19, Méchain to Delambre, 7 brumaire VII [28 October 1798].
9. KM, Delambre, *Base*, 1, title page.

Chapter 1 The North-Going Astronomer

1. Stendhal, *La chartreuse de Parme* (Paris: Garnier, 1962), p. 31, emphasis in original.

2. AML, Conseil Municipal, 'Déliberations', 21 August 1792. See also Georges Darney, *Histoire de Lagny* (Paris: Office d'Edition et Diffusion du Livre d'Histoire, 1994), p. 179. Jean Alexandre conversed with a gendarme near Lagny on 4 September 1792; see Pierre Caron, *Les massacres de septembre* (Paris: Maison du Livre Français, 1935), pp. 160–1.

3. Pétion, Mayor of Paris, read into the municipal records of Lagny, August 1792, in Darney, *Lagny*, p. 178.

4. For the actions of the Lagny militia, see AML, Conseil Municipal, 'Déliberations', August–September 1792. Also, Darney, *Lagny*, p. 180. On Petit-Jean, see Delambre to Madame d'Assy [August 1792], in Guillaume Bigourdan, *Le système métrique des poids et mesures* (Paris: Gauthier-Villars, 1901), pp. 118–19; also Petit-Jean to Delambre, received on 12 August 1792, in AOP E2-6, Delambre, 'Registre', p. 12.

5. For Delambre's passport, see Municipalité de Bruyères-Libre, 'Certificat', 17 prairial III [5 June 1795], in Bigourdan, *Système métrique*, p. 134.

6. On Bellet, see AAS Lavoisier 1967, Lavoisier, 'Etat des ouvriers et coopérateurs . . .' [1792–3].

7. Delambre to Mme d'Assy, 5 September 1792, in Bigourdan, *Système métrique*, pp. 119–22.

8. AAS Fonds Lavoisier, Nouvelle Ac. 30, Lavoisier to Delambre, 28 August 1792.

9. Delambre to Mme d'Assy, 5 September 1792, in Bigourdan, *Système métrique*, pp. 119–22.

10. Delambre to Mme d'Assy, 5 September 1792, in Bigourdan, *Système métrique*, pp. 119–22.

11. Mayor Aublan in municipal council, 16 August 1792, in Darney, *Lagny*, p. 176.

12. Delambre to Mme d'Assy, 5 September 1792, in Bigourdan, *Système métrique*, pp. 119–22, emphasis in original. See AML, Conseil Municipal, 'Déliberations', 4 September 1792.

13. Delambre to Mme d'Assy, 5 September 1792, in Bigourdan, *Système métrique*, pp. 119–22.

14. AOP E2-6, Delambre, 'Registre' 4 September 1792, p. 49.

15. Delambre to Mme d'Assy, 5 September 1792, in Bigourdan, *Système métrique*, pp. 119–22.

16. AML, Conseil Municipal, 'Déliberations', 5 September 1792.

17. Delambre to Mme d'Assy, 5 September 1792, in Bigourdan, *Système métrique*, pp. 119–22.

18. For Delambre's autobiography, see BI MS2042 fols 408–14, Delambre, 'Delambre par lui-même' [1821]; cited hereafter as Delambre, 'Lui-même'. Delambre's student Claude-Louis Mathieu drew on this unpublished autobiography for his biography in Mathieu, 'Delambre', *Biographie universelle*, ed. Michaud, new edn (Paris: Desplaces, n.d.), pp. 304–8. Vulfran Warmé, *Eloge historique de M. Delambre* (Amiens: Caron-Duquenne, 1824). ADSo 2E21/25, 'Acte de mariage de Jean-Nicolas-Joseph Delambre et Marie-Elisabeth Devisme', 27 January 1749; 'Baptême de Jean-Baptiste-Joseph Delambre', 19 September 1749.

19. BMA MS568(18), Delambre, 'Règles ou méthode facile pour apprendre la langue anglaise', n.d. On the literary club, see AAS Dossier Delambre, Résumé of letters from Delambre to Favart *fils*, sold by Cabinet Henri Saffroy in June 1943. See also AAS Col. Bertrand 9, Delambre to Favart *fils*, 20 October 1769.

20. On the d'Assy family, see Jean-Pierre Babelon, 'L'hôtel d'Assy, 58 bis, rue des Francs-Bourgeois au Marais', *Paris et Ile-de-France, Mémoires* 14 (1964), pp. 169–96; 16/17 (1965–6), pp. 231–40.

21. Delambre, 'Lui-même'.

22. BVCS MS99, Lalande, 'Journal', 1783. Lalande first noticed Delambre on 10 December 1782; see Joseph-Jérôme Le Français de Lalande, *Bibliographie astronomique; avec l'histoire de l'astronomie depuis 1781 jusqu'à 1802* (Paris: Imprimerie de la République, XI, 1803), p. 597.

23. Delambre, 'Lui-même'. On Delambre's assistance for Lalande, see AOP Z151(4), Lalande to [Delambre], 17 December 1783. Also 'Lalande au rédacteur', *Moniteur* 2 (1 December 1789), p. 273.

24. For Delambre's observatory, see CUS, Delambre to Cagnoli, 3 September [1788] and 23 November 1789. Also Guillaume Bigourdan, *Histoire de l'astronomie d'observation et des observatoires en France, seconde partie* (Paris: Gauthier-Villars, 1930), pp. 155–65.

25. For Geoffroy d'Assy's lawsuit to block a butcher from opening a shop in his territory in the 1780s, see AN AB XIX 322, d'Assy, 3, 13 September 1785.

26. Talleyrand, *AP* 24 (26 March 1791), p. 397.

27. For the election of the first expedition team, see AAS, 'Procès-verbaux de l'Académie' 109 (13 April 1791), p. 321.

28. BNRC Ge DD 2066 (3), Cassini IV, '2ème dialogue', pp. 43–5. Said by Cassini to be a 'word-for-word' recitation of the exchange.

29. BNRC Ge DD 2066 (3), Cassini IV, 'Mémoires', pp. 72–3; '2ème dialogue', pp. 25–6.

30. For Roland's threat, see *AP* 41 (3 April 1792), p. 110. Roland pressed the Academy again in May; see AN F12 1288, Roland to Comité d'Agriculture et Commerce, 19 May 1792. The academicians discussed their fear of Roland's goals in AAS Chabrol 1/71, Borda to Condorcet [May 1792].

31. For Delambre's election to the Commission, see AAS, 'Procès-verbaux de l'Académie' 110 (2, 5 May 1792), pp. 138–9, 142. For Delambre's account of this period, see Delambre, 'Lui-même'. Also Delambre, *Grandeur et figure de la terre* (Paris: Gauthier-Villars, 1912), p. 205. The division of sectors was provisional; see ENPC MS726, Delambre, 'Mesure du méridien', 19 messidor II [7 July 1794].

32. For the arrival of king's proclamation on 24 June, see Delambre, *Grandeur*, p. 203. For his start date of 24 June, see AOP E2-6, Delambre, 'Registre', p. 2. Also see chapter 2, below.

33. On Montmartre, see Delambre, *Base*, 1, p. 23.

34. On the sale of the Collégiale, see ADSM 1Q1047/2, 'Une église charpent . . .', 27 November, 11 December 1792. On the Saint-Martin-du-Tertre station, see AOP E2-6, Delambre, 'Registre', p. 51.

35. For Delambre's account of 10 August, see AOP E2-6, Delambre, 'Registre', p. 38. Also Delambre, *Base*, 1, pp. 29–30. On Montmartre on 10 August, see AN F7 4426, 'Section du Faubourg-Montmartre', 10 August 1792. F. Braesch, *La commune de dix août 1792* (Paris: Hachette, 1911), pp. 190–214, 335–46; Marcel Reinhard, *Chute de la royauté, 10 août, 1792* (Paris: Gallimard, 1969), pp. 39, 388.

36. On the local council and Petit-Jean, see Delambre to Mme d'Assy [August 1792], in Bigourdan, *Système métrique*, pp. 118–19.

37. On anti-royalism in Saint-Denis, see Anne Lombard-Jourdan, 'Traque et abolition des marques de religion, de royauté et de féodalité à Saint-Denis après 1789'; Bruno Hacquemand, 'F. A. Gautier, organiste de l'abbaye royale, et le vandalisme révolutionnaire à Saint-Denis'; and Philippe Weyl, 'Destruction des tombeaux et l'exhumation des rois à Saint-Denis'; in *Saint-Denis, ou, Le jugement dernier des rois*, ed. Roger Bourderon (Saint-Denis: PSD, 1993), pp. 209–64.

38. Delambre, *Base*, 1, p. 32. On the events of 6 September, see Delambre to Mme d'Assy, 8 September 1792, in Bigourdan, *Système métrique*, pp. 122–4. On Epinay, see André Clipet, *Epinay-sur-Seine: Son histoire* (Paris: Boudin, 1970), pp. 46, 165, 262–3, 268–9.

39. For the official record of the events of that day in Saint-Denis, see AMSD CT762, 'M. Delambre, deux voitures . . . arrêtées', 6 September 1792. For Delambre's account, see Delambre, *Base*, 1, pp. 32–4. For the history of Saint-Denis during the Revolution, see the eyewitness report in AHAP R4, Ferdinand-Albert Gautier, 'Supplément à l'histoire de l'abbaye de Saint-Denis', 1808; partially reprinted in J. Guiffrey, ed., 'La ville de Saint-Denis pendant la Révolution', *Cabinet historique* 20 (1874), pp. 280–303; 21 (1875), pp. 36–53, 118–34; especially pp. 285, 293. On Saint-Denis and the town hall, see Michaël Wyss, *Atlas historique de Saint-Denis: Des origines au XVIIIe siècle* (Paris: Éditions de la Maison des Sciences de l'Homme, 1996), pp. 299–300.

40. Delambre to Mme d'Assy, 8 September 1792, in Bigourdan, *Système métrique*, pp. 122–4.

41. Louis XVI, *Proclamation du roi, concernant les observations et expériences à faire par les Commissaires de l'Académie des Sciences, pour l'exécution de la loi du 22 août 1790*, 10 June 1792 (Paris: Imprimerie Royale, 1792), pp. 2–3.

42. The Saint-Denis *pinte* equalled 1.46 litres and a Paris *pinte* equalled 0.93 litres; see Alexis-Jean-Pierre Paucton, *Métrologie, ou traité des mesures, poids et monnoies des anciens peuples et des moderns* (Paris: Veuve Desaint, 1780), p. 808. For the Saint-Denis measures in the basilica, see Wyss, *Saint-Denis*, p. 62. For Amiens, see Léon Gaudefroy, *Rapport des mesures anciennes en usage à Amiens* (Paris: Camber, 1904), pp. 7–8.

43. Paucton, *Métrologie*, p. 7. For the efforts of royal administrators, see Nicolas de La Mare, *Traité de police* (Paris: Brunet, 1719), 2, p. 743. In 1321, 1510, 1576 and 1614, the Estates-General endorsed the king's measures to little effect; see Georges Picot, ed., *Histoire des Etats généraux* (Paris: Hachette, 1872), 2, pp. 256–7; 3, pp. 30, 204; 4, p. 130.

44. Anon., 'Poids', *Encyclopédie*, ed. Denis Diderot and Jean le Rond d'Alembert (Paris: Briasson, 1751–72), 12, p. 855. Jacques Necker, *Compte rendu au roi* (Paris: Imprimerie Royale, 1781), p. 121.
45. For the *Cahiers* at the national level, see Beatrice Fry Hyslop, *French Nationalism in 1789, According to the General Cahiers* (New York: Columbia University Press, 1934), p. 56. At the local level, the demand for 'one law, one king, one weight and one measure' figured in the complaints of eighteen of the surviving parish *Cahiers* in the Forez region alone; see Etienne Fournial and Jean-Pierre Gutton, eds, *Cahiers de doléances de la province de Forez* (Saint-Etienne: Centre d'Etudes Foréziennes, 1974), pp. 57, 106, 122, 127, 141, 149, 151, 160, 170, 179, 182, 217, 263, 311, 314, 319, 334, 353. Saint-Denis made its request as part of the Paris region; see Charles-Louis Chassin, *Les cahiers de 1789 et les cahiers du Sénat* (Paris: Suffrage Universel, 1875), 4, pp. 263–4. For Epinay, see 'Epinay-sur-Seine', *AP* 4, p. 517.
46. Delambre, *Base*, 1, pp. 33–4.
47. For the evening of 6 September 1792, see Delambre to Mme d'Assy, 8 September 1792, in Bigourdan, *Système Metrique*, pp. 122–4.
48. For the proclamation of the National Convention, see *Moniteur* 13 (12 September 1792), p. 666. For the release of Delambre's carriages, see AMSD 1D1-1, Registres municipales, 8 September 1792.
49. For the volunteers and subsequent history of Saint-Denis, see AHAP R4, Gautier, 'Supplément à l'histoire de l'abbaye de Saint-Denis' (1808), p. 218.
50. Alexandre Lenoir, 'Notes historiques sur les exhumations faites en 1793 dans l'abbaye de Saint-Denis', in *Description historique et chronologique des monuments*, 6th edn (Paris: L'auteur, X [1802]), pp. 338–56; see p. 241 for one version of this quotation. For another, see Weyl, 'Destruction', p. 245.
51. BMSD S10, David et al., 'Rapport des Commissaires nommés par la Commission Temporaire des Arts pour conférer . . .', 3 ventôse III [21 February 1794].

Chapter 2 The South-Going Astronomer

1. Miguel de Cervantes, *The History and Adventures of the Renowned Don Quixote*, trans. Tobias Smollett (Chatham, England: Mackays, 1998), p. 862.
2. For the official departure date, see Delambre, *Base*, 1, p. 21. Needless to say, all accounts of this expedition have taken this date to be accurate. For the series of Méchain's predicted dates of departure, see Méchain to Cotte, 28 January 1792, in Joseph Laissus, 'Un astronome français en Espagne: Pierre-François-André Méchain (1744–1804)', *Comptes rendues du 94e Congrès National des Sociétés Savantes*, Pau, 1969 (Paris: Bibliothèque Nationale, 1970), pp. 36–59, especially pp. 48–50. AMNM Leg. 2294, J. Gonzales to A. Valdez, 4 July 1792. AOP MS1058III, Méchain to Flaugergues, 9 June 1792; KBD NKS1304, Méchain to Bugge, 23 June 1792.

3. AN M.C. Etude XXIII, 'Procuration . . . fait et passé à Paris en la demure du Sr Méchain', signed by Méchain, and the notaries François Brichard and [Antoine-Marie] Boulard, 28 June 1792.

4. On the Méchain family dwelling, see BL, 'Registres', 17 nivôse II [6 January 1794]; see also AAS Dossier Méchain, 'Certificat de mariage', 4 November 1777.

5. On the Marjou family, see AN O1 682, 'Marjou, valet de garde robe de Monsieur', 6 January 1780. For Méchain's pay, see AAS Lavoisier 1042, 'Quittances', 10 March, 15 June 1792, 1 July 1793. The move of the king from Versailles to Paris cost Mme Méchain her 'patrimoine'; see Méchain to Cotte, 7 January 1790, in Laissus, 'Astronome', pp. 45-6. For Mme Méchain's promise to conduct astronomical work, see AAS 1J4, Lalande, 'Journal', 28 April [1793], p. 63. See also KBD NKS1304, Mme Méchain to Bugge, 19 March 1793. Guillaume Bigourdan, 'Le Bureau des Longitudes, Son histoire et ses travaux, de l'origine (1795) à ce jour', *Annuaire du Bureau des Longitudes* (1928), pp. A1–72; (1929), pp. C1–92; (1930), pp. A1–110; (1931), pp. A1–151; (1932), pp. A1–117; see especially (1928), p. A34, and (1932), pp. A26–7. For her duties as substitute *concierge*, see Charles Wolf, *Histoire de l'Observatoire de Paris de sa fondation à 1793* (Paris: Gauthier-Villars, 1902), pp. 326–8.

6. For Delambre's subtle phrasing of his own departure date, and his assertion that he had been working 'since' 26 June, see Delambre, *Base*, 1, p. 23. For his logbook account indicating a start date of 24 June, see AOP E2-6, Delambre, 'Registre', p. 2.

7. I have carefully considered whether I am mistaken about the date of Méchain's departure. In general, a notarized document is the most certain proof one can have as to the presence of a particular person at a particular time and place. This notarized document was entered into the bound register of the notary François Brichard on 28 June 1792. There are, however, a few possible grounds upon which one might challenge this proof – although none seems convincing. First, the handwriting of the place and date on the document does differ from the rest of the text and was almost certainly filled in after the rest of the document. The most plausible explanation for this, however, is that the power of attorney, which covers three long pages of dense legal boilerplate, was prepared in the notary's office before he went to the home of Méchain for his signature, with blanks left to fill in the place and date on that occasion. So the question is whether the place, date and Méchain's signature were all affixed on that same occasion. Second, it is possible that Méchain signed the boilerplate in advance of his departure, allowing his wife to fill in the date and place at a time of her choosing. However, a signature must be witnessed by notaries to be valid, and the two notaries who signed the document attested to the date and place where they signed. Finally, Méchain may in fact have left Paris on 25 June, only to return almost immediately, signing the power of attorney on 28 June, and then proceeding on his mission. This would make Delambre's claim (narrowly) truthful, while preserving the authenticity

of the notarized document. A plausible scenario for this quick return is imaginable. Méchain had difficulty at a barricade at Essonne, one day's ride out of Paris, and this may have led him to return to Paris to get some new passports. However, Delambre himself noted that this obstacle at Essonne did not cause Méchain to turn around and call a halt to his mission; see Delambre, *Notice historique sur M. Méchain, lue, le 5 messidor XIII* [24 June 1805], (Paris: Baudouin, January 1806), p. 13. Moreover, it seems far-fetched to think Méchain would have chosen this occasion to sign a power of attorney. In sum, it would seem almost certain that Méchain did not leave Paris until sometime on or shortly after 28 June. Moreover, I have not found any place where Méchain himself mentioned his departure date. In a private letter of 1793, Madame Méchain recollected that her husband left Paris on 24 June 1792, which was clearly mistaken, and was instead the date on which the king's proclamation was finally delivered; see KBD NKS 1304, Mme Méchain to Bugge, 19 March 1793. Lalande recorded in his journal that Méchain left for Barcelona on 25 June; however, he intended to (and actually did) publish his journal as an official history of astronomy in his own time, so it too should be read as an official statement, and hence protective of Méchain's precedence. AAS 1J4, Lalande, 'Journal', (25 June 1792), p. 57; Lalande, *Bibliographie astronomique*, p. 717.

8. For the seven-month estimate, see KBD NKS1304, Méchain to Bugge, 23 June 1792. Méchain expected the entire operation to be completed in two years; see Méchain to Cotte, 28 January 1792, in Laissus, 'Astronome', pp. 48–50.

9. AOP MS1058III, Méchain to Flaugergues, 10 August 1789.

10. Cassini IV to Marquis de B***, in J.-F.-S. Devic, *Histoire de la vie et des travaux scientifiques et littéraires de J.-D. Cassini IV* (Clermont: Daix, 1851), p. 123. On the break-in at the Méchain home at the Observatory, see BML 26CA6, Méchain to Cotte, 21 July 1789. See also Wolf, *Observatoire*, pp. 319–23; Bigourdan, 'Bureau des Longitudes' (1928), p. A66.

11. AOP MS1058III, Méchain to Flaugergues, 22 October 1789.

12. The main source for Méchain's early career, presumably on the basis of Méchain's own notes, is an article by his friend, F.-X. Zach, 'Méchain', *MC* 2 (1800), pp. 96–117. For doubts about the story that Lalande purchased the instruments, see chapter 12 below.

13. Méchain took over the *Connaissance des temps* in 1785. For Méchain's sense of the importance of the journal, see 'Avertissement', *CT pour 1793* (1791).

14. D'Angiviller to Cassini IV, 7 March 1783, in Wolf, *Observatoire*, pp. 45–9.

15. For a physical description of Méchain, see Jacques-François-Laurent Devisme, 'Méchain', *Histoire de la ville de Laon* (Laon: Le Blan-Courtois, 1822), pp. 360–7.

16. For the 1788 survey, see Jean-Dominique Cassini IV, Pierre-François-André Méchain and Adrien-Marie Legendre, *Exposé des opérations faites en France en 1787, pour la jonction des observatoires de Paris et de Greenwich* (Paris: Institution des Sourds-Muets, 1790), pp. 34–6, 59. Also Sven

Widmalm, 'Accuracy, Rhetoric, and Technology, The Paris–Greenwich Triangulation, 1784–88', in *The Quantifying Spirit in the Eighteenth Century*, eds Tore Frängsmyr, John L. Heilbron and Robin E. Rider (Berkeley: University of California Press, 1990), pp. 179–206. Suzanne Débarbat, 'Coopération géodesique', *Echanges d'influences scientifiques et techniques entre pays européens de 1780 à 1830*, Actes du 114e Congrès National des Sociétés Savantes, Paris, 3–9 April 1989 (Paris: CTHS, 1990), pp. 47–76.

17. AOP D5-7, William Roy to Cassini IV, 29 January 1789.

18. Cassini IV, 'Application du cercle', *MAS* (1789; pub. II [1794]), p. 617.

19. For their confidence in Méchain, see Cassini IV, Méchain and Legendre, *Opérations Paris et Greenwich*, pp. 58–62.

20. For Méchain's work with the repeating circle, see Cassini IV, 'Application du cercle', *MAS* (1789; pub. II [1794]), pp. 617–23.

21. Méchain to Cotte, 7 January 1790, in Laissus, 'Astronome', pp. 45–6.

22. For Tranchot's assignment to the meridian project on 27 May 1792, see SHAT Xem 176, Tranchot, 'Etat des services de feu Monsieur Tranchot', 31 October 1815. For Méchain's previous work with Tranchot, see Puissant, 'Notice', *CT* (1822), pp. 293–7.

23. Young, *Travels*, 1, p. 59. Also Henry Swindburne, *Travels Through Spain in the Years 1775 and 1776* (London: Davis, 1787), 1, pp. 3–11.

24. For the ban on the *Journal de physique*, see Richard Herr, *The Eighteenth-Century Revolution in Spain* (Princeton: Princeton University Press, 1958), p. 255. On this period in Barcelona, see Jaume Carrera Pujal, *La Barcelona del segle XVIII* (Barcelona: Casa Editorial, 1951). For a contemporary witness, see Rafel d'Amat i de Cortada, Barón de Maldà, *Calaix de Sastre, 1792–1794* (Barcelona: Curial Edicions Catalanes, 1987–). Also Arthur Young, 'Tour in Catalonia', *Annals of Agriculture* 8 (1787), pp. 235–41.

25. For the Franco-Spanish frontier survey, see Josef Konvitz, *Cartography in France, 1660–1848: Science, Engineering, and Statecraft* (Chicago: University of Chicago Press, 1987), p. 37. On Spanish interest in French geodetic techniques, see Antonio Ten, 'Scientifiques et Francisés, Dépendances intellectuelles des scientifiques espagnols à la fin du XVIIIe siècle et au début du XIXe', *Echanges d'influences scientifiques et techniques entre pays européens de 1780 à 1830*, Actes du 114e Congrès National des Sociétés Savantes, Paris, 1989 (Paris: Comité des Travaux Historiques et Scientifiques, 1990), pp. 27–8; Antonio Ten, *Medir el metro: La historia de la prolongación del arco de meridiano Dunkerque–Barcelona, base del Sistemo Métrico Decimal* (Valencia: Universitat de València, 1996), pp. 107–19.

26. On Méchain's prior acquaintance with Gonzales, see AOP MS1058III, Méchain to Flaugergues, 23 April 1791. For the advance Franco-Spanish coordination, see, on the French end: AMAE Corr. Pol. Espagne 632, B.-C. Cahier [Min. Int.] to Delessart [Min. Aff. Etr.], 9 March 1792; Borda to [Min. Aff. Etr.], 28 March 1792. On the Spanish end, see AMNM Leg. 2294/53, Gonzales to Valdez, 11 July 1792; Antonio Ten, 'El sistema métrico decimal y España', *Arbor* 134 (1989), pp. 101–21.

27. For a description of the operations, see AOP E2-20, Méchain, 'Détails des opérations faites en Catalogne', April 1793; also, AOP E2-19, Méchain to Delambre, 14 vendémiaire IV [4 October 1795]. See diagram of signal in ADPO L1128, Méchain to Lucia [sic], 6 October 1793. For rumours in Barcelona, see Barón de Maldà, in Enric Moreu-Rey, *El naixement del metre* (Palma de Mallorca: Moll, 1956), pp. 70–1.

28. For a brilliant account of the ambiguities of national and geographic demarcations, see Peter Sahlins, *Boundaries: The Making of France and Spain in the Pyrénées* (Berkeley: University of California Press, 1989).

29. For Méchain's account of their procedures, see AMNM Leg. 2294, Méchain, 'Francisco Planez', 7 February 1793.

30. Méchain, in Delambre, *Base*, 1, p. 486.

31. On the local inns, see Joseph Townsend, *A Journey Through Spain in the Years 1786 and 1787* (Bath: Longman, 1814), 1, p. 78. A description of the Pyrénées in this area is in Young, *Travels*, 1, pp. 56–9, 626.

32. On the effect of the border tensions on the mission, see AOP E2-19, Méchain to Lalande, 11 ventôse IV (1 March 1796).

33. AOP B4-11, Méchain to Cassini IV, 8 September 1792.

34. For Méchain's complaints about the popular disregard for his mission, see Méchain to Admin. de Pyrénées-Orientales, 28 August 1792, in Pierre Vidal, *Histoire de la Révolution dans le département des Pyrénées-Orientales* (Perpignan: Indépendant, 1889), 2, pp. 373–4; AAS Fonds Lavoisier 1257, Méchain to Lavoisier, 4 September 1792; AOP B4-11, Méchain to Cassini IV, 8 September 1792.

35. CNAM C8, Jumelin to Bureau de Consultation, 12 September 1792.

36. Jean-Charles de Borda, *Description et usage du cercle de réflection* (Paris: Didot, 1787), p. 5. On Borda, see Jean Mascart, *La vie et les travaux du chevalier Jean-Charles de Borda, 1733–1799* (Lyon: Rey, 1919). On the relationship between Borda and Lenoir, see A. J. Turner, *From Pleasure and Profit to Science and Security: Etienne Lenoir and the Transformation of Precision Instrument-Making in France, 1760–1830* (Cambridge, England: Whipple Museum, 1989).

37. On Puig Rodos, see AOP E2-19, Méchain to Lalande, 3 brumaire IV [25 October 1795]. On the lights in the city, see Barón de Maldà, in Moreu-Rey, *Naixement del metre*, pp. 71–2. On the use of mirrors, see AMNM Leg. 2294/74, Gonzales to Alvarez, 27 October 1792.

38. AOP E2-19, Méchain to Delambre, 14 vendémiaire IV [4 October 1795].

39. For fossils and tombstones, see Townsend, *Journey Through Spain*, 1, pp. 127–9.

40. Townsend, *Journey Through Spain*, 1, p. 128. Also, on the Mont-Jouy fortress, see Swindburne, *Travels*, 1, pp. 71–5; Young, 'Tour in Catalonia', pp. 235–41; *Military Museum: Montjuich Castle* (Barcelona: Graistudio, 1997); Pedro Voltes Bou, *Historia de Montjuich y su castillo* (Barcelona: Ayuntamiento de Barcelona, 1960), pp. 129–43; Estanislau Roca, *Montjuïc, La muntanya de la ciutat*, 2nd edn (Barcelona: Institut d'Estudis Catalans, 2000).

41. The extension to the Balearic Islands had been mooted before Méchain left Paris; see AOP MS1058III, Méchain to Flaugergues, 9 June 1792. The idea had been first proposed by José de Mendoza, a Spanish savant in Paris who had helped the French academicians plan the expedition; see Ten, 'Sistemo métrico'. Borda reported on the plan in *AP* 53 (25 November 1792), p. 583. Méchain reported his need for better scopes in AOP B4-11, Méchain to Cassini IV, 8 September 1792. For Gonzales' account of his trip to Mallorca, see AMNM Leg. 2294, Gonzales to Valdez, 16 January 1793; also AOP B4-10, Gonzales to Méchain, 21 September 1803. For Méchain's priorities, see ENPC MS1504, Méchain to Borda, 13 February 1793.

42. For Méchain's observatory and methods, see AOP E2-19, Méchain to Delambre, 12 vendémiaire IV [4 October 1795]. Also see logbooks such as AOP E2-20.

43. On the importance of latitude, see Delambre, *Base*, 2, p. 158.

44. AOP E2-20, Méchain, 'Détails des opérations faites en Catalogne', April 1793. For Méchain's boast to Borda about his success so far, see AOP E2-19, Méchain to Borda, 10 January 1794.

45. AMNM Leg. 2294, Planez to [Unknown], 16 February 1793. See also AMNM Leg. 2294, Gonzales to Valdez, 16, 29 January, 16 February, 23 March 1793. Méchain promised to let his Spanish hosts try the instrument, but only after he was done with all his own measurements; see AMNM Leg. 2294, Méchain, 'Planez', 7 February 1793.

46. For a description of the winter, see ENPC MS504, Méchain to Borda, 13 February 1793.

47. Delambre, *Notice historique sur M. Méchain*, p. 16.

48. *Diario de Barcelona* 2 (17 January 1793), pp. 66–7. A new comet was international news in those days and this one was reported (via Lalande) ten months later in *The Times* (London), 3 October 1793. See Méchain's belated published report in Méchain, 'Comète de 1793 . . . lu le 10 nivôse, an XIII [31 December 1804]', *MI* 6 (1806), pp. 290–309.

49. AMAE Barcelone 20, Aubert to Cit. Min., 20, 23 February 1793. For Méchain's account of the corsairs, see AOP E2-19, Méchain to Delambre, 23 February 1793. For another eyewitness to the event, see Maldà, *Calaix de Sastre*, 2, pp. 67–8. For the incident and the cause of the war, see Philippe Torreilles, *Perpignan pendant la révolution (1789–1800)*, (Perpignan: Schrauben, 1989), 2, p. 3.

50. See AOP E2-19, Méchain to Delambre, 23 February 1793.

51. AOP E2-19, Delambre to Méchain, 31 March 1793.

52. For Gonzales' complaints, see AMNM Leg. 2294, Gonzales to Valdez, 19 February 1793. Méchain mailed a summary report of his data; see ENPC MS504, Méchain, 'Correspondance et observations', January–February 1793.

53. For the recall of Méchain's Spanish collaborators, see BMCF Fonds Chazelle, Méchain to Lavoisier, 6 March 1793.

54. For the war on the southern front, see Alain Degage, 'Les principaux aspects de la stratégie des armées françaises des Pyrénées-Orientales durant la guerre franco-espagnole (1793–1795)', *L'Espagne et la France à*

l'époque de la Révolution française (1793–1807), ed. Jean Sagnes (Perpignan: Presses Universitaires de Perpignan, 1993), pp. 11–31; Lluís Roura i Aulinas, *Guerra Gran a la Ratlla de França, Catalunya dins la guerra contra la Revolució Francesa, 1793–1795* (Barcelona: Curial, 1993). For the demand that Méchain leave the fort, see AN F17 1326 dos. 18, Méchain to Lavoisier, 11 May 1793.

55. For Méchain's retrospective account of the accident, see AOP E2-19, Méchain to Lalande, 12 floréal IV [1 May 1796]. For his immediate, vague account, see AN F17 1326 dos. 18, Méchain to Lavoisier, 11 May 1793. Méchain dated the accident to 1 May 1793; see ENPC MS1504, Méchain to Borda, 10 January 1794. On the pump, see Townsend, *Journey Through Spain*, 1, pp. 134–5.

56. AOP E2-19, Méchain to Lalande, 19 ventôse III [9 March 1795].

57. For the Spanish decision to detain Méchain in Spain because war had broken out, see AMNM Leg. 2294/116–117, Consular note, n.d.

58. AN F17 1326 dos. 18, Méchain to Lavoisier, 11 May 1793. The letter is not in Méchain's hand.

59. For Mme Méchain's expectations of her husband's return, see KBD NKS1304, Mme Méchain to Bugge, 19 March 1793. For her worries, see AAS Fonds Lavoisier 1229(2), Lavoisier to Méchain [mid-June 1793]. Lavoisier to Méchain, 15 June 1793, in CUL MS4712++, Lavoisier, 'Registres de l'Académie des Sciences', 1792–3. See also AAS Lavoisier 1228, Lavoisier to Mme Méchain [mid-June 1793].

60. Lavoisier to Méchain, 29 June 1793, in CUL MS4712++, Lavoisier, 'Registres de l'Académie des Sciences', 1792–3.

61. For Lavoisier's guarantee of the funds, see AAS Fonds Lavoisier 972, Lavoisier, 'Au nom de Antoine-Laurent Lavoisier' [June 1793].

62. AAS Fonds Lavoisier, 1229(2), Lavoisier to Méchain [mid-June 1793]. See also, Lavoisier to Méchain, 15 June 1793, in CUL MS4712++, Lavoisier, 'Registres de l'Académie des Sciences', 1792–3.

63. AAS Fonds Lavoisier 1229(2), Lavoisier to Méchain [mid-June 1793].

64. AAS Fonds Lavoisier 1228(45), Lavoisier to Lakanal [May–June 1793].

Chapter 3 The Measure of Revolution

1. Thomas Pynchon, *Mason & Dixon* (New York: Holt, 1997), p. 669.

2. Louis-Sébastien Mercier, *Le nouveau Paris* ([Brunswick]: n.p., 1800), 3, p. 44.

3. For the station at Saint-Martin, see AOP E2-6, Delambre, 'Registre', p. 57.

4. For the history of Belle-Assise, see ADSM 30Z216, Froment, 'Monographie de la commune de Joissigny', 1888.

5. For the view of Bruyères, see Delambre, *Base*, 1, p. 137. For his *ancien régime* observations there, see Bigourdan, *Astronomie d'observation*, pp. 170–2.

6. For the old measures stored at the Panthéon, see Lavoisier, 'Rapport sur le local destiné . . . pour les poids et mesures', September 1793, in

Antoine-Laurent Lavoisier ,*Œuvres de Lavoisier* (Paris, Imprimerie Impériale, 1862–93), 6, p. 690.

7. *Journal de la Montagne* (28 messidor II [16 July 1794]), p. 647, quoted in Centre Canadien d'Architecture, *Le Panthéon: Symbole des révolutions* (Tardy, France: Picard, 1989), pp. 133–6. On alterations to the Panthéon, see Antoine Quatremère de Quincy, *Rapport fait au directoire du département de Paris, 13 novembre 1792, sur l'état actuel du Panthéon* (Paris: Ballard, 1792).

8. Antoine Quatremère de Quincy, *Rapport fait au directoire du département de Paris sur les travaux entrepris, continués ou achevés au Panthéon français* (Paris: Ballard, II [1794]), pp. 6–7, 17–18. For Delambre's proposed observatory, see AOP MS1033B, Delambre, 'Plan de la lumière du dome du Panthéon française' [January 1793].

9. AOP E2-19, Delambre to Méchain, 31 March 1793; received in Barcelona in June 1793.

10. BN Rés Ye 3641, Lalande, *Article pour les cahiers dont les 36 rédacteurs sont prier instament et requis expressément de faire usage* ([Paris]: n.p., 1789).

11. On Lalande's youth, see Julien Raspail, 'Papiers de Lalande', *La Révolution française* 74 (1921), pp. 236–54. On his science, see Jean-Claude Pecker, 'L'œuvre scientifique de Lalande', *Jérôme de Lalande (1732–1807)*, ed. Pecker (Bourg-en-Bresse: Société d'Emulation de l'Ain, 1985), pp. 12–16.

12. For the predicted comet, see Lalande, *Réflexions sur les comètes qui peuvent approcher de la terre* (Paris: Gibert, 1773). For the wag's comment, see BN Vz 1695, [Pierre Hourcastrenne], *Dissertation sur les causes qui ont produit* (Paris: Pougin [1791]), p. 29. For the effects on women, countryfolk and sales of unleavened bread, see Condorcet to Voltaire, 16 May 1773, in Voltaire, *Voltaire's Correspondence*, ed. Theodore Besterman (Geneva: Institut et Musée Voltaire), D18372. Simon Schaffer, 'Authorized Prophets: Comets and Astronomers after 1759', *Studies in Eighteenth-Century Culture* 17 (1987), pp. 45–74.

13. Friedrich Melchior Grimm, *Correspondance littéraire*, ed. Maurice Tourneux (Paris: Garnier, 1877–82), 10 (April 1773), pp. 235–8.

14. For Lalande's first voyage to England in 1763, see Lalande, 'Journal d'un voyage en Angleterre', ed. Helène Monod-Cassidy, *Studies on Voltaire and the Eighteenth Century* 184 (1980), pp. 5–116. On Lalande's ballooning, see Seymour Chapin, 'What Better Way to Go to Gotha: Lalande, Blanchard, and the Balloon', *Griffith Observer* 47 (November 1983), pp. 2–19. On Lalande and the masons, see Louis Aimable, *Le franc-maçon, Jérôme Lalande* (Paris: Charavay, 1889), pp. 22–38.

15. Constance de Salm-Reifferscheid-Dyck, *Eloge historique de M. De La Lande* (Paris: Sajou, 1810), p. 33.

16. Lalande, 'Testament moral', in Aimable, *Lalande*, p. 52.

17. Voltaire to Lalande, 6 February 1775, in Voltaire, *Correspondence*, D19323.

18. Voltaire to Lalande, 11 June 1770, in Voltaire, *Correspondence*, D16404: 'Dans l'universe entier ta gloire est répandue, / Et ce n'est qu'avec lui que périra ton nom.'

19. Denis-Bernard Quatremère-Disjonval, *De l'aranéologie* (Paris: Fuchs, 1797), pp. 141–2.

20. For Lalande's height, see CUS, Lalande, 'Passeport', 19 prairial VI [7 June 1798]. For Piery, see the correspondence between Lalande and Louise-Elizabeth-Félicité Pourra de la Madeleine du Piery in BYU Lalande papers. Jérôme Lalande, *Astronomie des dames* (1st edn, 1785; 2nd edn, Paris: Cuchet, 1795; later editions in 1806, 1817, 1820, 1841, 1900). For English edition, see Lalande, *Ladies' Astronomy*, trans. W. Pengree (London: Dutton, 1815).

21. Lalande, 'Journal', 1763, in Jules Claretie, *L'empire, les Bonapartes & la cour* (Paris: Dentu, 1871), p. 231.

22. Lalande, 'Mémoires de Lalande', in Salm-Reifferscheid-Dyck, *La Lande*, pp. 43–4. For his apology, see BNR NAF4073, Lalande to Sophie Germain, 4 November 1797.

23. For enrolments in Lalande's course, see BVCS MS99, Lalande, 'Journal', 1777–1807.

24. Jérôme Lalande, *Abrégé de navigation* (Paris: Chez l'auteur, 1793), p. 3. For his family feeling, see Lalande, 'Testament moral', in Aimable, *Lalande*, p. 52.

25. BYU, Lalande to Piery, 1 messidor II [19 June 1794].

26. BYU, Lalande to Piery, 12 August 1788.

27. On Lalande's star catalogue, see Lalande, *Bibliographie astronomique*, pp. 681–2, 690–4.

28. For Méchain's prize-winning work, see Méchain, 'Recherches sur les comètes de 1532 et 1661', *Mémoires de mathématique et de physique* (Paris: Moutard [1782], 1785), 10, pp. 330–96.

29. KBD NKS1304, Lalande to Bugge, 16 June 1788.

30. For the Mercury sighting see AAS, 'Procès-verbaux de l'Académie des Sciences' 105 (1786), pp. 276–8; Delambre, *Histoire de l'astronomie au dix-huitième siècle*, ed. Claude-Louis Mathieu (Paris: Bachelier, 1827), pp. 564–5; Curtis Wilson, 'Perturbations and Solar Tables from Lacaille to Delambre: The Rapprochement of Observation and Theory', *Archive for the History of the Exact Sciences* 22 (1980), pp. 54–304, especially pp. 268–96.

31. Delambre, 'Joseph-Jérôme Lefrançais de Lalande', *Biographie universelle* (Paris: Michaud, n.d.), p. 612.

32. For Lalande's report on Mercury, see Lalande, 'Sur la théorie de Mercure', *MAS* (1786), p. 272. For the report of the other observer, see Messier, 'Observation du passage de Mercure', *MAS* (1786), pp. 121–4.

33. For Méchain's missed observation of Mercury, see BML 26CA6, Méchain to Cotte, 8 May 1786.

34. CUS, Delambre to Cagnoli, 23 November 1789. On Laplace, see Charles Coulson Gillispie, *Pierre-Simon Laplace, 1749–1827: A Life in Exact Science* (Princeton: Princeton University Press, 1997). For the report on Delambre's prize submission, see 'Rapport de Cassini IV, Lalande et Méchain', 5 July 1788, in AAS 'Procès-verbaux de l'Académie des Sciences' 107 (1788), pp. 176–9.

35. Lalande, *Bibliographie astronomique* [1792], p. 703.

36. For evidence that Delambre and Méchain observed together, see AAS Dossier Méchain, Méchain to Delambre, 29 September [1786]. Delambre asked a friend to intercede for him with Méchain; see CUS, Delambre to Cagnoli, 23 November 1789.

37. AAS Dossier Méchain, Méchain to M. l'abbé de Lambre [Delambre], 29 September [1785/86].

38. Delambre, 'Lui-même'.

39. For pre-Revolutionary discussion of metric reform, see Henri Pigeonneau and Alfred de Foville, eds, *L'Administration de l'Agriculture au Controle Générale des Finances, 1785–1787, Procès-verbaux et rapports* (Paris: Guillaumin, 1882), pp. 127–8, 324–6, 378, 404. For the first proposals within the Academy, see AAS, J.-B. Le Roy, 'Procès-verbaux de l'Académie des Sciences' 108 (27 June, 14 August 1789), pp. 171, 207. For Prieur's pre-Revolutionary proposals, see BEP Prieur 4.2.4.1, Prieur, 'Mesure universelle, Extrait du *Journal encyclopédique*' (1785), p. 491. For his post-Revolution proposals, see Prieur to Louis XVI, April 1790, in Georges Bouchard, *Un organisateur de la victoire: Prieur de la Côte-d'Or* (Paris: Clavreuil, 1946), pp. 458–9; also, Claude-Antoine Prieur-Duvernois, *Mémoire sur la nécessité et les moyens de rendre uniformes dans le royaume toutes les mesures* (Dijon: Causse, 1790). For a survey of various *ancien régime* proposals, see Ronald Zupko, *Revolution in Measurement: Western European Weights and Measures Since the Age of Science* (Philadelphia: American Philosophical Society, 1990).

40. Tillet and Abeille, *AP* 11 (6 February 1790), pp. 466–86. Tillet and Abeille cited Lalande's proposal and reproduced a portion of his article from the *Tribut de la Société Nationale de Neuf-Soeurs* 1 (1790), pp. 7–16, in *AP* 11 (6 February 1790), pp. 486–7. The only surviving copy of this volume of the *Tribut* has apparently been lost from the Bibliothèque Nationale.

41. Talleyrand, *AP* 12 (9 March 1790), p. 106. Identical phrasing can be found in Condorcet's proposal in BI MS833. Also, Talleyrand admitted that he had consulted 'men of art'; see Talleyrand, *Mémoires* (Paris: Bonnot, 1989), 1, p. 134.

42. For the old Paris *toise*, see Charles Wolf, 'Recherches historiques sur les étalons de l'Observatoire', *Annales de chimie et de physique*, 5th series, 25 (1882), pp. 5–112.

43. On units of weight, see A. Birembaut, 'Les deux déterminations de l'unité de masse du système métrique', *Revue d'histoire des sciences* 12 (1959), pp. 25–54.

44. On Stevin, see René Taton, 'La tentative de Stevin pour la décimalisation de la métrologie', in *Acta Metrologiae Historicae*, ed. Gustav Otruba (Linz: IIIe Congrès International de la Métrologie Historique, 1983), pp. 39–56. On Vauban, see Sébastien Le Prestre de Vauban, 'Description géographique de l'élection de Vezalay', *Projet d'une dixme royale* (Paris: Alcan, 1933). On Lavoisier, see Lavoisier, *Traité élémentaire de chimie* [1789], in ,*Œuvres*, 1, pp. 248–51. On the ease of calculation, see Laplace, 'Mathématiques', *Séances des Ecoles Normales [de l'an III], Leçons* (Paris: Reynier [1795]), 1, pp. 10–23. On the naturalness of the

decimal system, see [René-Just Haüy], *Instructions sur les mesures déduites de la grandeur de la terre*, 1st edn (Paris: Imprimerie Nationale, II [1794]), pp. xxvii–xviii.

45. On base 12, see A. G. Leblond, *Sur la fixation d'une mesure et d'un poid – lu à l'Académie des Sciences, 12 mai 1790* (Paris: Demonville, 1791); also Rollin, 12 frimaire II [22 November 1793], in *PVCIP*, 3, pp. 88, 90–1. For base 8, see Gueroult, *Observations sur la proposition faite par le Cit. Prieur* (Paris: Guerin [1796–7]), p. 5. For base 2 (and 4), see CNAM R14, Leturc to Citoyens agents, 4 fructidor III [21 August 1795]. For base 11, see the views of Lagrange, in Laplace, 'Mathématiques', *Ecoles Normales*, 1, p. 23. For attacks on those who preferred base 7 and 11, see Charles-Etienne Coquebert [de Montbret], 'An Account of a New System of Measures Established in France', *Journal of Natural Philosophy* 1 (August 1797), p. 195.

46. Leblond, *Sur la fixation d'une mesure*, 10.

47. Mercier, *Le nouveau Paris*, 3, p. 44.

48. For various seventeenth-century pendulum proposals, see Isaac Beeckman to Marin Mersenne, 7 October 1631, in Mersenne, *Correspondance du P. Marin Mersenne* (Paris: Beauchesne, 1932–88), 3, pp. 209–10; Christian Huygens, *Horologium Oscillatorium* (Paris: Muguet, 1673), sect. 4, prop. 25. For eighteenth-century attempts, see Turgot to Messier, 3 October 1775, in Etienne-François Turgot, *Œuvres* (Glashütten im Taunus: Auvermann, 1972), 5, pp. 31–3; Turgot to Condorcet [1775], in *Correspondance inédite de Condorcet et de Turgot 1770–1779*, ed. Charles Henry (Paris: Charavay, 1883), pp. 234–5; Keith Michael Baker, 'Science and Politics at the End of the Old Regime', in *Inventing the French Revolution: Essays on French Political Culture in the Eighteenth Century* (Cambridge: Cambridge University Press, 1990), pp. 153–66.

49. Talleyrand, *AP* 12 (9 March 1790), pp. 104–8. Talleyrand and Miller were in regular communication; see Yves Noël and René Taton, 'La réforme des poids et mesures, Origines et premières étapes (1789–91)', in *Œuvres de Lavoisier*, vol. 7, *Correspondance*, ed. Patrice Bret (Paris: Académie des Sciences, 1997), 6, pp. 339–65. John Riggs Miller, *Speeches in the House of Commons upon the Equalization of the Weights and Measures of Great Britain* (London: Debrett, 1790).

50. BI MS883 fol. 34, Condorcet, 'Sur une mesure commune', n.d. Jefferson worked closely with David Rittenhouse (a mathematician) and Robert Leslie (a watchmaker). For an overview of early American efforts for metric reform, see Julian P. Boyd, 'Report on Weights and Measures', in Thomas Jefferson, *The Papers of Thomas Jefferson*, ed. Julian P. Boyd (Princeton: Princeton University Press, 1950–), 16, pp. 602–17. See also C. Doris Hellman, 'Jefferson's Efforts toward the Decimalization of United States Weights and Measures', *Isis* 16 (1931), pp. 265–314. In 1785 James Madison had also suggested basing a standard of measure on the pendulum; see Madison to James Monroe, 28 April 1785, in *The Writings of James Madison*, ed. Gaillard Hunt (New York: Putnam, 1901), 2, pp. 142–3. For the early involvement of the US government in

metric matters, see Sarah Ann Jones, *Weights and Measures in Congress: Historical Summary*, National Bureau of Standards, Miscellaneous Publication M122 (Washington, DC: US GPO, 1936).

51. *AP* 15 (8 May 1790), p. 439. This proposal was passed by the king in *Proclamation du roi sur le décret de l'Assemblée Nationale du 8 mai 1790*, 22 August 1790 (Paris: Imprimerie Royale, 1790).

52. For Miller's praise, see Miller, *Speeches* (1790), pp. viii–iv, xiv. Jefferson to Speaker of the House, 4 July 1790, in Jefferson, *Papers*, 16, p. 653. See his First Report (April 1790), Second Report (20 May 1790) and Final Report (4 July 1790), in Jefferson, *Papers*, 16, pp. 623–48. For Jefferson's earlier preferences, see Jefferson to Leslie, 27 June 1790, in Jefferson, *Papers*, 16, p. 576. For the preference of one Paris savant, see Brisson, 'Essai sur l'uniformité de mesures' (14 April 1790), *MAS* (1790), pp. 722–6.

53. [Laplace], 'Discours', 4 messidor VII [22 June 1799], in Delambre, *Base*, 3, p. 585. The speech is unsigned, but was delivered by Laplace; see Bigourdan, *Système métrique*, pp. 160–6.

54. Borda, 'Rapport à l'Académie des Sciences', 19 March 1791, *AP* 24 (26 March 1791), pp. 379, 394–7.

55. On the debates over geodesy, see Jean-Jacques Levallois, *Mesurer la terre: 300 ans de géodésie française* (Paris: Presses de l'Ecole Nationale des Ponts et Chaussées, 1988); Mary Terrall, *The Man Who Flattened the Earth: Maupertuis and the Sciences in the Enlightenment* (Chicago: University of Chicago Press, 2002). Also see Mary Terrall, 'Representing the Earth's Shape: The Polemics Surrounding Maupertuis's Expedition to Lapland', *Isis* 83 (1992), pp. 218–37; Rob Iliffe, 'Aplatiseur du monde et de Cassini', *History of Science* 31 (1993), pp. 335–75; John L. Greenberg, *The Problem of the Earth's Shape from Newton to Clairaut: The Rise of Mathematical Science in Eighteenth-Century Paris and the Fall of 'Normal' Science* (Cambridge: Cambridge University Press, 1995).

56. Fontenelle, 'Physique', in *HAS* 1 (1674, pub. 1733), p. 178. See also d'Alembert, 'Pendule', *Encyclopédie*, 12, p. 294.

57. Condorcet to President of the National Assembly, and Borda, 'Rapport', in *AP* 24 (26 March 1791), pp. 379, 394–7.

58. BLL Add MS33272, ff. 97–8, Charles Blagden to Joseph Banks, 8 September 1791, emphasis in original.

59. Jefferson, 'Memorandum to James Monroe', before 4 April 1792, in Jefferson, *Papers*, 27, pp. 818–22. Jefferson received a copy of the French meridian law from Condorcet, translated it himself and wrote back 'to confess that it is not what I would have approved'; see Condorcet to Jefferson, 3 May 1791, Jefferson to Short, 28 July 1791, Jefferson to Condorcet, 30 August 1791, in Jefferson, *Papers*, 20, pp. 353–60; 20, pp. 686–91; 22, pp. 98–9.

60. Borda's new budget came to 300,000 *livres*; see AN F12 1289, [Académie des Sciences], 19 March 1791. The decree of 20 August 1790 had set the Academy's total budget at 93,458 *livres*; see PVCIP, 1, p. 260n. For contracts with instrument-makers, see AN F12 1289, Borda to Pavé (Min. Int.), 12 brumaire II [2 November 1793]. For a

description of Lenoir – he was four foot ten inches tall – see CNAM R6, 'Etienne Lenoir, Certificat de residence', 22 floréal II [11 May 1794]. On the personnel in Lenoir's shop, see AAS Lavoisier 1967, Lavoisier, 'Etat des ouvriers et coopérateurs' [1793].

61. Mercier, *Le nouveau Paris*, 3, p. 44.

62. Jean-Paul Marat, *Les charlatans modernes; ou, Lettres sur le charlatanisme académique* ([Paris]: n.p., 1791), p. 40.

63. On Borda's motives, see Delambre, *Grandeur*, pp. 202–3, 213; also Lalande, *Bibliographie astronomique*, p. 718. On Laplace's motives, see Laplace, 'Mathématiques', *Ecoles Normales*, 5, pp. 203–14.

64. BN Vz 1695, [Pierre Hourcastrenne], *Dissertation sur les causes qui ont produit* (Paris: Pougin, [1791]), pp. 18–19. For speculation about change in the shape of the earth over time, see Laplace, 'Mathématiques', *Ecoles Normales*, 5, p. 212. For other objections to the meridian, see AN A34 1037, 'Observations sur le rapport relatif au choix d'une unité de mesure', n.d. AN AFII 67 plaq. 496, Gueroult, *Observations sur le nouveau système des poids et mesures* (Paris: Rousyef, II [1794]).

65. For Lalande's objections, see *AP* 11 (6 February 1790), pp. 486–7.

66. Delambre, 14 April 1793, in Bigourdan, *Système métrique*, pp. 128–30.

67. For Delambre's passport, see ADSe 3AZ259, Destournelles and Coulenbeau, 'Extrait de registres des déliberations du Conseil Général', 15 April 1793. The passport itself was granted on 24 April; see BMA Arch Rev 2K10, Delambre, 'Pétition', 9 August 1793. See Delambre's account in Delambre, *Base*, 1, pp. 41–2. For one of Delambre's cautionary letters, see BMA 1953(17), Delambre to [?], 31 March [1793].

68. For Delambre's dossier, see BMA Arch Rev 2K10, Delambre, 'Pétition', 9 August 1793.

69. AOP E2-6, Delambre, 'Registre', p. 127. See also AOP MS1033c, Garcia to Delambre, 7 June 1793.

70. For Delambre's progress from Watten to Bayonvilles, see AOP E2-6, Delambre, 'Registre', pp. 132–98; AAS Cabrol 120/121, Delambre to Lavoisier, 11, 16 July 1793; Delambre, *Base*, 1, p. 44.

71. CUS, Delambre to Amélie Lefrançais, 22 August 1793.

72. For Lalande's hopes for his nephew's election, see CUS, Lalande to Delambre, 28 July [1793].

73. AAS Lavoisier 1128(41), Lavoisier to Delambre, [August 1793]. See the advance warning of the Academy's demise, which Lavoisier provided to Delambre in AAS Bertrand Col. 4, Lavoisier to Delambre, 23 July 1793.

74. BNRC Ge DD 2066 (3), Cassini IV, 'Mémoires, 2ème dialogue', p. 31. For attempts to save the Academy, see *PVCIP*, 2, pp. 240–60. For the Academy in general, see Roger Hahn, *The Anatomy of a Scientific Institution: The Paris Academy of Sciences, 1666–1803* (Berkeley: University of California Press, 1971).

75. The demand for a 'provisional' metre had been first made by Roland; see *AP* 41 (3 April 1792), p. 100. First presented to the CIP by Arbogast, 'Rapport', July 1793, in *PVCIP*, 2, pp. 9–20. Based on a report by Borda, Lagrange and Monge to the Academy of Sciences

and sent to the committee on 29 May 1793. For the legislation, see *AP* 70 (1 August 1793), pp. 70–4, 112–18. BN Le38 2501, Arbogast, *Sur l'uniformité et le système général des poids et mesures* (Paris: Imprimerie National, 1793).

76. For Borda's estimate, see AAS Chabrol 1/71, Borda to Condorcet [May 1792].

77. For the metric needs of the *cadastre*, see Jean-Baptiste Jollivet, *Rapport et projet de décret sur une nouvelle et complète organisation de la contribution foncière*, 21 August 1792 (Paris: Imprimerie Nationale, 1792), pp. 151–81. For the calculation of the provisional metre, see AAS, 'Procès-verbaux de l'Académie' 110 (19 January 1793), pp. 327–35; published as Borda, Lagrange and Laplace, 'Fait à l'Académie des Sciences, sur l'unité des poids et mesures et sur la nomenclature de ses divisions', in CUL QC89. F8A16, *Recueil de pièces relatives à l'uniformité des poids et mesures* ([Paris]: n.p., 1793).

78. CUS, Lalande to Delambre, 2 frimaire II [23 November 1793].

79. For Amiens, see AAS Dossier Delambre, Delambre to Lavoisier, 18 August 1793. AOP MS1033c, Delambre, 'Méridien de France (Partie du Nord)', [1793–4].

80. [Delambre], 'Proposition d'un citoyen', *Affiches du Département de la Somme* 28 (9 July 1791), pp. 120–1. For Delambre's participation in the Société, see BMA Arch. Rév. 315(1), 'Société des Amis de la Constitution', 16 April 1791. Delambre did not visit Amiens once between 1782 and 1790; see Bigourdan, *Astronomie d'observation*, p. 169.

81. [François-Noël Babeuf], *Affiches du Département de la Somme* 29 (16 July 1791), p. 125; 30 (23 July 1791), p. 128, in M.-J. Foucart, 'Un aspect inconnu de Babeuf à l'été charnière de 1791', *Société des Antiquaires de Picardie* 62 (1989), pp. 371–9.

82. For salaries on mission, see AAS Dossier Delambre, Delambre to Lavoisier, 18 August 1793. Delambre, 'Mesure du méridien, dépense faite par Delambre', 26 floréal III [15 May 1795], in Observatoire de Paris, *Longueur et temps* (Paris: Observatoire de Paris, 1984). Delambre, *Grandeur*, p. 281.

83. Delambre to Lavoisier, 13 frimaire II [3 December 1793], in Bigourdan, *Système métrique*, p. 132.

84. Jacques Soyer, 'Un acte de vandalisme dans le département du Loiret en l'an II', *Bulletin de la Société Archéologique et Historique de l'Orléanais* 18 (1917), pp. 99–106.

85. At Châtillon, see AOP E2-6, Delambre, 'Registre', p. 270.

86. Lagrange [CPM] to Delambre, 8 nivôse II [28 December 1793], in Bouchard, *Prieur*, p. 461. See also Lagrange to Comité de Salut Public, 8 nivôse II [28 December 1793]; Carnot and Lindet, 9 nivôse II [29 December 1793], in Bouchard, *Prieur*, pp. 461–2.

87. For Delambre's request for a delay, see ENPC MS724, Delambre to [Lagrange, CPM], 13 nivôse II [2 January 1794]. Delambre, *Base*, 1, p. 48.

88. Delambre noted in the margin that the commissioner was Prony in KM, Delambre, *Base*, 1, p. 48. See also Bigourdan, *Système métrique*,

p. 133; ENPC MS724, CPM to Commission de Subsistances, 15 nivôse II [4 January 1794].

89. Denis Lottin, *Recherches historique sur Orléans* (Orléans: Jacob, 1838), 4, pp. 195, 427–8.

90. For Delambre's estimate of 1200 *lieues* (2400 miles), see Delambre, 'Mesure du méridien, dépense faite par Delambre', 26 floréal III [15 May 1795], in Observatoire de Paris, *Longueur et temps*; see also ENPC MS726, Delambre, 'Mesure du méridien', 19 messidor II [7 July 1794].

91. AOP E2-6, Delambre, 'Registre', 3 pluviôse II [22 January 1794].

92. Delambre, *Base*, 1, p. 49. See also ENPC MS724, Min. Int. to Prony, 18 nivôse II [7 January 1794].

93. AN AF II 67 plaq. 496, Prieur, Barère, Carnot, Lindet and Billaud-Varenne, 'Le Comité de Salut Public, considérant combien . . .', 3 nivôse II [23 December 1793]. For the published version, see *RACSP*, 9, p. 600. Note that in Delambre, *Base*, 1, p. 50, Delambre gives the names of the signatories as the three (dead) radicals Robespierre, Couthon and Collot-d'Herbois, rather than the three (living) engineers-moderates who actually signed the order: Prieur, Carnot and Lindet. However, he corrects this deliberate change in a marginal note in KM, Delambre, *Base*, 1, p. 50.

94. AOP E2-19, Delambre to CPM, 4 pluviôse II [23 January 1794].

95. For d'Assy's arrest, see AN F7 4722, Comité de Sûreté Générale 'Geoffroy d'Assy', 6 pluviôse II [25 January 1794].

Chapter 4 The Castle of Mont-Jouy

1. Blaise Pascal, *Pensées sur la religion*, ed. Louis Lafuma (Paris: Luxembourg, 1951), pp. 52–3, (fragment 60–108).

2. AOP E2-19, Méchain to Lalande, 19 ventôse III [9 March 1796]. News of Méchain is found in CUS, Lalande to [Delambre], 28 July 1793. For the effect of heat on the circle, see AOAB Cart. 88, Méchain to Oriani, 2 April 1795.

3. Jaume Carrera Pujal, *La Barcelona del segle XVIII* (Barcelona: Casa Editorial, 1951), v. 1. Maldà, *Calaix de Sastre*, 2, pp. 91–3. Torreilles, *Perpignan*, 2, pp. 62–3.

4. On Méchain's pay, see AAS Lavoisier 1228 (36), Lavoisier to Méchain, 6 October 1793. For Méchain's deference to the defunct Academy, see AOP E2-19, Méchain to Borda, 10 January 1794. For Méchain's fears of replacement, see AOP E2-19, Méchain to Borda, 10 January 1794. For Delambre's speculation as to why he was purged while Méchain was spared, see Delambre, *Grandeur*, p. 214.

5. For rumours in Paris that Méchain had received outside (foreign) offers of employment, see KM, Delambre, *Base*, 1, p. 52. For the correspondence regarding Méchain's funds, see CUL MS4712++, Lavoisier, 'Registres de l'Académie des Sciences', pp. 1792–3.

6. For Delambre's offer to help, see ENPC MS724, Delambre, 'Mesure du méridien', 19 messidor II [7 July 1794].

7. ADPO L1128, Gauderique Costaseca et al., 'Extrait des registres de la municipalité de Valmagnne [sic]', 7 October 1793.

8. For the help of Llucía, see ADPO L1128, Méchain to Lucia [sic], 6 October 1793.

9. Lalande, Pingré, Chabert, Cassini IV and Méchain, 'Rapport', 4 June 1791, in Antoine Albitreccia, *Le plan terrier de la Corse au XVIIIe siècle* (Paris: Presses Universitaires de France, 1942), p. 74.

10. Méchain, as quoted in Delambre, *Base*, 1, p. 430. Delambre notes that the data for this site were recorded in Tranchot's hand; see KM, Delambre, *Base*, 1, pp. 429–35. In AOP E2-19 Delambre notes that Méchain supplied Tranchot with correction factors without explaining how or why to use them. Also see Delambre, *Grandeur*, p. 280.

11. For Tranchot's supply of the latitude data of the Spanish forts to the French army, see Henri-Marie-Auguste Berthaut, *Les ingénieurs géographes militaires, 1624–1831: Étude historique* (Paris: Imprimerie du Service Géographique, 1902), 1, p. 172. I have not been able to locate the original sources for this claim. Unfortunately, Berthaut almost never cites his sources, and there is something contradictory about the story as he tells it. He says Tranchot got these plans to Dugommier in the year IV, but Dugommier died in the year II; and the plans would not have been any use in the year IV, since the French had already occupied most of these forts and the peace had been signed in the year III. Despite this slip, however, Berthaut was an extremely reliable and knowledgeable historian, who worked exclusively with archival materials from the Dépôt de la Guerre, to which he had complete access for many years. Many of the documents in that holding have since been misplaced. So the gist of his account is likely to be correct, though the timing and manner of the espionage are unclear.

12. For Méchain at Puig Camellas, see Méchain, in Delambre, *Base*, 1, pp. 429–45.

13. For Lalande's news of Méchain and Tranchot, see CUS, Lalande to Delambre, 3 frimaire II [23 November 1793].

14. ADPO L1128, Tranchot (at Puig de l'Estelle) to [Llucía], 3 November 1793. Tranchot is quoting from Méchain's letter to him, now lost.

15. ADPO L1128, Tranchot to [Llucía], 3 November 1793.

16. Llucía, in Vidal, *Pyrénées-Orientales*, 3, p. 46.

17. For the prohibition against sending data, see AOP E2-19, Méchain to Borda, 10 January 1794. For the orders of Ricardos, see Delambre, *Notice historique sur Méchain*, 18.

18. On Dugommier, see Arthur Chuquet, *Dugommier, 1738–1794* (Paris: Fontemoing, 1904). On the battles of 1793–4, see Joseph-Napoléon Fervel, *Campagnes de la Révolution française dans les Pyrénées-Orientales* (Paris: Dumaine, 1861), 2, pp. 210–11.

19. AOP E2-19, Méchain to Borda, 10 January 1794. There are two copies of the letter from Méchain to Borda of 10 January 1794; neither is complete and each contains elements that the other omits: 1) AOP E2-19, Méchain to Borda, 10 January 1794; and 2) ENPC MS1504, Méchain to Borda, 10 January 1794.

20. For the Barcelona measurements, see Méchain's later admission that he was crippled in his right arm and had to operate 'with the help of others' in AOP E2-19, Méchain to Lalande, 4 messidor IV [22 June 1796].

21. For the Mont-Jouy triangulations, see Delambre, *Base*, 1, p. 503. For the weather on 16 March, see *Diario de Barcelona 77* (18 March 1794), p. 305.

22. For the tensions in Barcelona, see Roura i Aulinas, *Guerra Gran*, p. 75. Herr, *Eighteenth-Century Spain*, pp. 311–12.

23. For Méchain's departure from Barcelona, see BUP MS168.1, Méchain to Slop, 6 September 1794. For the lightning strike, see AOAB Cart. 88, Méchain to Oriani, 2 April 1795.

24. SHAT B4 112, Dugommier to La Unión, 28 prairial II [16 June 1794].

25. SHAT B4 112, Dugommier to La Unión, 15 messidor II [3 July 1794].

26. La Unión to Dugommier [June–July 1794], in J. Delbar, 'La comte de La Union', *Etudes réligieuses* 47 (1889), pp. 235–54; 48 (1889), pp. 57–85, 278–98, 428–50; quotations on 48, p. 287.

27. Delbrel, 'Notes du Conventionel Delbrel', 27 brumaire III [17 November 1794], *Revue de la Révolution* 5 (1885), p. 53.

Chapter 5 A Calculating People

1. Charles de Secondat de Montesquieu, *Esprit des lois* [1750], in Montesquieu, ,*Œuvres complètes* (Paris: Garnier, 1875), 5, pp. 412–13.

2. Marie-Jean-Antoine-Nicolas Caritat de Condorcet, *Observations . . . sur le vingt-neuvième livre de l'Esprit des lois* [1793], in Condorcet, *Œuvres* (Paris: Didot, 1847), 1, pp. 376–81.

3. For the Committee of Public Safety's pleas to use the new measures, see *RACSP* 9 (30 November 1793), p. 63. For the persistence of the metric diversity into 1794, see AN F12* 208, Min. Aff. Etr. to Comité des Subsistances, 30 nivôse II [19 January 1794].

4. Daryl Hafter, 'Measuring Cloth by the Elbow and Thumb: Resistance to Numbers in France of the 1780s', *Cultures of Control*, ed. Miriam Levin (Amsterdam: Harwood, 2000), pp. 69–79.

5. For an example of the regional diversity of anthropometric measures at the end of the *ancien régime*, see Jean-Baptiste Galley, *Le régime féodal dans le pays de Saint-Etienne* (Paris: Imprimerie de la Loire Républicaine, 1927), pp. 315–16, 326. For examples of anthropometric measures hidden behind seemingly abstract units, see Alfred Antoine Gandilhon, ed., *Département du Cher: Cahiers de doléances du bailliage de Bourges* (Bourges: Tardy-Pigelet, 1910), pp. 768, 770.

6. Witold Kula, *Measures and Men*, trans. R. Szreter (Princeton: Princeton University Press, 1986). The word 'anthropometric' can also mean the measurement of the human body; here, following Kula, it is used to denote measures derived from human needs. See also the prescient article by Marc Bloch, 'Le témoinage des mesures agraires', *Annales d'histoire économique et sociale* 6 (1934), pp. 280–2.

7. Galley, *Régime féodal*, p. 282.

8. For a typical claim of how the creation of uniform and abstract units of land area would improve agricultural productivity, see Bureaux de

Pusy, *AP* 15 (8 May 1790), p. 440. For the most sophisticated recent treatment of agricultural productivity under the *ancien régime*, see Philip T. Hoffman, *Growth in a Traditional Society: The French Countryside, 1450–1815* (Princeton: Princeton University Press, 1996). This study, for all its quantitative virtues, tells us little about agricultural productivity outside 'modern' farms where records were kept, or about the exchanges which dominated early modern transactions. For a contrasting view, see Jean Peltre, *Recherches métrologiques sur les finages lorrains* (Lille: Atelier Reproduction des Thèses, 1977).

9. Young, *Travels*, 2, p. 44–6.

10. Galley, *Régime féodal*, p. 307.

11. For scientific definitions of 'natural', see Maurice Crosland, '"Nature" and Measurement in Eighteenth-Century France', *Studies on Voltaire and the Eighteenth Century* 87 (1972), pp. 277–309.

12. On the theory of the just price, see Raymond de Roover, 'The Concept of the Just Price: Theory and Economic Policy', *Journal of the History of Economic Thought* 18 (1958), pp. 418–34.

13. For Notre-Damme-de-Lisque and such practices throughout *ancien régime* Europe, see Kula, *Measures and Men*, pp. 194–5.

14. Charles Porée, ed., *Département de l'Yonne, Cahiers de doléances du Bailliage de Sens* (Auxerre: Imprimerie coopérative ouvrière 'l'Universelle', 1906), pp. 177–8. For the acknowledgement by the official of the role played by metrical diversity, see Robert Vivier, 'Contribution à l'étude des anciennes mesures du département d'Indre-et-Loire', *Revue d'histoire économique et sociale* 14 (1926), pp. 180–99; 16 (1928), pp. 182–227, especially 14, p. 196.

15. On the *ancien régime* economy, see Judith Miller, *Mastering the Market: The State and Grain Trade in Northern France, 1700–1860* (Cambridge: Cambridge University Press, 1999), pp. 34–6. Also see Steven L. Kaplan, *The Bakers of Paris and the Bread Question, 1700–1775* (Durham, NC: Duke University Press, 1996). On the distinction between the market principle and the marketplace, see Steven L. Kaplan, *Provisioning Paris: Merchants and Millers in the Grain and Flour Trade during the Eighteenth Century* (Ithaca: Cornell University Press, 1984), pp. 47–8, 68–9. See also Kula, *Measures and Men*, pp. 71–8.

16. Borda, *AP* 70 (1 August 1793), pp. 117–18. See also my 'Note on Measures'.

17. Lavoisier, 'Eclaircissements historiques sur les mesures des anciens', n.d., *Œuvres*, 6, p. 703.

18. Condorcet, *Sketch for a Historical Picture of the Progress of the Human Mind*, trans. J. Barraclough (London: Weidenfield & Nicolson, 1955), p. 199. K. M. Baker, 'An Unpublished Essay of Condorcet on Technical Methods of Classification', *Annals of Science* 18 (1962), pp. 99–123; Gilles-Gaston Granger, 'Langage universelle et formalisation des sciences: Un fragment inédit de Condorcet', *Revue d'histoire des sciences* 7 (1954), pp. 197–219. Louis Marquet, 'Condorcet et la création du système métrique décimal', in *Condorcet, mathématicien, économiste, philosophe, homme politique*, eds Pierre Crépel and Christian Gilian (Paris: Minerve,

1989), pp. 52–62. On Condorcet as a political economist, see Emma Rothschild, *Economic Sentiments: Adam Smith, Condorcet, and the Enlightenment* (Cambridge: Harvard University Press, 2001). On the Enlightenment view of rational language, see Michel Foucault, *The Order of Things: An Archeology of the Human Sciences* (New York: Random House, 1970), pp. 78–124.

19. Condorcet, *Mémoires sur les monnaies* (Paris: Baudouin, 1790), p. 3.

20. [Prieur], ATPM, *Notions élémentaires sur les nouvelles mesures*, 1st edn (Paris: Imprimerie de la République, IV [1795]), pp. 1, 3–4.

21. BEP Prieur 4.2.3.2, Prieur, 'Motifs,' n.d. [1794].

22. BEP Prieur 4.2.3.3, Prieur, 'Chez une peuple', n.d. Also Prieur, *Nouvelle instruction sur les poids et mesures*, 2nd edn (Paris: Du Pont, IV [1795–6]), pp. 26–31.

23. ATPM, *Aux citoyens rédacteurs*, pp. 9–10.

24. For the elimination of patois, see Patrice Higonnet, 'The Politics of Linguistic Terrorism and Grammatical Hegemony during the French Revolution', *Social History* 5 (1980), pp. 41–69. The abbé Henri Grégoire, who led the campaign against patois, segued directly from metric uniformity to linguistic uniformity in his famous denunciation of vandalism; see Grégoire, *AP* 96 (31 August 1794), p. 154.

25. Romme, 'Rapport sur l'ère de la République', *PVCIP* 2 (20 September 1793), p. 442. Romme initially named the months after the republican virtues. The seasonal names were proposed by the poet Fabre d'Elgantine; see Gilbert Romme et al., *Calendrier de la République française* (Paris: Imprimerie Nationale, II [1793]).

26. For Lalande's doubts about the ten-day week, see NYPL *KVR 756, Lalande, preface to Cubières, *Le calendrier républicain, poème en deux chants*, 3rd edn (Paris: Mérigot, VII [1798–9]), p. 5.

27. The first proposal for decimal time was read to the National Assembly by Borda, *AP* 53 (5 November 1792), pp. 583–5. See AN F17 1135, Anon., *Rapport sur les questions relatives au nouveau système horaire* (Paris: Baudelot et Eberhart, pluviôse II [January 1794]). J. de Rey Pailhade, 'La montre décimal de Laplace', *Revue chronométrique* 60 (1914), pp. 34–7, 51–6.

28. For metric angles, see BN Le38 2501, Borda et al., *Sur le système général des poids et mesures* ([Paris]: n.p., 1793), sent to CIP on 29 May 1793. Some navigators were enthusiasts for the reform; see Charles-Pierre Claret de Fleurieu, *Voyage autour du monde* (Paris: Imprimerie de la République, VIII [1798]), 4, pp. iii–viii, 1–130. Other navigators were dubious about the changes and doubted sailors would ever accept them; see Jean-François de Galaup La Pérouse, *Voyage de La Pérouse*, ed. L. A. Milet-Mureau (Paris: Imprimerie de la République, 1797), pp. xxx–xxxi.

29. For the decimal log tables, see BI MS883 fol. 38, Condorcet, no title, n.d.; Prony, 'Eclaircissements sur un point d'histoire des tables trigonométriques', *MI* 5 (XII [1803–4]), pp. 67–93; I. Grattan-Guinness, 'Work for the Hairdressers: The Production of Prony's Logarithmic and Trigonometric Tables', *Annals of the History of Computing*

12 (1990), pp. 177–85; Lorraine Daston, 'Enlightenment Calculators', *Critical Inquiry* 21 (1994), pp. 182–202. Proposals were also heard to rationalize the measurement of temperature, which was generally measured on the 80° Réaumur scale at this time. Some proposed that it be decimalized according to the plan of Anders Celsius of Sweden; see AN F17 1135, Cotty to CPM, 30 prairial II [18 June 1794].

30. For military exemptions, see Lavoisier, 'Rapport et projet de décret sur la réquisition des ouvriers', 1793, in *Œuvres*, 6, pp. 665–9. See also Delambre's request to exempt Bellet in *PVCIP* 2 (21 September, 1 October 1793), pp. 452, 520, 527–9.

31. For Lavoisier's attempt to save the Academy, see Lavoisier to Lakanal, 'Observations sur l'Académie des Sciences', 17 July 1793, in *Œuvres*, 6, pp. 621–2. For Borda's attempt to save Lavoisier, see AN F17 4770, Borda and Haüy, 'Extrait de registre des délibérations de la CPM', 28 frimaire II [28 December 1793]; received by the Comité de Sûreté Générale on 2 nivôse II [22 December 1793]; and ignored on 29 frimaire II [18 January 1794]. For the purge, see *PVCIP*, 3, pp. 233–42.

32. Delambre, *Grandeur*, p. 213.

33. BEP Prieur 4.2.1.3, Prieur, 'L'uniformité des poids et mesures', [1794].

34. AN F7 4722, Chalandon and Wibert, 'Aujourd'hui . . . d'Assy', 12 pluviôse II [31 January 1794]. For the sealing of the d'Assy home, see AN F7 4722, Chalandon and Wibert, 'd'Assy', 19 frimaire II [9 December 1793]. For d'Assy's arrest, see AN F7 4722, Comité de Sûreté Générale, 'Geoffroy d'Assy', 6 pluviôse II [25 January 1794].

35. For Delambre's return to Paris, see Delambre, 'Mesure du méridien, dépense faite par Delambre', 26 floréal III [15 May 1795] in Observatoire de Paris, *Longueur et temps*. For Delambre's proof of residence, see AN F12 1289, Borda and Haüy (CPM) to Paré (Min. Int.), 23 frimaire II [13 December 1793]; Min. Int., 'Certificat', 27 frimaire II [17 December 1793]; Min. Int. to Borda, 29 frimaire II [19 December 1793].

36. AN F7 4666, Hevaud and Wibert, 'Delambre', 11 ventôse II [1 March 1794]; also, 13 pluviôse II [1 February 1794].

37. Delambre, *Base*, 1, p. 52.

38. Lagrange's comment reported in Delambre, 'Notice sur la vie et les oeuvres de M. le comte J.-L. Lagrange', in Lagrange, *Œuvres* (Paris: Gauthier-Villars, 1867), 1, p. xl.

39. AN W410, Fouquier, 'Jugement qui condamne', 4 messidor II [22 June 1794]; 'Procès-verbal d'exécution de mort', 21 messidor II [9 July 1794]. Ciffinat, 'Noms', 21 messidor II [9 July 1794]. For an account of the accusations against the Luxembourg prisoners, see Henri Wallon, *Le tribunal révolutionnaire*, 2nd edn (Paris: Plon, 1900), 2, pp. 332–64, esp. pp. 347–9.

40. For further visits of Delambre to Paris, see AN F7 4722, d'Assy, 17 prairial II [5 June 1794]; 23 prairial II [11 June 1794]. Delambre 'Réclamation', 21 nivôse III [10 January 1795], from catalogue of Terry Bodin, *Autographes*, September 2000. For the fast pace of the tragedy, see CUS, Delambre to Cit. Charles (Prof. Physique), 27 vendémiaire III [18 October 1794].

41. Devic, *Cassini IV*, p. 187. See also Wolf, *Observatoire*, pp. 338–46.
42. AN F17 1065, 'Les astronomes de l'Observatoire de la République au CIP', 26 vendémiaire II [17 October 1793]. For their astronomical collaboration, see Cassini IV, Nouet, Villeneuve and Ruelle, 'Extrait des observations', *MA* (1786), pp. 314–17.
43. AAS 1J4, Lalande, 'Journal', p. 64. See also AN F17 1065, Professeurs de l'Observatoire to Convention Nationale, 1 ventôse II [19 February 1794].
44. On the Observatory during the Revolution, see Wolf, *Observatoire*, pp. 353–9. See also Devic, *Cassini IV*, pp. 170–221. For the order evicting Cassini, see Perny, 'Billet', 4 October 1793, in Claude Teillet, 'Cassini IV, Témoin de la Révolution', *Actes du colloque de Clermont: La Révolution dans le Clermontois et dans l'Oise*, 7, 8 October 1989 ([Clermont]: GEMOB, 1990), p. 72.
45. The original accusation against Ruelle has been lost, but is summarized in *PVCIP 4* (27 thermidor II [14 August 1794]), p. 941. For the invitation to Delambre, see *PVCIP 4* (9 fructidor III [26 August 1794]), p. 984.
46. For the Bureau, see Henri Grégoire, *Rapport sur l'établissement du Bureau des Longitudes, séance du 7 messidor III* [25 June 1795], (Paris: Imprimerie Nationale, messidor III [June 1795]).
47. Ruelle begged for release to visit his wife, a longtime servant to the Cassinis, and to see his dying six-year-old son; see AN F7 4775(4), Ruelle to Commission des Admin. Civile, 24 vendémiaire III [15 October 1794]; AN F17 1065, Lalande to CIP, 29 vendémiaire IV [21 October 1795].
48. Jean-Dominique Cassini IV, 'Mon apologie à un de mes confrères' [1795–6], in *Riens qui vaillent* ([Paris]: n.p., 1842), p. 70. Cassini IV did show up for one early meeting of the Bureau on 25 July 1795; see Bigourdan, 'Bureau des Longitudes', (1928), pp. A16–18. For Delambre's efforts on Cassini's behalf, see Delambre, 'Réponse à la note' [1796], in Bigourdan, 'Bureau des Longitudes', (1928), p. A41.
49. Cassini IV, 'Mon apologie' [1795–6], pp. 63–76.
50. Lalande, 'Testament moral', in Aimable, *Lalande*, pp. 50, 53. For the admission of women to the Collège and teaching, see BVCS MS99, Lalande, 'Journal', 1790 and 5 May 1791. For his purple waistcoat, see BYU, Lalande to Piery, 23 July [1788]. For his umbrella, see BYU Delambre folder 8, Delambre to Lalande, 7 May 1792. Lalande stopped writing in his personal diary in 1793–4 in case the Revolutionary police seized his papers; see AAS Dossier Lalande, Synopsis of Lalande, 'Journal, 1756–1807'.
51. Lalande to Amélie, 9 October 1793, in Raspail, 'Lalande', p. 243. On the protection afforded by his atheism, see Lalande in Sylvain Maréchal, *Dictionnaire des athées* (Paris: Grabit, VIII [1799–1800]), p. 227.
52. Delambre, 'Lalande', *Biographie universelle*, p. 613. Du Pont, 'Discours prononcé . . . aux obsèques de Joseph-Jérôme de Lalande', *Moniteur* 103 (6 April 1807), pp. 11–12; see also [Du Pont] to Lalande, 24 August 1793, in AAS Dossier Lalande, Synopsis of Lalande, 'Journal, 1756–1807'. For his risky publications, see BNR Fr 12273, Lalande,

marginal notes in his own 'Eloge de Bailly', *Décade philosophique*, 30 pluviôse III [18 February 1795].

53. Raspail, 'Lalande', p. 243. See also BN Rés 3640, [Lalande], *L'ether, ou l'être suprême élémentaire [et pneumatique]* (Paris: Petits Augustins, 1796).

54. CUS, Lalande to Président du Tribunal Criminel, 8 vendémiaire IV [30 September 1795].

55. CUS, Lalande to [unknown], 25 March 1797. For his progress on his stars, see BYU, Lalande to Piery, 1 messidor II [19 June 1794]; APS, Lalande to Hassler, 12 October 1794. For 41,000 stars, see Lalande, *Bibliographie astronomique*, p. 781.

56. For Delambre's passport, see Municipalité de Bruyères-Libre, 'Certificat . . . Delambre', 17 prairial III [5 June 1795], in Bigourdan, *Système métrique*, p. 134. Delambre apparently had a cousin who emigrated, and this in itself could have been risky for him; see CUS, J.-B.-J. Delambre to [cousin] Delambre, notary, 7 floréal [c. 1800]. For his observations in April–May 1795, see Bigourdan, *Astronomie d'observation*, p. 170.

57. *PVCIP* 6 (19 floréal III [8 May 1795]), p. 187. For Delambre's participation in the calendar reform, see *PVCIP* 6 (20, 29 germinal, 19 floréal III [9, 18 April, 8 May 1795]), pp. 180–8. For Delambre's opinion as to the impossibility of reconciling the two dates, see AOP Z137(2), Delambre (from Bruyères-Libre) to Romme [1795]; Delambre, *Astronomie théorique et pratique* (Paris: Courcier, 1814), 3, p. 696. For background, see Michel Froeschlé, 'Le calendrier républicain correspondait-il à une nécessité scientifique?', *Scientifiques et sociétés pendant la Révolution*, Actes du 114e Congrès National des Sociétés Savantes (Paris: Comité des Travaux Historiques et Scientifiques, 1990), pp. 454–65; Bruno Morando et al., 'Le calendrier républicain', *Astronomie* (June 1989), pp. 269–74.

58. SHAT 3M4, Calon to Méchain, 13 pluviôse III [1 February 1795]. Numa Broc, 'Un musée de géographie en 1795', *Revue d'histoire des sciences* 27 (1974), pp. 37–43; Patrice Bret, 'Le Dépôt Général de la Guerre et la formation scientifique des ingénieurs-géographes militaires en France (1789–1830)', *Annals of Sciences* 48 (1991), pp. 113–57. Berthaut, *Ingénieurs géographes*.

59. For Calon's solicitation of Delambre, see SHAT 3M4, Calon to Delambre (at Bruyères-Libre), 12 brumaire III [2 November 1794]. For Calon's assumption that Delambre was in prison, see Delambre, *Grandeur*, p. 216.

60. For the development of the metric law, see Prieur, *Rapport sur les moyens*, read to the National Convention on 25 fructidor III [11 September 1794], (Paris: Imprimerie Nationale, vendémiaire IV [September–October 1794]). Prieur, *PVCIP* 5 (11 ventôse III [1 March 1795]), pp. 551–63. The law of 1 vendémiaire IV [23 September 1795] declared that the use of the metre would become obligatory throughout Paris in three months, on 1 nivôse IV [22 December 1795]. For a lukewarm defence of decimal time, see Lagrange, *PVCIP* 3 (29 ventôse II [19 March 1794]), pp. 605–6.

61. AN AFII 67 plaq. 496, Prieur, 'Arrêté relatif aux poids et mesures', 12 prairial III [31 May 1795].

62. AOP E2-19, 'Extrait de registre des déliberations du CIP', 18 floréal III [7 May 1795]. See also PVCIP 6 (12 prairial III [31 May 1795]), pp. 244–5, 247.

63. AOP E2-19, Prieur to Delambre, 14 prairial III [2 May 1795].

64. AOP E2-19, Delambre to Méchain, 12 frimaire IV [3 December 1795]. See also, Delambre, *Grandeur*, p. 217.

65. ENPC MS726, Delambre, 'Mesure du méridien', 19 messidor II [9 July 1794]. PVCIP 5 (8 prairial III [27 May 1795]), p. 236. Delambre's colleagues on the various committees spoke up on his behalf. AN F17 1135, ATPM, 'Rapport présenté au CIP', 29 floréal III [18 May 1795]; 'Projet d'arrêté du CIP', 8, 12 prairial III [27, 31 May 1795].

66. On Delambre's equipment, see BMR Tarbé XXI/137, Delambre, 'Etat de dépense pour la mesure de la méridienne', 1 ventôse V [19 February 1797]. His new assistant was named Plessis and had little astronomical experience.

67. On Bourges, see AOP E2-1, Delambre, 'Reprise des opérations', messidor II [June 1795]; KM, Delambre, *Base*, 1, p. 210.

68. Emile Mesle, *Histoire de Bourges* (Roanne: Horvath, 1988), pp. 247–9.

69. On the Bourges pelican, see AOP E2-6, Delambre, 'Registre', p. 279; AOP E2-19, Delambre to Méchain, 12 frimaire IV [3 December 1795].

70. For Delambre's expenses, see BMR Tarbé XXI/137, Delambre, 'Etat de dépense pour la mesure de la méridienne', 1 ventôse V [19 February 1797].

71. For Delambre's plea to Calon for funds, see SHAT 3M4, Calon to Delambre, 26 thermidor III, 14, 22 vendémiaire, 29 brumaire IV [13 August, 6, 15 October, 20 November 1795]. For Delambre's military rank, see AOP MS1033C, Delambre to Méchain, 17 fructidor III [3 September 1795]. For Méchain's military rank, see AOP E2-19, Méchain to Delambre, 8 thermidor III [26 July 1795].

72. AOP E2-19, Delambre to Méchain, 12 frimaire IV [3 December 1795].

73. Christophe Sauvageon, c. 1700, in Bernard Edeine, *La Sologne: Contribution aux études d'ethnologie métropolitaine* (Paris: Mouton, 1974), p. 193.

74. Légier, 'Traditions et usages de la Sologne', *Mémoires de l'Académie Celtique* 2 (1808), p. 206. On the *dindon*, see Sauvageon, c. 1700, in Edeine, *Sologne*, pp. 681–3.

75. For a description of his work from Vouzon to Sainte-Montaine, see Delambre, *Base*, 1, pp. 180–204. See also AOP E2-6, Delambre, 'Registre', pp. 336–61.

76. Delambre, *Base*, 1, pp. 203–6. See also AOP E2-6, Delambre, 'Registre', pp. 312–35.

77. Delambre to [Lavoisier?], quoted in Hélène Richard, 'Dans le cadre de la "meridienne verte"', *Bulletin de la Société Archéologique de Puiseaux* 29 (1999), p. 40.

78. For his accuracy in Sologne, see Delambre, *Grandeur*, p. 235.
79. For Delambre's request for information, see AOP MS1033C, Delambre to [Méchain], 13 fructidor III [3 September 1795].

Chapter 6 *Fear of France*

1. Johann Wolfgang von Goethe, *The Sorrows of Young Werther*, trans. Victor Lange (New York: Rinehart [1774], 1959), p. 53.
2. BUP MS168.1, Méchain to G. Slop, 14 June 1794. For prior contacts between Slop and the French, see BUP MS167.11, Lalande to G. Slop, 21 October 1792; see also BUP MS168.1, Dépôt de la Marine to G. Slop, 2 March 1778. Josepho Slopio de Cadenberg [G. Slop], *Theoriae cometarum* (Pisa: Pizzornius, 1771).
3. BUP MS168.1, Branacci to G. Slop, 20 June 1794. For accounts of this period, see BUP MS168.1, Méchain to G. Slop, 21 June 1794; see also Oriani to Piazzi, 12 November 1794, in *Correspondenza astronomica fra Giuseppe Piazzi e Barnaba Oriani* (Milan: Hoepli, 1874), pp. 31–2.
4. For the Slop family, see F. Menestrina, 'L'astronomo Giuseppe G. Slop e la sua famiglia', *Studi Trentini di scienze storiche* 26 (1947), pp. 3–24, 127–50.
5. BUP MS168.1, Méchain to G. Slop, 2 August 1794.
6. BUP MS168.1, Méchain to G. Slop, 4 October 1794. We can infer that Méchain told G. Slop all this because his subsequent letters assume G. Slop's prior knowledge, including knowledge of the Mont-Jouy results; see BUP MS168.1, Méchain to G. Slop [27 September 1794].
7. BUP MS168.1, Méchain to G. Slop, 4 October 1794. We can infer that Méchain swore G. Slop to silence based on his demand that G. Slop burn his letters and keep them secret. It is unlikely, however, that G. Slop burned any of the letters, since there are almost no breaks in the weekly correspondence.
8. For Méchain's arrival in Genoa, see BUP MS168.1, Méchain to G. Slop, 12 July 1794. Méchain was still lodged at the Albergo del Leon d'Oro on the Piazza de Scolopi in late October; see AOP B4-10, Oriani to Méchain, 22 October 1794.
9. On Genoa in the run-up to the Revolution, see Lalande, *Voyage en Italie*, 2nd edn (Paris: Desaint, 1786), 8, pp. 292–390; René Boudard, *Gênes et la France dans la deuxième moitié du XVIIIe siècle* (Paris: Mouton, 1962), pp. 77–80, 173–6; Pietro Nurra, 'Genova durante la rivoluzione francese: un conspiratore: il patrizio Luca Gentile', *Giornale storico della Liguria*, new series, 4 (1928), pp. 124–31.
10. On Napoleon's visit, see Napoleon, *Correspondance de Napoléon I* (Paris: Plon, 1858), 1, pp. 54–5, 61–4. Théodore Jung, *Bonaparte et son temps, 1769–1799* (Paris: Charpentier, 1883), 2, pp. 438–40; Pietro Nurra, 'La missione del Generale Bonaparte à Genova nel 1794', in *La Liguria nel Risorgimento* (Genoa: Dalla sede del Comitato, 1925); Ramsay Weston Phipps, *The Armies of the First French Republic* (London: Oxford University Press, 1931), 3, pp. 231–2.

11. BUP MS169.15, Tranchot to F. Slop, 16 August 1794. For Méchain's view of Robespierre's fall, see BUP MS168.1, Méchain to G. Slop, 9 August 1794.
12. BUP MS169.15, Tranchot to F. Slop, 25 October 1794.
13. BUP MS169.15, Tranchot to F. Slop, 11 October 1794.
14. BUP MS168.1, Méchain to G. Slop, 2 August 1794.
15. On Tranchot's frustration, see BUP MS169.15, Tranchot to F. Slop [November 1794].
16. BUP MS168.1, Méchain to G. Slop, 9 August 1794.
17. BUP MS168.1, Méchain to G. Slop, 2 August 1794.
18. BUP MS168.1, Méchain to G. Slop, 10 September 1794.
19. BUP MS168.1, Méchain to G. Slop, 25 August 1794.
20. BUP MS168.1, Méchain to G. Slop, 25 August 1794.
21. BUP MS168.1, Méchain to G. Slop, 20 September 1794.
22. BUP MS168.1, Méchain to G. Slop, 20 September 1794.
23. BUP MS168.1, Méchain to G. Slop [27 September 1794].
24. For Méchain's discussion of his appointment, see BUP MS168.1, Méchain to G. Slop, 4 October 1794. Méchain heard from Calon in SHAT 3M4, Calon to Méchain, 13 vendémiaire II [4 October 1794]. For further discussion, see BUP MS168.1, Méchain to G. Slop, 18 October 1794.
25. BUP MS168.1, Méchain to G. Slop, 18 October 1794.
26. BUP MS168.1, Méchain to G. Slop, 4 October 1794.
27. For Méchain's relay of the news from Lalande, see BUP MS168.1, Méchain to G. Slop, 4 October 1794. For Lalande's expectation that Méchain would soon return, see AAS 1J4, Lalande, 'Journal', pp. 69–70.
28. BUP MS168.1, Méchain to G. Slop, 18 October 1794.
29. BUP MS168.1, Méchain to G. Slop, 20 September 1794.
30. For Tranchot's preparations, see BUP MS169.15, Tranchot to F. Slop, 25 October 1794.
31. AOAB Cart. 88, Méchain to Oriani, 2 October 1794. On Oriani's trip to Paris in 1786, see A. Mandrino, G. Tagliaferri and P. Tucci, eds, *Un viaggio in Europa nel 1786: Diario di Barnaba Oriani* (Milan: Olschki, 1994), pp. 145, 147, 152, 154.
32. Boudard, *Gênes*, p. 308. AOP C6-6, Méchain, 'A Gênes sur la terrasse de l'hôtel du grand Cerf', 16 October 1794. Méchain was aware that Zach had observed from the same hotel terrace in 1787; see AOAB Cart. 88, Méchain to Oriani, 27 December 1794. Zach also visited Méchain while he was Italy; see BI MS2042 fol. 323, Delambre, 'Zach, Journal de Gotha' [1810s].
33. AOP B4-10, Oriani to Méchain, 22 October 1794.
34. SHAT 3M4, Calon to Méchain, 13 pluviôse III [1 February 1795].
35. For Méchain's proposals for delay, including his pendulum experiment; see BUP MS168.1, Méchain to G. Slop, 2, 9, 16 August 1794. Méchain worried that the refraction project would be read as a 'pretext' for deferring his return to France; see AOAB Cart. 88, Méchain to Oriani, 5 November 1794.
36. BUP MS168.1, Méchain to F. Slop, 25 October 1794.

37. BUP MS168.1, Méchain to G. Slop, 22 November 1794.
38. AOAB Cart. 88, Méchain to Oriani, 20 November 1794.
39. BUP MS168.1, Méchain to G. Slop, 6 December 1794.
40. BUP MS168.1, Méchain to G. Slop, 17 January 1795.
41. BUP MS168.1, Méchain to G. Slop, 7 February 1795.
42. For a contemporary nosology of melancholy, see Philippe Pinel, *A Treatise on Insanity*, trans. D. D. Davis (Sheffield: Todd [1801], 1806), pp. 136–49, 222–34.
43. For Méchain's claim that one of his Paris colleagues – perhaps Lalande – had 'moved heaven and earth' to prevent him from getting the pendulum, see AOAB Cart. 88, Méchain to Oriani, 12 February 1795. For the sale of the circle, see AOAB Cart. 88, Méchain to Oriani, 12, 28 February, 5, 14 March, 2 April 1795.
44. On refraction, see AOAB Cart. 88, Méchain to Oriani, 27 December 1794; see also BUP MS168.1, Méchain to G. Slop [27 September 1794]. Méchain conducted these observations in the company of Francesco Pezzi, captain of the engineering corps and mathematics professor at the University of Genoa. Pezzi was also involved in anti-oligarchy politics. Nurra, 'Genova durante la rivoluzione', pp. 124–31. The Barcelona results were published as Méchain, 'Eclipses de soleil et occultations d'étoiles', *Ephemerides astronomicae anni 1795* (Milan: Galeatrium, 1794), pp. 81–3. Méchain discovered a typographical error in the published Mont-Jouy latitude, which should have read 41°21'45", not 41°21'25". See AOAB Cart. 88, Méchain to Oriani, 29 November 1794, 2 April 1795.
45. Esteveny left Genoa on 22 December 1794 and arrived in Paris on 25 January 1795; see AOAB Cart. 88, Méchain to Oriani, 27 December 1794; BUP MS168.1, Méchain to G. Slop, 17 January 1795; SHAT 3M5, Calon to Méchain, 13 pluviôse III [1 February 1795]; AN F12 1288, Ginguieni, 'Rapport', 22 pluviôse III [10 February 1795].
46. AOAB Cart. 88, Tranchot to Oriani, 27 December 1794, 2 April 1795.
47. BUP MS169.15, Tranchot to F. Slop, 4 April 1794.
48. BUP MS168.1, Méchain to G. Slop, 4 October 1794.
49. SHAT 3M5, Calon to Méchain, 13 pluviôse III [1 February 1795].
50. BUP MS168.1, Méchain to G. Slop, 4 April 1795.
51. For Tranchot's tracking of the war, see BUP MS169.15, Tranchot to F. Slop, 8 November 1794, 21 February 1795, 24 ventôse 1795 [14 March 1795], 4 April 1795. Phipps, *Armies of the First French Republic*, 3, pp. 234–6.
52. Napoleon to Oriani, 5 prairial IV [24 May 1796], in Napoleon, *Correspondance*, 1, pp. 491–2. On Slop's son, see Menestrina, 'Slop'.
53. On the assumption that Méchain was coming north, see Delambre, *Base*, 1, p. 61.
54. BUP MS168.1, Méchain to G. Slop, 4 October 1794. On Calon's plans for Tranchot, see Berthaut, *Ingénieurs géographes*, p. 164.
55. AOP E2-19, Méchain to Lalande, 13 thermidor III [31 July 1795].
56. SHAT 3M4, Calon to Méchain, 28 thermidor III [15 August 1795]. For Méchain's complaints to Delambre, see AOP E2-19, Méchain to Delambre, 8 thermidor III [26 July 1795].

57. For Méchain's arrival in Perpignan, see BML, Méchain to Calon, 13 fructidor III [30 August 1795]. For the story of the carriage, see Delambre, *Grandeur*, p. 216.

Chapter 7 Convergence

1. William Blake, *Milton: Book the First*, in *The Portable Blake*, ed. Alfred Kazin (New York: Penguin Books, 1974), p. 433.
2. AOP E2-19, Méchain to Delambre, 8 thermidor III [26 July 1795].
3. AOP MS1033c, Delambre to [Méchain], 13 fructidor III [3 September 1795].
4. AOP E2-19, Méchain (Perpignan, Estagel) to Delambre, 12–23 vendémiaire IV [4–15 October 1795].
5. For Méchain's horror of publication, see Delambre, *Notice historique sur M. Méchain*, p. 30.
6. AOP E2-19, Delambre to Méchain, 12 frimaire IV [3 December 1795].
7. AOP E2-19, Delambre to Méchain, 12 frimaire IV [3 December 1795].
8. AOP E2-19, Delambre to Méchain, 12 frimaire IV [3 December 1795].
9. On Delambre's passage through Paris, see AAS 1J4, Lalande, 'Journal' (22 frimaire IV [13 December 1795]), p. 73; see also, AN F17 3702, Delambre to Min. Int., 28 frimaire IV [19 December 1795]. On the Academy elections, see *PVCIP*, 6, pp. 832–6. Méchain was appointed to the astronomy section, along with Lalande, in the first round, but only after some backroom jockeying; the four other places in astronomy, elected in a second round of voting by those appointed in the first round, were taken by Le Monnier, Pingré, Messier and Cassini IV. This meant that Delambre was elected to the mathematics section in the second round, though by his own admission this was not the appropriate section for him. See Delambre, 'Lui-même'.
10. CUS, Lalande to Delambre, 2 pluviôse IV [22 January 1796].
11. For Delambre's observations at Dunkerque, see Delambre, *Base*, 2, pp. 249–96. AOP E2-16, Delambre, 'Registre: Méridien de France, Partie du nord', vol. 2. AOP MS1033c, Delambre, 'Hauteur du pole à Dunkerque', [1795–6]. Delambre to [Janvier?], 2 pluviôse IV [22 January 1796], in Adolphe Desboves, *Delambre et Ampère* (Amiens: Hecquet, 1881), pp. 32–3.
12. Delambre, *Grandeur*, pp. 239–40.
13. Delambre, *Grandeur*, p. 279.
14. AOP MS1033c, Delambre to Méchain, 17 fructidor III [3 September 1795].
15. AOP E2-1, Delambre, 'Reprise des opérations', 9 fructidor III [26 August 1795].
16. AOP E2-19, Méchain to Delambre, 12 floréal IV [1 May 1796]. Méchain asked Lalande about Delambre's observations in March, especially for his data on other stars; see AOP E2-19, Méchain to Lalande, 11 ventôse IV [1 March 1796].
17. AOP E2-19, Delambre to Méchain, floréal IV [mid-May 1796]. In this letter, Delambre admits to having received Méchain's letter of 12 floréal IV [1 May 1796].

18. AOP E2-19, Delambre to Méchain, floréal IV [mid-May 1796].
19. SHAT 3M4, Calon to Méchain, 26 thermidor III [13 August 1795]. For costs, see AOP E2-19, Méchain to Delambre, 12–23 vendémiaire IV [4–15 October 1795].
20. AOP E2-19, Méchain to Lalande, 1 ventôse IV [1 March 1796]. On the delicate subject of Tranchot's pay, see SHAT 3M4, Calon to Méchain, 9 nivôse IV [30 December 1795]. On the inflated salaries, see BL, 'Procès-verbaux', 22 pluviôse IV [11 February 1796].
21. BML, Méchain to Calon, 13 fructidor III [30 August 1795].
22. Méchain got rid of Décuve first, and Bouvet that winter; see BA MS P2100, Méchain to Calon, 5 ventôse IV [24 February 1796].
23. For Forceral, see AOP E2-19, Méchain to Delambre, 12–23 vendémiaire IV [4–15 October 1795].
24. AOP E2-19, Méchain to Lalande, 3 brumaire IV [25 October 1795].
25. Gaston Jourdanne, *Contribution au folk-lore de l'Aude* (Paris: Maisonneuve, 1973), p. 28; for *sinagries*, see p. 22. On modern local folklore, see ADAu D°2441, Roger Antoni, *Rennes-le-château ou la mystification biblique* (Montreal: Chez l'auteur, n.d.).
26. This description of typical French inns of the region and time comes from Young, *Travels*, 1, pp. 40–50. Original quote reads 'nidus' instead of 'nest'.
27. On mountain travel in the region in this period, see Serge Biffaud, *Naissance d'un paysage: La montagne pyrénéenne à la croisée des regards, XVIe-XIXe siècle* (Tarbes: Mauran, 1994).
28. For Alaric and Tauch, see AOP E2-19, Méchain to Lalande, 1 ventôse IV [1 March 1796].
29. For the preparations for the Perpignan baseline, see AOP E2-19, Méchain to Lalande, 1 frimaire IV [22 November 1795], 28 prairial IV [16 June 1796]. For permission to have the army engineers build the pyramids, see SHAT 3M4, Calon to Méchain, 26 messidor IV [14 July 1796]; 3M5, Calon to Méchain, 5 thermidor IV [23 July 1796]. Nor would his colleagues let Méchain conduct the pendulum measurements, as he still wished to do; see SHAT 3M4, Calon to Méchain, 28 fructidor III [14 September 1795]. On Tranchot's use of the chain and his role in setting up the pyramid markers, see AOP E2-19, Méchain to Delambre, 9 germinal VI [29 March 1798].
30. AOP E2-19, Méchain to Lalande, 1 ventôse IV [1 March 1796].
31. AOP E2-19, Méchain to Lalande, 12 floréal IV [1 May 1796].
32. For Méchain's plea to do extra work, see AOP E2-19, Méchain to Delambre, 21 frimaire V [11 December 1796].
33. On the intermediate latitudes, see AOP E2-19, Delambre to Méchain, 8 fructidor IV [25 August 1796]. AOP E2-19, Borda to Méchain, 14 messidor IV [2 July 1796].
34. AOP E2-19, Méchain to Delambre, 3 ventôse V [21 February 1797].
35. AOP E2-19, Borda to Méchain, 14 messidor IV [2 July 1796].
36. AOP E2-19, Borda to Méchain, 12 frimaire VI [2 December 1797].
37. AOP E2-19, Delambre to Méchain, 17 floréal V [6 May 1797]. See also AOP E2-19, Delambre to Méchain, 5 nivôse V [25 December 1796].

38. AOP E2-19, Delambre to Méchain, 12 frimaire IV [3 December 1795].
39. AOP E2-19, Méchain to Lalande, 12 floréal IV [1 May 1796].
40. AOP E2-19, Delambre to Méchain, 8 fructidor IV [25 August 1796]. See also AOP E2-6, Delambre, 'Registre', p. 20. For the tower at Morlac, see Delambre, *Base*, 1, p. 74; also ADC 1L629, Delambre, 'Morlac', 26 fructidor IV [12 September 1796]; Baudat, 'Aux citoyen président' [1796–7]; Delambre to Baudat, 16 vendémiaire VI [7 October 1797]. ADC 1L646, 'Renseignements sur les édifices non aliénés servant à l'exercice des cultes', X [1801–2]; 'Cher: Morlac', in *Cahiers de doléances, région Centre* (Dijon: Coloradoc, 1995), 3, pp. 32–3; BMR Tarbé XXI/137, Delambre, 'Etat de dépense pour la mesure du méridienne', 1 ventôse V [19 February 1797].
41. For Arpheuille, see Delambre, *Grandeur*, p. 236.
42. For Delambre's observations at Evaux, see AOP E2-19, Delambre to Borda, 4 pluviôse V [23 January 1797]; AOP MS1033c, Delambre, 'Evaux', V [winter 1796–7]; AOP E2-6, Delambre, 'Registre', pp. 55–98. Delambre, *Base*, 1, pp. 250–1.
43. On Cassini III's error, see DLSI MSS420A, Delambre to [Calon], 27 frimaire V [17 December 1796]. For Delambre's willingness to share data with Méchain, see AOP E2-19, Delambre to Méchain, 27 ventôse V [17 March 1797].
44. AOP E2-19, Delambre to Méchain, 27 ventôse V [17 March 1797].
45. AOP E2-19, Delambre to Borda, 4 pluviôse V [23 January 1797].
46. SHAT 3M5, Calon to Méchain, 14 ventôse V [4 March 1797]. For Delambre's pleas for funds, see DLSI MSS420A, Delambre to [Calon], 27 frimaire V [17 December 1796]; SHAT 3M5, Calon to Delambre, 19 frimaire, 5 nivôse, 10, 17 pluviôse, 2 ventôse V [9, 25 December 1796, 29 January, 5, 20 February 1797]. Granting military rank to the expedition team had not eased their burden. Officers could only collect army rations when operating near the frontiers, and Delambre had been obliged to admit that he was as far from the frontiers as it was to possible to be. AOP E2-19, Delambre to Méchain, 5 nivôse V [25 December 1796]. CUS, Raynal Rouby to Delambre, 25 ventôse V [13 March 1797].
47. CUS, Lalande to Delambre, 17 March 1797. See also CUS, Lalande to Delambre, 1 March 1797.
48. AOP E2-19, Delambre to Méchain, 5 nivôse V [25 December 1796]. For Delambre's use of his own funds, see AOP E2-19, Delambre to Borda, 4 pluviôse V [23 January 1797].
49. On theories of geology in this period regarding the Auvergne, see Harladur Sigurdsson, *Melting the Earth: The History of Ideas on Volcanic Eruptions* (New York: Oxford University Press, 1999); Carl Gustaf Bernhard, *Through France with Berzelius: Live Scholars and Dead Volcanoes* (Oxford: Pergamon Press, 1985); G. Poulett Scrope, *Memoir on the Geology of Central France* (London: Longman, 1827), pp. 120–3.
50. Johannel (Commissionnaire de pouvoir exécutif du canton d'Herment) to Boutarel, 9 messidor V [27 June 1797], in Philippe Bourdin, *Le*

Puy-de-Dôme sous le Directoire: Vie politique et esprit public (Clermont-Ferrand: Mémoires de l'Académie des Sciences et Belles-Lettres de Clermont-Ferrand, 1990), p. 268. On Delambre's stay in Herment, see Delambre, *Base*, 1, p. 79; AOP E2-6, Delambre, 'Registre', pp. 115–17; Ambroise Tardieu, *Histoire de la ville, du pays et de la baronnie d'Herment* (Marseille: Laffitte, [1866]).

51. Frances Gostling, *Auvergne and Its People* (New York: Macmillan, 1911), pp. 112–14. On the debate over Delambre's signals in Bort, see AOP E2-19, Delambre to CPM, 8 messidor V [26 June 1797]; Delambre, *Base*, 1, pp. 79–80; Bort-les-Orgues, *Histoire et tourisme* (Bort-les-Orgues: OTSI, 1985), pp. 45, 85–97.

52. Delambre to Lalande [thermidor V; August 1797], in Lalande, *Bibliographie astronomique*, pp. 780–1.

53. On farming in the Cantal region in the eighteenth century, see Albert Rigaudière, *Etudes d'histoire économique rurale au XVIIIe siècle* (Paris: Presses Universitaires de France, 1965), pp. 12–14; Jonathan R. Dalby, *Les paysans cataliens et la Révolution française, 1789–1794* (Clermont-Ferrand: Université de Clermont-Ferrand, 1989), p. 74.

54. On mistaking Delambre for a sorcerer, see Lakanie, 'Mémoires', 22 floréal V [11 May 1797], from Archives Départementales du Cantal, Aurillac, reprinted on the website of Gilbert Coudon, http://perso.infonie.fr/gilbert.coudon/delambre.htm. For general views of sorcery in the region, see Gostling, *Auvergne*, pp. 239–47. Cit. Legrand, *Voyage fait en 1787 et 1788 dans la ci-devant haute et basse Auvergne* (Paris: Imprimerie des Sciences et Arts, III [1795]), 1, pp. 30–6.

55. Legrand, *Voyage*, 1, p. 100. On the request that Delambre report on provincial views of the metric system, see AN F12* 2103, ATPM to Delambre, 3 fructidor III [20 August 1795]. Delambre could not supply receipts to Calon because the workmen were illiterate; see DLSI MSS420A, Delambre to [Calon], 27 frimaire V [17 December 1796].

56. On Delambre in the rain storm, see Delambre, *Base*, 1, p. 86. On the roads of the Auvergne, see Franck Imberdis, *Le réseau routier de l'Auvergne au XVIIIe siècle* (Paris: Presses Universitaires de France, 1967).

57. On lizards in the region, see Young, *Travels*, 1, p. 28.

58. On Delambre's sighting of 'Méchain's' signal, see Delambre, *Base*, 1, p. 82. Delambre said he had not heard from Méchain since spring; see AOP E2-19, Delambre to CPM, 8 messidor V [26 June 1797]. For Méchain's instructions to Tranchot, see AOP E2-19, Méchain to Tranchot, 21 messidor V [9 July 1797].

59. AOP E2-6, Delambre, 'Registre', p. 157, quoting from Virgil, *Aeneid*, 3, line 714.

60. On the tower of Rodez, see Louis Bousquet, 'Contribution à l'histoire du clocher de la cathédrale de Rodez', *Procès-verbaux de la Société des Lettres, Sciences, et Arts de l'Aveyron* 29 (1924), pp. 146–50; Jacques Bousquet, 'La cathédrale de Rodez sous la Révolution: Philosophie du vandalisme', *Revue du Rouergue* (1989), pp. 177–205.

61. On Delambre's departure for Paris, see BL, 'Procès-verbaux', 4 jour comp. V, 9 vendémiaire VI [20, 30 September 1797]. Tranchot left Rodez for Paris one day after Delambre.

Chapter 8 Triangulation

1. Laurence Sterne, *The Life and Opinions of Tristram Shandy*, ed. Graham Petrie (London, Penguin Books [1759], 1987), p. 161.
2. For the history of Mme Méchain's move into the Observatory apartments, see Méchain to Jeaurat, 17 floréal VIII [7 May 1800], in Bigourdan, 'Bureau des Longitudes' (1930), pp. A86–8, A96–7. See also BL, 'Procès-verbaux', 27 messidor III [15 July 1795], 17 nivôse IV [7 January 1796], 2 floréal IV [6 April 1796].
3. For Mme Méchain's surprise that her husband had not written to his colleagues, see AOP E2-19, Mme Méchain to Delambre, 20 thermidor V [7 August 1797].
4. AOP E2-19, Méchain to Delambre, 25 germinal V [14 April 1797].
5. AOP E2-19, Méchain (Pradelles) to Delambre, 20 brumaire IV [10 November 1797].
6. AOP E2-19, Delambre to Borda, 4 frimaire VI [24 November 1797].
7. On the amateur astronomers of Carcassonne, see ADAu 3K3, Anon., 'Eloge pour Fabre', *Mémorial du département de l'Aude* 58 (1810), pp. 37–45. [Isidore] Dougados, *Le dernier Juge-Mage . . . de Carcassonne: Raymond de Rolland* (Carcassonne: Pomiés, 1856). Gilbert Larguier et al., *Cahiers de doléances audois* (Carcassonne: Association des Amis des Archives de l'Aude [1989]). On the Saint-Vincent church, see Juliette Costeplane, 'L'église Saint-Vincent de Carcassonne au XVIIIe siècle', *Bulletin de la Société d'Etudes Scientifiques de l'Aude* 81 (1981), pp. 95–102. Swindburne, *Travels*, 2, pp. 367–72.
8. For Méchain's mixed recollection of this period of his life, see Méchain to Rolland, 9 nivôse X [30 December 1801], in [Isidore] Dougados, 'Lettres de l'astronome Méchain à M. Rolland', *Mémoires de la Société des Arts et des Sciences de Carcassonne* 2 (1856), pp. 74–130, especially p. 123. Méchain lodged with 'Cit. Comelerand, traiteur, près la Comédie ou la porte des Cazernes'; see AOP E2-19, Méchain to Delambre, 13 vendémiaire VII [4 October 1798].
9. For Méchain's request to extend his portion, see AOP E2-19, Méchain to Delambre, 27 ventôse V [17 March 1797].
10. AOP E2-19, Méchain to Delambre, 25 germinal V [14 April 1797].
11. For Calon's promise not to show the data to anyone else, see SHAT 3M4, Calon to Méchain, 28 thermidor III [15 August 1795]. For Méchain's plea for more time to work on his data, see AOP E2-19, Méchain to Delambre, 3 ventôse V [21 February 1797].
12. AOP E2-19, Méchain to Delambre, 3 ventôse V [21 February 1797].
13. AOP E2-19, Méchain to Delambre, 27 ventôse V [17 March 1797].
14. AOP E2-19, Méchain to Delambre, 3 ventôse V [21 February 1797].
15. Larguier, *Cahiers audois*, pp. 185–9. Jean-Louis Bonnet, 'Le Cabardes: Population et ressources aux XVIIe et XVIIIe siècles', *Bulletin de la*

Société d'Etudes Scientifiques de l'Aude 79 (1979), pp. 89–96; André David, *Les Montagnes Noires* (Carcassonne: Bonnafous, 1924).

16. Méchain to Rolland, 2 brumaire VI [23 October 1797], in Dougados 'Lettres de Méchain', p. 74.

17. Méchain to Rolland, 4 frimaire VI [24 November 1797], in Dougados, 'Lettres de Méchain', p. 78. On the storm, see Claude-Joseph Trouvé, *Description générale et statistique du Département de l'Aude* (Paris: Didot, 1818), p. 148.

18. Dr Astruc, in Claude Ignace Barante, *Essai sur le département de l'Aude* (Carcassonne: Gareng, brumaire XI [October–November 1802]), 100–2.

19. For the preparations for the northern baseline, see AN F17 1135, CIP, 23 thermidor III [10 August 1795]; CIP to Commission des Travaux Publiques, 10 vendémiaire IV [2 October 1795]. Also ENPC MS724, ATPM to Prony, 27 thermidor III [14 August 1795]. CUS, Lalande to [CPM], 4 fructidor III [21 August 1795]. For the tree pruning, see Delambre, *Base*, 1, pp. 84–5; AOP E2-6, Delambre, 'Registre', 2, pp. 168–92. For Delambre's final measurements, see AOP E2-6, Delambre, 'Registre', 2, pp. 168–92; E2-19, Delambre to Buache, 29 pluviôse VI [17 February 1798]. See the write-up of the baseline measurement in *Décade philosophique* 27 (30 prairial VI [18 June 1798]).

20. For the new rulers, see Borda, 'Expériences sur les règles qui ont servi à la mesure des bases', in Delambre, *Base*, 3, pp. 313–36. For the various logbooks of the baseline measurement, see AOP E2-3, Tranchot, 'Base de Melun', floréal–prairial VI [May–June 1798]; AOP E2-2, Pommard, 'Base de Melun', floréal–prairial VI [May–June 1798]. Anon., 'Method Employed Between Melun and Lieusaint in France', *Philosophical Magazine* 1 (1798), pp. 269–74.

21. Humboldt to Zach, 3 June 1798, in E.-T. Hamy, ed., *Lettres américaines d'Alexandre de Humboldt, 1798–1807* (Paris: Guilmoto [1905]), pp. 2–4.

22. AOP E2-19, Méchain to Delambre, 9 germinal VI [29 March 1798].

23. For Mme Méchain's promise to Delambre, see AOP E2-19, Mme Méchain to Delambre, 12 floréal VI [1 May 1798].

24. AOP E2-19, Mme Méchain to Delambre, 11 prairial VI [30 May 1798].

25. Note appended by Delambre to AOP E2-19, Méchain to Delambre, 7 brumaire VII [28 October 1798]. Delambre confessed that he never understood the enmity Méchain had for Tranchot; although he noted that Tranchot had admitted to once having uttered a few *vivacités* to Méchain.

26. For Delambre's claim that the Méchain family finances had improved with the Revolution, see Delambre, *Notice historique sur M. Méchain*, p. 31.

27. AOP E2-19, Méchain to Delambre, 19 fructidor VI [5 September 1798].

28. AOP E2-19, Mme Méchain to Delambre, 15 fructidor VI [1 September 1798].

29. Méchain did not leave Carcassonne for Rodez until early messidor [mid-June], well after he would have heard of his wife's plans to come and fetch him; see AOP E2-19, Fabre to Delambre, 3 thermidor VI [21

July 1798]. Mme Méchain met her husband in Rodez on 7 July, and left him on 18 August in Rieupeyroux, during which interval he conducted his observations at those two stations. Méchain was also accompanied by Agoustenc for these stations. See Méchain's logbook in AOP E2-10.

30. Méchain to Rolland, 5 jour comp. VI [21 September 1798], in Dougados, 'Lettres de Méchain', p. 85.

31. For Delambre's promise to bring Méchain back to Paris, see Delambre, 'Méchain', *Astronomie au dix-huitième*, p. 761.

32. Méchain to Rolland, 19 fructidor VI [5 September 1798], in Dougados, 'Lettres de Méchain', p. 82.

33. AOP E2-19, Méchain to Delambre, 19 fructidor VI [5 September 1798]. On the mistaking of a signal for a guillotine, see AOP E2-19, Fabre to Delambre, 15 frimaire VI [1 September 1798]. For the suspicions of local administrators about the signals, see AN F17 1135, Commissaire de Directoire de Lacaune to Admin. du dépt. du Tarn, 2 fructidor V [19 August 1797]; Commissaire de Directoire de Montredon to Admin. du dépt. du Tarn, 5 fructidor V [22 August 1797]. Méchain in Delambre, *Base*, 1, pp. 306–7.

34. Jeanne Bardou (p. 18), in Remy Cazals, *Autour de la Montagne Noire au temps de la Révolution, 1774–1799* (Carcassonne: CLEF, 1989), pp. 11–20. Bardou was recounting what her two servants had told her about the surveyor Pierre de Lalande (no known relation to Jérôme Lalande).

35. Méchain to Rolland, 5 jour comp. VI [21 September 1798], in Dougados, 'Lettres de Méchain', p. 87. On the militia to guard at the Montalet site, see ADT L210, Commissaire de Lacaune, 29 thermidor VI [16 August 1798]; L266, idem, 11, 14, 17, fructidor 2 jour comp. VI [28, 31 August, 3, 18 September 1798].

36. AOP E2-19, Méchain to Delambre, 19 fructidor VI [5 September 1798].

37. On the Perpignan road, see ADPO L1105, Saussine, 'Dépt. de Pyrénées-Orientales, Ponts et Chaussées', 18 nivôse VI [7 January 1798]; Jacques Freixe, 'Tracé de la voie Domitienne de Narbonne à Gerona', *Revue d'histoire et d'archéologie du Roussillon* 2 (1901), pp. 387–405; 3 (1902), pp. 202–16, 285–317; Pierre Ponsich, 'Les voies antiques du Roussillon et de la Cerdagne', in *Les routes du sud de la France: De l'antiquité à l'époque contemporaine*, Colloque Montpellier, 1985 (Paris: CTHS, 1985), pp. 91–105.

38. Swindburne, *Travels*, 1, p. 2.

39. For the Perpignan baseline, see AOP E2-6, Delambre, 'Registre', pp. 255–318; AOP E2-4, Pommard, 'Base de Perpignan', thermidor–fructidor VI [August–September 1798]; AOP E2-5, Tranchot, 'Base de Perpignan', thermidor–fructidor VI [August–September 1798]. For the results, see *ASPV* 21 (21 brumaire VII [11 November 1798]), p. 492. The match of the two measurements was mostly due to compensating errors; see Levallois, *Mesurer la terre*, p. 64.

40. ENPC MS724, Delambre to Prony, 4 jour comp. VI [20 September 1798].

41. CUS, Lalande (Gotha) to Delambre, 19 August 1798.

42. AOP E2-19, Méchain to Delambre, 19 fructidor VI [5 September 1798].

43. For the mountain road to Saint-Pons, described as 'extremely difficult' in the eighteenth century, see J. Sahuc, *Saint-Pons: Dictionnaire topographique et historique* (Paris: Res Universis, 1993), pp. 12–13. But Méchain admitted he was located only a three-hour ride up from town; see AOP E2-19, Méchain to Delambre, 29 fructidor VI [15 September 1798].

44. AOP E2-19, Méchain to Delambre, 9 vendémiaire VII [30 September 1798].

45. While Delambre waited for Méchain's answer, he worked with Bellet and Tranchot to take geodetic measurements of the angle between Mont Alaric and Saint-Pons from Narbonne on 17–18 vendémiaire VI [8–9 October 1798]; see AOP E2-6, Delambre, 'Registre', p. 319.

46. AOP E2-19, Méchain to Delambre, 13 vendémiaire VII [4 October 1798].

47. For Méchain's procrastinations, see AOP E2-19, Méchain to Delambre, 19, 29 fructidor, 4 jour comp. VI, 9, 11, 13, 15, 22, 28 vendémiaire, 1, 7 brumaire VII [5, 15, 20, 30 September, 2, 4, 6, 13, 19, 22, 28 October 1798].

48. AOP E2-19, Méchain to Delambre, 19 fructidor VI [5 September 1798].

49. AOP E2-19, Fabre to Delambre, 15 fructidor VI [1 September 1798].

50. AOP E2-19, Méchain to Delambre, 29 fructidor VI [15 September 1798].

51. AOP E2-19, Méchain to Delambre, 4 jour comp. VI [20 September 1798].

52. AOP E2-19, Méchain to Delambre, 19 fructidor VI [5 September 1798].

53. Méchain to Rolland, 19 fructidor VI [5 September 1798], in Dougados, 'Lettres de Méchain', p. 83.

54. AOP E2-19, Méchain to Delambre, 4 jour comp. VI [20 September 1798].

55. AOP E2-19, Méchain to Delambre, 11 vendémiaire VII [2 October 1798].

56. For the offer of the directorship of the Observatory, see Delambre, 'Méchain', *Astronomie au dix-huitième*, p. 763.

57. Though Lalande expected their arrival on 15 November, they did not in fact arrive in Paris until 17 November; see BL, 'Procès-verbaux', 24 brumaire VI [14 November 1798].

Chapter 9 The Empire of Science

1. Louis-Sébastien Mercier, *Satires contre les astronomes* (Paris: Terrelonge, XI, 1803), pp. 16–17: 'On réserve sur-tout des cadeaux magnifiques / Pour ceux qui s'embrouillent dans les mathématiques, / Aux triangles liés avec d'énormes frais, / Qu'ils soient faux ou trompeurs, ne renoncent

jamais . . . / Qu'ont fait les nouveaux poids, les nouvelles *mesures*? / A tous nos bons vieillards apporter des tortures. / Pour boire une chopine, auner un long ruban, / Ou réduire et changer les heures d'un cadran, / L'arc du *méridien* était-il nécessaire? / On peut très-bien auner sans mesurer la terre; / Et si ce haut calcul n'est point exempt d'erreur, / Briser longue habitude est mauvaise rigueur.'

2. Méchain to Rolland, 6 frimaire VII [26 November 1798], in Dougados, 'Lettres de Méchain', pp. 88, 90. The President of the Directory was the nation's chief executive.

3. Méchain to Rolland, 6 frimaire VII [26 November 1798], in Dougados, 'Lettres de Méchain', p. 90.

4. For the claim to be the first international scientific meeting, see Maurice Crosland, 'The Congress on Definitive Metric Standards, 1798–1799: The First International Scientific Conference?' *Isis* 60 (1969), pp. 226–31. The best account of the meeting is by Thomas Bugge, published in Danish and translated immediately into German under his direction as *Reise nach Paris in den Jahren 1798 und 1799*, trans. Johann Nicolaus Tilemann (Copenhagen: Brummer, 1801). An English version appeared soon after, with the sections on the metric system excised, as *Travels in the French Republic*, trans. John Jones (London: Phillips, 1801). A recent re-edition of this translation has been issued, with some of the metric material reinserted as *Science in France in the Revolutionary Era*, ed. Maurice Crosland (Cambridge: MIT Press, 1969).

5. CUS, Laplace to [Delambre], 5 pluviôse VI [24 January 1798].

6. Laplace to Delambre, 10 pluviôse VI [29 January 1798] in Yves Laissus, 'Deux lettres de Laplace', *Revue d'histoire des sciences* 14 (1961), pp. 285–96. For oblique references to the debate over the proposition of Laplace and the objections of Borda, see *ASPV* 1 (1, 5 pluviôse VI [20, 24 January 1798]), pp. 334–5.

7. Talleyrand to Cisalpine Republic [Piedmont], 5 July 1798, in Kula, *Measures and Men*, p. 271.

8. For Napoleon's election to the Academy on 25 December 1797, see M. E. Maindron, 'Bonaparte, Membre de l'Institut Nationale', *Revue scientifique de la France* 1 (1881–2), pp. 321–38. For Napoleon and science, see the excellent study by Joachim Fischer, *Napoleon und die Naturwissenschaften* (Stuttgart: Steiner, 1988).

9. *Clef du Cabinet*, 3 floréal VI [22 April 1798], in François-Alphonse Aulard, ed., *Paris pendant la réaction thermidorienne et sous le Directoire* (Paris: Cerf, 1898–1902), 4, p. 596.

10. *Narrateur universel*, 24 frimaire IV [14 December 1797], in Aulard, *Thermidorienne*, 4, pp. 490–1.

11. For Jefferson's switch of latitude preference, see Jefferson to William Short, 26 July, 26 September 1790, in Jefferson, *Papers*, 17, pp. 281, 528. For the early French optimism regarding Jefferson and American participation, see *Loi relatif à l'établissement de nouvelles mesures pour les grains* (Paris: Imprimerie Nationale, 1790), p. 9. For Jefferson's views on the British-French coordination, see Jefferson to Rittenhouse, 20, 30 June 1790, in Jefferson, *Papers*, 16, pp. 542–3, 587–8. For the US

Senate, see Committee on Weights and Measures, 5 April, 18 December 1792, in Joseph Gales, ed., *Debates and Proceedings in the Congress of the United States* (Washington, DC: Gales and Seaton, 1834–56), 3, pp. 117–18, 621–2. As time went on, Jefferson became even more convinced that the French had acted selfishly in choosing the meridian. See Jefferson to Doctor Patterson, 11 September, 10 November 1811, in Thomas Jefferson, *The Writings of Thomas Jefferson*, ed. H. A. Washington (New York: Derby and Jackson, 1859), 6, p. 11; 13, pp. 95–108.

12. For Dombey's mission, see *PVCIP* 3 (11, 13, 21, 29 frimaire, 5 nivôse II [1, 3, 11, 19, 25 December 1793]), pp. 54, 64, 136, 197, 211. *RACSP* 9 (21, 26 frimaire II [11, 16 December 1793]), pp. 321, 436–7. Yves Laissus, 'Note sur le deuxième voyage et la mort de Joseph Dombey', *Comptes rendus du 94e Congrès National des Sociétés Savantes, Histoire des sciences* (Paris: Bibliothèque Nationale, 1970), pp. 61–79.

13. [Joseph Fauchet], *Joseph Fauchet, Minister Plenipotentiary of the French Republic near the United States to Mr. Randolph, Secretary of the United States, 15 Thermidor 2nd year of the French Republic (2d August 1794, Old Style)* (Philadelphia: Fenno [1794]). See also APS, Fauchet to Rittenhouse, 10 September 1794.

14. Fauchet to Randolph, 15 thermidor II [2 August 1794], in 3rd Congress, 2nd Session, 8 January 1795, no. 60, in Walter Lowrie and Walter S. Franklin, eds, *American State Papers: Documents* (Washington, DC: Gales and Seaton, 1834), 1, pp. 115–16. For public accounts in the US, see Anon., 'Foreign Literature, France', *American Monthly Review* (February 1795), pp. 195–8.

15. Fauchet to Commissaire du Dépt. des Relations Extérieures, 26 nivôse II [15 January 1795], in *Annual Report of the American Historical Association* 2 (1903), pp. 544–6. For Washington's addresses, see Washington to Congress, 8 January 1790 and 25 October 1791, in Gales, *Congress of the United States*, 1, pp. 968–72; 3, pp. 11–16. For Washington's actions in favour of Fauchet's proposals, see Washington to Congress, 8 January 1795, in Gales, *Congress of the United States*, 4, p. 809.

16. For the end of the Congressional debate, see Harrison, 'Weights and Measures', House of Representatives, 14, 19 May 1796, in Gales, *Congress of the United States*, 5, pp. 1376–83, 1405. For Fauchet's views, see 'Mémoire sur les Etats-Unis d'Amérique', 24 frimaire IV [15 December 1795], in *Annual Report of the American Historical Association* 1 (1936), pp. 85–131. John J. Reardon, *Edmund Randolph: A Biography* (New York: Macmillan, 1974), pp. 307–15.

17. On Jefferson's deference to Congress on this matter, see Jefferson to John Rutherford, 25 December 1792, in Jefferson, *Papers*, 24, p. 783.

18. Cuthbert Clarke, *A New Complete System of Weights and Measures* (Edinburgh: Author, 1789), p. 6. See the long list of British attempts at measurement standardization between 1200 and 1730 appended to Miller, *Speeches*, pp. 29–40. For one eighteenth-century attempt, see House of Commons, *A Report from the Committee Appointed to Enquire into the Original Standards of Weights and Measures in this Kingdom* (London:

Whiston, 1758). For Scotland, see Lord John Swinton, *A Proposal for Uniformity of Weights and Measures in Scotland* (Edinburgh: Elliot, 1779). For other complaints about fraud, see Hubert Hall and Freida Nichols, *Select Tracts and Tables: Books Relating to English Weights and Measures, 1100–1742* (London: Offices of the Society, 1929), pp. 47–51. For an overview of the history of English weights and measures, see R. D. Connor, *The Weights and Measures of England* (London: Her Majesty's Stationery Office, 1987).

19. Miller, 6 February 1790, *Speeches*, pp. 17–18.

20. For the pendulum 'yard', see John Whitehurst, 'An Attempt Toward Obtaining Invariable Measures', n.d., in *The Works of John Whitehurst* (London: Dent, 1792), p. iv. For an economist, see James Stueart, 'A Plan for Introducing an Uniformity of Measures over the World' [written c. 1760, first printed in 1790], in Stueart, *Works* (London: Cadell, 1805), 5, pp. 379–415.

21. Renerus Budelius, *De Monetis* (1591), quoted in Miller, *Speeches*, p. 52. At Miller's request, Talleyrand inserted language allowing for 'a location to be determined' into the law passed by the National Assembly; see Miller, *Speeches*, pp. xiv–xv.

22. John Rotheram, *Observations on the Proposed Plan for an Universal Standard of Weights and Measures in a Letter to Sir John Sinclair, M.P.* (Edinburgh: Creech, 1791), p. 10. In the same fashion, the British expected the Americans to follow their lead in measurement standards; see Rotheram, *Observations*, pp. 35–6. George Skene Keith, *Tracts on Weights, Measures, and Coins* (London: Murray, 1791).

23. For Blagden, see BLL Add MS33272, ff. 97–8, Charles Blagden to Joseph Banks, 8 September 1791. For British mockery, see 'New System of Weights and Measures', *The Times* (London), 1 October 1798. For the parliamentary debate, see T. C. Hansard, ed., *The Parliamentary History of England from the Earliest Period to the Year 1803*, vol. 28 (1789–91), cols 297, 315–23, 874–5, 876–9.

24. For a German sceptical of the reform, see Frederich Johann Lorenz Meyer, *Fragmente aus Paris* (Hamburg: Bohn, 1798), 2, pp. 265–83. The CPM sent sample metre sticks to German commercial cities like Hamburg so that they might draw up conversion tables, see p. 279. On the German savants' objections to the meridian project, see pp. 268–9.

25. For Delambre's recommendation of Tranchot for the German map project, see CUS, Delambre to [Prony?], 7 fructidor IX [25 August 1801]. Within days, Tranchot had the job and was on his way; SHAT 3M401, Tranchot to Gen. Andeossy (Dépôt de la Guerre), 8 fructidor IX [15 September 1801].

26. Delambre, *Rapport historique*, 9, pp. 77–8.

27. For Jérôme-Isaac Méchain's role as aide to Nicolas-Auguste Nouet, the former monk who had served as Cassini IV's assistant under the *ancien régime*, see Jean-Joseph Marcel et al., *Histoire scientifique et militaire de l'expédition française en Egypte* (Paris: Dénain, 1830–6), 4, p. 57. For P.-F.-A. Méchain's connection to Nouet, see KBD NKS1304, Méchain to Bugge, 1 vendémiaire X [23 September 1801]. For the geographic

aspect of the expedition, see Anne Godlewska, 'The Napoleonic Survey of Egypt: A Masterpiece of Cartographic Compilation and Early Nineteenth-Century Fieldwork', *Cartographia* 25 (1988), pp. i–xiii, 1–171, especially pp. 17–22; Anne Godlewska, 'Map, Text and Image: The Mentality of Enlightened Conquerors. A New Look at the *Description de l'Egypte*', *Transactions of the Institute of British Geographers* 20 (1995), pp. 5–28; Ghislain Alleaume, 'Entre l'inventaire du territoire et la construction de la mémoire: L'oeuvre cartographique de l'expédition d'Egypte', in *L'expédition d'Egypte, Une entreprise des lumières, 1798–1801*, ed. Patrice Bret (Paris: Hachette, 1998), pp. 279–94; Antoine Tramoni, 'Du plan terrier de la Corse à la carte de l'Egypte: La géographie des militaires', in *Bonaparte et les îles méditerranéennes et l'appel de l'Orient*, Actes du Colloque d'Ajaccio, 29–30 May 1998, *Cahiers de la Méditerranée* 57 (1998), pp. 87–99. Yves Laissus, *L'Egypte, Une adventure savante* (Paris: Fayard, 1998); Muséum National d'Histoire Naturelle, *Il y a 200 ans, Les savants en Egypte* (Paris: NATHAN, 1998).

28. Fourier, in E. F. Jomard, 'Description de Syène et de ses cataractes', *Description de l'Egypte* 1 (Paris, 1821), p. 121. See also Nouet, 'Observations . . . haute Egypte', *Décade égyptienne* 3 (IX [1800–1]), pp. 15–16.

29. Marcel, *Expédition française en Egypte*, 5, p. 53. For Nouet's reports, see 'Rapport . . . Alexandrie', 'Mémoire . . . du Kaire, 11 messidor VII [29 June 1799]', 'Rapport . . . styles, 21 messidor VII [9 July 1799]', 'Position . . . points de l'Egypte', 'Observations . . . haute Egypte', 'Position . . . des pyramides', *Décade égyptienne* 1 (VII [1798–9]), pp. 165–82; 2 (VIII [1799–1800]), pp. 129–58, 226–31, 267–71; 3 (IX [1800–1]), pp. 7–27, 101–10.

30. For the Nilometre, see Girard to Le Père, 30 thermidor VII [17 August 1799], in *Courier de l'Egypte* 37 (29 fructidor VII [15 September 1799]), p. 3; Pierre-Simon Girard, 'Résumé des deux mémoires sur le nilomètre de l'Ile d'Eléphantine et l'ancien coudée des égyptiens', *Mémoires de l'Institut des Sciences Morales et Politiques*, 7 vendémiaire X [29 September 1801], pp. 63–74. Marcel, *Expédition française en Egypte*, 4, pp. 494–7. For the *ancien régime* suppositions about a connection between the pyramids and measures, see Jean-Sylvain Bailly, *Histoire de l'astronomie ancienne* (Paris: De Bure, 1781), pp. 77–85, 167–76. Also Paucton, *Métrologie*, pp. 6–7; and Paucton, *Explication de l'hiéroglyphie du grand principe de la nature consacré dans les pyramides d'Egypte* (Paris: Desaint, 1781), pp. 345–7. Laplace himself had endorsed the idea; see Laplace, 'Mathématiques', *Ecoles Normales*, 5, p. 203. For new speculations about the connection based on data gathered 1798–1801, see E. Jomard, *Mémoire sur le système métrique des anciens égyptiens* (Paris: Imprimerie Royale, 1817).

31. AN Colonies C8A 59, Thomassin de Farret, 'Projet pour le commerce des colonies', 1752. See also, AN Colonies A23, Conseil Supérieur de Louisiane, 1 April 1715; Arrêté du Conseil Supérieur, 19 July 1725; Arrêté du Conseil d'Etat de Roy, 1 March 1744. AN Colonies C88 12, Jean-François Pierre, 'Mémoires pour l'établissement d'un poids public', 1767.

32. For La Pérouse, see La Pérouse, *Voyage de La Pérouse*, pp. 286–7. Fleurieu, ed., *Voyage autour du monde*, 4, pp. iii–viii, 1–130. For the Borda circle on the Entrecasteaux expedition, see Elisabeth-Paul-Edouard Rossel, ed., *Voyage d'Entrecasteaux* (Paris: Imprimerie Impériale, 1808), 1, pp. 33, 594–9; and volume 2.

33. Méchain to Rolland, 6 frimaire VII [26 November 1798], in Dougados, 'Lettres de Méchain', p. 90.

34. On the Panthéon at this moment, see [W. F. Blagdon], *Paris as It Was and as It Is* (London: Baldwin, 1803), 2, p. 140; see also Meyer, *Fragmente aus Paris*, 1, pp. 166–82.

35. Warmé, *Delambre*, p. 29.

36. AOP E2-8, Delambre, 'Méridienne, Partie du nord, Observations', [1798–9]. Delambre tried (and failed) to hire young Pomar [sic] as an assistant at the Observatory on 23 April 1798; see Bigourdan, 'Bureau des Longitudes' (1928), pp. A17–18. Delambre to Humboldt, 22 January 1801, in Humboldt, *Briefe aus Amerika*, p. 120.

37. AOP E2-19, Méchain to Delambre, 6 nivôse VII [26 December 1798]. Méchain believed that the Paris latitude measurements should belong to him alone, to compensate him for Delambre's greater contributions elsewhere. But Delambre quietly insisted on conducting observations in parallel; see Delambre, *Grandeur*, p. 222.

38. AOP E2-19, Méchain to Borda, 7 nivôse VII [27 December 1798].

39. AOP E2-19, Méchain to Borda, 7 nivôse VII [27 December 1798].

40. On Méchain's avoidance of his colleagues, see Delambre, 'Méchain', *Astronomie au dix-huitième*, p. 762.

41. Danish State Archives F6 1087, Bugge to Danish Secretary of State, 17 November 1798, translation from the Danish by Arne Hessenbruch. This letter was written the day before Delambre and Méchain returned to Paris. Bugge also considered Lalande the greatest egotist and charlatan in all astronomy; see BLL Add MS8099, Bugge to Banks, 19 November 1798. Bugge was a long-time correspondent of Méchain, and respected him. He was also impressed by the precision of the repeating circle, when Méchain finally demonstrated its use to him. Bugge, *Science in France*, pp. 205–6. The Academy noted the arrival of the foreign savants, but did not meet with them in formal session until 28 November 1798, two weeks after Delambre and Méchain returned; *ASPV* 1 (16 vendémiaire, 6 frimaire VII [7 October, 26 November 1798]), pp. 476, 496–7. For the Academy's ban on publication, see *ASPV* 1 (16 prairial VI [4 June 1798]), p. 403.

42. Zach to Lalande, 28 May 1799, in Bigourdan, *Système métrique*, pp. 240–1.

43. Zach to Lalande, 3 December 1798, in Bigourdan, *Système métrique*, p. 240.

44. AAS Dossier Delambre, Margelay (Monthéçon, dépt. de l'Allier) to Delambre, 18 floréal VII [7 May 1799].

45. *Décade philosophique* 15 (30 pluviôse VII [18 February 1799]), p. 372.

46. For Delambre's presentation, see ENPC MS726, Trallès, Van Swinden, Laplace, Legendre and Delambre, 'Observations rélatives à la mesure

de la méridienne . . . du Citoyen Delambre', 14 pluviôse VII [2 February 1799].

47. On Laplace's deal with Méchain, see AOP E2-19, Méchain to Delambre, 18 pluviôse VII [6 February 1799].

48. For Méchain's Paris results, see AOP E2-19, Méchain to Delambre, 17 pluviôse VII [15 February 1799]; see also, Delambre, 'Méchain', *Astronomie au dix-huitième*, p. 762.

49. Méchain to Rolland, 6 germinal VII [26 March 1799], in Dougados, 'Lettres de Méchain', p. 93. For Méchain's presentation, see ENPC MS726, Legendre, Van Swinden, Prony, Méchain, Ciscar, Laplace and Trallès, 'Tableau des observations pour la calcul de la méridien . . . du citoyen Méchain', 21 ventôse VII [11 March 1799]. The International Commission did reject a few of Méchain's observations; see KM, Delambre, *Base*, 1, p. 501.

50. For the setting aside of Méchain's latitude data for the Fontana de Oro, see Delambre's note [c. 1805–10] at the end of AOP E2-9, Méchain, 'Registre des observations astronomiques fait au Mont-Jouy et Barcelone en 1792 et 1793'.

51. For the criticism of Delambre's results, see Delambre, *Grandeur*, p. 223. The Commissioner was Bugge, so the comment may have come earlier.

52. For Delambre's new calculation methods, see Delambre, *Méthodes analytiques pour la détermination d'un arc du méridien*, 11 germinal VII [31 March 1799]; with Legendre, *Méthode pour déterminer la longueur exacte du quart du méridien*, 9 nivôse VII [29 December 1798] (Paris: Crapelet, VII [1799]).

53. On Borda's funeral, see Bougainville, 'Borda', *Décade philosophique* 16 (10 ventôse VII [28 February 1799]), pp. 434–8.

54. Méchain to Rolland, 22 floréal VII [11 May 1799], in Dougados, 'Lettres de Méchain', p. 101.

55. Méchain to Rolland, 22 floréal VII [11 May 1799], in Dougados, 'Lettres de Méchain', p. 101.

56. [Blagdon], *Paris as It Was*, 2, p. 141. This comment almost certainly comes from Lalande. For Laplace's suppositions about the figure of the earth, see Laplace, 'Mémoire sur la figure de la terre' (1783, pub. 1786), in ,*Œuvres complètes de Laplace* (Paris: Gauthier-Villars, 1878–1912); 11, pp. 3–32. Laplace, 'Mathématiques', *Ecoles Normales*, 5, p. 213. For the tentative justification for taking the intermediate latitudes, see the subtle change between the draft and final version of the letter sent by Lavoisier in AAS Lavoisier 1228(36), Lavoisier to Méchain, 6 October 1793. In this letter Lavoisier first claims that the intermediate latitudes will be 'more useful' for identifying 'irregularities in the figure of the world' than for 'establishing the new measures', and then amends this to note simply that they will 'serve to identify' the irregularities. Lavoisier also censored a reference to the fact that it was Laplace who had insisted on the intermediate latitude measurements. On Boscovich's expedition, see John L. Heilbron, *Weighing Imponderables and Other Quantitative Science around 1800* (Berkeley: University of California Press, 1993), pp. 226–9.

57. For the eccentricity data, see Laplace, *Traité de mécanique céleste* (Paris: Crapelet, VII [1799]), 2, pp. 138–45; Laplace in BL, 'Procès-verbaux', 19 frimaire VIII [10 December 1799].

58. The French first learned of this precious metal through Lalande; see Lalande, 'Lettre sur un métal appellé platine', *Journal des sçavans* (January 1758), pp. 46–59. Donald McDonald and Leslie B. Hunt, *History of Platinum and Its Allied Metals* (London: Johnson Matthey, 1982), pp. 179–93. W. A. Smeaton, 'Platinum Sales Problems in the French Revolution: Janety Writes to Sir Joseph Banks', and 'Bertrand Pelletier, Master Pharmacist: His Report on Janety's Preparation of Malleable Platinum', *Platinum Metals Review* 12 (1968), pp. 64–6; 41 (1997), pp. 86–8. On the budget for platinum, see AAS Lav. 167, Lavoisier, 'Etat des ouvriers', [1793]. For discovery of the short-weighting – the shipment was 78 *marcs* short of the contracted 500 *marcs* – see Guyton, 'Rapport', *ASPV* 1 (21 thermidor VII [8 August 1799]), pp. 610–13. For the secret scramble for platinum in the final days before the conference, see Danish State Archives F6 1087, Bugge to Danish Secretary of State, 17 November 1798.

59. Méchain to Rolland, 22 floréal VII [11 May 1799], in Dougados, 'Lettres de Méchain', p. 103. This was Lenoir's second, more accurate, comparator; see Delambre, 'Mètre définitif', *Base*, 3, pp. 691–8; Taylor, *Pleasure to Profit*. For the physical standards, see C. Wolf, 'Recherches historiques sur les étalons de l'Observatoire', *Annales de chimie et de physique*, 5th series, 25 (1882), pp. 5–112. The other three platinum bars were given to the Academy of Sciences, the Conservatoire des Arts et Métiers and the *cadastre* survey. Copper bars of the same length were distributed to all the different *départements* and the major ministries. Iron bars were distributed to the visiting savants.

60. [Laplace], 'Discours', 4 messidor VII [22 June 1799], in Delambre, *Base*, 3, pp. 581–9.

61. Van Swinden, in Delambre, *Base*, 3, p. 648. The members of the International Commission had been assured that the leading role at the ceremony would be given to the foreign savants. Méchain, as senior expedition leader, had expected to enjoy the honour of presenting results, but he bowed to political exigencies. Besides, he had never been one for public speeches. KBD NKS1304, Méchain to Bugge, 10 brumaire VIII [1 November 1799]. Méchain to Rolland, 22 floréal VII [11 May 1799], in Dougados, 'Lettres de Méchain', p. 103.

62. On the delay in returning the finished bar, see *ASPV* 2 (16 nivôse, 6 germinal VIII [6 January, 27 March 1800]), pp. 76, 128.

63. Baudin, 'Réponse', 4 messidor VII [22 June 1799], in Delambre, *Base*, 3, p. 651, quotation taken from J.-J. Rousseau.

64. Coquebert de Montbret to Alexandre Brongniart, 1794, in Isabelle Laboulais-Lesage, *Lectures et pratiques de l'espace: L'itinéraire de Coquebert de Montbret, savant et grand commis d'Etat, 1755–1831* (Paris: Champion, 1999), p. 299.

65. For a list of government pamphlets, see AN F12 1237, ATPM, 'Etat des différens ouvrages', 10 vendémiaire IV [2 October 1795]. For graphs,

see [Prieur], *Echelles graphiques pour la comparaison de l'aune de Paris avec le mètre* (Paris: Imprimerie de la République, thermidor III [July–August 1795]). Scores of privately printed guides helped citizens learn the new measures; see Cit. Bonnin, *Vocabulaire étymologique des poids et mesures de la République française* (Paris: Fournier, VII [1799]); Pierre Periaux, *Tableaux comparatifs des mesures républicaines avec les anciennes* ([Rouen]: n.p., VII [1799]); C. F. Martin, *Le régulateur universel des poids et mesures* (Paris: Guyot, 1807). For a quasi-official national almanac, see *Le manuel républicain* (Paris: Didot, VII [1799]), pp. 76–82. For a paper dial-up calculator, see BNR Estampes IA mat 3a, Leblond, 'Cadrans logarithmiques adaptés aux poids et mesures', 16 pluviôse VII [4 February 1799]. For playing cards, see BNR Estampes Kh383 no. 227, Bézu, 'Jeu de 52 cartes historiques' ([Egalité-sur-Marne]: n.p. [1792]). For the marble metre, see Fernand Gerbaux, 'Le mètre de marbre de la rue Vaugirard', *Bulletin de la Société Historique du VIe Arrondissement de Paris* (1904), pp. 1–72. For the blind instructor, see AN F17 1237, ATPM, 'Tableaux', III–IV [1794–6].

66. François Gattey, *Tables des rapports des anciennes mesures agraires avec les nouvelles*, 2nd edn (Paris: Michaud, 1810), p. 6. For a conversion table for Paris, see Min. Int., *Tables de comparaison entre les mesures* (Paris: Imprimerie de la République, IX [1800–1]). For the frustrations of local administrators translating local measures, see ADSM L259, 'Rapport au département par la CPM', 26 ventôse VI [16 March 1798]; ADSM L260, Seine-et-Marne, 'Registre des séances du Commission Temporaire des Poids et Mesures', 30 pluviôse VI–25 pluviôse VII [18 February 1798–13 February 1799]; ADSM, François de Neufchâteau (Min. Int.), *Instruction sur les nouvelles mesures pour les terrains* (Paris: Imprimerie de la République, fructidor VI [August–September 1798]).

67. ATPM, *Avis instructif sur la fabrication des mesures de longueur à l'usage des ouvriers* (Paris: Imprimerie de la République, III [1795]). For the shortage of rulers, see AN F12 7637, 'Rapport sur le nombre de mètres à envoyer dans chaque des sections de Paris' [1795]; AN F17 1237, ATPM, 'Etat des mesures linéaires entrées en magasin', 14 brumaire IV [5 November 1795]. Ultimately, the cost of making enough rulers for all France was expected to come to 11 million *francs*. The government budgeted less than 2 per cent of that; see BEP Prieur 4.4.6.2, [Prieur], 'Aperçu des dépenses de l'établissement des nouvelles mesures' [1794–5]. The difference was supposed to be made up through the sale of the rulers by private sub-contractors. For attempts to spur production, see AN F12 1289, CPM to Paré (Min. Int.), 18 ventôse II [8 March 1794]; see also AN F17 1237, ATPM, 'Tableau sommaire des engagements contractés', 12 thermidor IV [30 July 1796]. For a request to use an abbé's home to manufacture metre sticks, see BEP Prieur [no number], Feras et Cornu to Prieur, 21 pluviôse II [9 February 1794].

68. For attempts to mass produce metre sticks, see AN F12 1310, 'Extrait des Registres du Comité de Salut Public', 27 floréal III [16 May 1795]; AN F12 1311, ATPM to [Atelier de Perfectionnement], 8 floréal IV [26

April 1796]. For more on the mass production of guns and metre sticks, see Ken Alder, *Engineering the Revolution: Arms and Enlightenment in France, 1763–1815* (Princeton: Princeton University Press, 1997), pp. 253–91. Etienne Lenoir was among those who received an award for the design for his machine; see AN F4 2556, ATPM, 'Comptes', IV [1795–6]. For the inaccuracy of the rulers, see Meyer, *Fragmente aus Paris*, 2, pp. 279–80.

69. For customer preferences, see Dupin, 'Rapport', in Aulard, *Thermidorienne*, 4 (25 February 1798), pp. 556–7; see also 5 (30 December 1798), pp. 98–9; also, 5, pp. 108–9, 287, 477–8, 576, 579, 632. For police frustration, see AN F12* 215, ATPM to Min. Police, 9 messidor IV [27 June 1796]; AN F17 1135, Min. de Police Générale to Min. Int., 21 vendémiaire VI [12 October 1797]; ADSe VD* 429, Bureau Central de Paris, *Avis mesures de capacité pour les liquides*, 22 brumaire VIII [13 November 1799] (Paris: Lottin, VIII [1799]). For the snafu over enforcement between the Min. Police and various administrative units, see ADSe VD* 2421, 2486, 2073, 2075, 4037, 4065, which date from the years IV through VI [1795–8].

70. Meyer, *Fragmente aus Paris*, 2, pp. 282–3, emphasis added.

71. For law expanding the use of the new measures, see Min. Int., *Proclamation du directoire exécutif*, 28 messidor VII [16 July 1799]. For the frustrations of police inspectors, see François-Alphonse Aulard, ed., *Paris sous le Consulat* (Paris: Cerf, 1903), 1 (November–December 1799), p. 65; 2 (12 September 1801), p. 521. For the warnings about cheating in restaurants and fine grocers, see *Almanach des gourmands* 7 (1809), pp. 196–8.

72. ATPM, *Aux citoyens rédacteurs*, pp. 5–6. See the original complaint in *Feuille du cultivateur* 38 (9 messidor III [27 June 1795]), pp. 227–8.

73. For the Ecole Normale, see Laplace, 'Mathématiques', in *Ecoles Normales*, 5, pp. 201–19. For the mixed reception of the metric system in Revolutionary public schools, see Y. Marec, 'L'arithmétique révolutionnaire à Rouen (1789–99)', *Etudes normandes* 3 (1980), pp. 69–83.

74. ATPM, *Aux citoyens rédacteurs*, 1, pp. 18–19.

75. François de Neufchâteau (Min. Int.), 'Emploi des nouvelles mesures', 12 fructidor V [29 August 1797], in Nicolas-Louis François de Neufchâteau, *Recueil des lettres circulaires, instructions, discours et autres actes publics emanés du Cen. François de Neufchâteau* (Paris: Imprimerie de la République, VII [1798–9]), 1, pp. xlii–xliv.

76. Emmanuel Pérès, *Rapport . . . relative aux peseurs publics*, 21 vendémiaire VIII [12 October 1799] ([Paris]: n.p. [1799]). For the law on the Bureaux des Poids et Mesures of 27 brumaire VII [17 November 1798], see Bigourdan, *Système métrique*, pp. 186–7. The law was amplified and extended to all major market towns on 7 brumaire IX [29 October 1800], and again on 16 June 1808; see Désiré Dalloz, ed., *Jurisprudence générale* (Paris: Bureau de la Jurisprudence Générale, 1845–70), 35, pp. 983–5. For the analogy to poison, see Monseignut, *Opinion sur le projet . . . concernant l'établissement des peseurs publics*, 23 fructidor VII [9 September 1799] (Paris: Imprimerie de la République, VII [1799]).

77. On the use of troops to police the market, see A.-B.-J. Guffroy, *Avis civique contre un projet libérticide* (Paris: Everat, vendémiaire VII [September–October 1798]), p. 12. For a defence of the Paris Bureau, see Brillat, Binot and Pelletier, *Mémoire des citoyens nommés pour administrer les Bureaux de Poids Public du département de la Seine* (Paris: Bailleul [1799]). Also, Brillat, Binot and Pelletier, *Réponse des Administrateurs du poids public . . . aux calomnies de Joseph Guffroi* ([Paris]: n.p., fructidor VII [August–September 1799]). Brillat had been lobbying for the contract for several years; see also BEP Prieur 4.5.10, Chef de la 4e division du Min. Int. to Bureau des Poids et Mesures, 13 pluviôse V [1 February 1797]; Brillat to Min. Int., 16 frimaire VI [6 December 1797].

78. On physicians, see Vincent-Jean-Paul Biron, *Rapport fait à la Société de Médicine de Paris, 21 et 27 pluviôse X sur l'application des nouveaux poids et mesures* [10, 16 February 1802] (Paris: Brasdor et Pelletier, X [1801–2]). On notaries, surveyors and accountants, see Reveillière-Lépeaux et al., 7 pluviôse IV [27 January 1796], in Antonin Debidour, ed., *Recueil des actes du Directoire-Exécutif* (Paris: Imprimerie Nationale, 1910–17), 1, p. 492; Aulard, *Thermidorienne* 5 (August–September 1798), p. 99; Min. Int. to Admin. Centrales des Dépts., 21 brumaire VII [11 November 1798], in François de Neufchâteau, *Recueil des lettres*, 1, pp. 273–5. On the national legislators, see AN F12* 210, ATPM to Conseil des Anciens, 14 nivôse IV [4 January 1796]. For the package labelled in old units, see Louis Marquet, 'Anciens mesures, anciens poids', *Amis du vieux Saint-Etienne* 36 (1957), p. 9.

79. For the shifts within the artillery, see SHAT 4c3/2, F.-M. Aboville, 'Mémoire', 1 nivôse IV [21 December 1795]; Gen. Drouân, 'Mémoire', 3 vendémiaire V [24 September 1796]; Min. Guerre to Comité Central d'Artillerie, 19 vendémiaire X [10 October 1801]; Chief Inspector of Revenue to Gen. Songis, 25 fructidor XIII [11 September 1805]; Comité Central d'Artillerie, 'Observations', 29 March 1806. In 1822, the artillery had yet to take up metric conversion; see Anon., 'Mémoire', 1822.

80. On Napoleon's use of the old units, see BN Piece 8-D3 MON-36, Musée de l'Histoire de France (Paris), *Le mal de changer: Les français et la révolution métrique*, Exposition, Archives Nationales, 1 June–31 August 1995 (Paris: Presses Artistiques, 1995), p. 10.

81. APS, Lalande to [Fabroni], 16 December 1801. For the consultations with Delambre and Laplace, see *Moniteur* 41 (11 brumaire IX [2 November 1800]), p. 157; law of 13 brumaire IX [4 November 1800], in Bigourdan, *Système métrique*, pp. 190–1.

82. Sabatier et al., 5 brumaire VII [27 October 1799], in Maindron, 'Bonaparte', p. 326. See also, McDonald and Hunt, *Platinum*, pp. 181–2.

83. Lalande, in Claretie, *Empire*, p. 234. For Laplace's proclamation identifying the reform of weights and measures as proof of the republicanism of the new régime, see Laplace to Admin. Centrales et Municipales, 30 brumaire VIII [21 November 1799], in François de Neufchâteau, *Recueil des lettres*, 3, p. 103. For Laplace's willingness to abandon the

nomenclature, which he blamed on Prieur, see Laplace to Chaptal (Min. Int.), 3 February 1804, in Bigourdan, *Système métrique*, p. 192.

Chapter 10 The Broken Arc

1. George Sand, *Winter in Majorca*, trans. Robert Graves (Chicago: Cassandra [1956], 1978), p. 29. Graves's translation.
2. For Méchain's astronomy, see Méchain to Chaptal, in *Moniteur* 98 (28 messidor IX [17 July 1801]), p. 1232. Also Méchain, *MC* (May 1800), pp. 290–311. For his mood, see Méchain to Rolland, 18 fructidor VII [4 September 1799], 16 messidor VIII [5 July 1800], in Dougados, 'Lettres de Méchain', pp. 105–9, 115–17.
3. For the insults to Méchain behind his back, see Bugge, *Travels*, pp. 247–8; Delambre, 'Méchain', *Astronomie au dix-huitième*, p. 765.
4. For Méchain's retreat, see Delambre, 'Méchain', *Astronomie au dix-huitième*, p. 763.
5. For Delambre's expectation that the *Base* would be published in three volumes and finished within a year, see AAS Dossier Delambre, Delambre to Petit-Genest, 20 prairial VII [18 June 1799]. Méchain knew that his role in writing the *Base* would be minimal; see KBD NKS1304, Méchain to Bugge, 2 brumaire IX [24 October 1800], 1 vendémiaire X [23 September 1801].
6. Delambre and Méchain cooperated on commissions to judge astronomical work, see *ASPV* 2 (11 germinal VIII [1 April 1800], 21 brumaire X [12 November 1801]), pp. 129, 429–30. For Delambre and Napoleon, see Delambre, 'Lui-même'. For Méchain and Napoleon, see KBD NKS1304, Méchain to Bugge, 2 brumaire IX [24 October 1800].
7. Méchain to Rolland, 18 floréal IX [8 May 1801], in Dougados, 'Lettres de Méchain', p. 120. At one point the bureaucrats of the Bureau of Longitudes would not even forward letters to Méchain which were addressed to him as 'Director'; see KBD NKS1304, Méchain to Bugge, 2 brumaire IX [24 October 1800]. In fact, Delambre was appointed 'administrator' of the Bureau, rather than its president, so his stay may have been legitimate. And he served for a little over a year, not two years as Méchain alleged.
8. KBD NKS1304, Méchain to Bugge, 1 vendémiaire X [23 September 1801]. For squabbles over firewood and supplies, see BL, 'Procèsverbaux', 19 vendémiaire X [11 October 1800]. For Méchain's threat to resign, see Méchain to Rolland, 18 floréal IX [8 May 1801], in Dougados, 'Lettres de Méchain', p. 120.
9. For Méchain's sense of being wronged, see KBD NKS1304, Méchain to Bugge, 10 brumaire VIII [1 November 1799]. Delambre discovered Méchain's knowledge of these events when he in turn became Méchain's scientific executor in 1805; see Delambre's marginal note (c. 1810) on AOP E2-19, Delambre to Borda, 4 frimaire VI [24 November 1797].
10. For Delambre's initial election as (temporary) Secretary of the Academy on the same day that Napoleon was elected its President, see

ASPV 2 (1 germinal VIII [22 March 1800], pp. 11, 25 pluviôse XI [14 February 1803]), pp. 126, 625, 629. The presidency of the Academy had always been (and remained) a rotating and largely honorific office, although Napoleon used it to reorganize the Academy into two branches – one for the mathematical sciences (maths, physics, astronomy, geography and the mechanical arts) and one for the physical sciences (chemistry, the life sciences and medicine). Delambre was made Permanent Secretary for the mathematical sciences, with Cuvier made Permanent Secretary for the physical sciences. Napoleon's new regulations provided that the Permanent Secretary be named by the branch itself, but with 'the approbation' of the First Consul, meaning in effect that the position was in Napoleon's gift. Delambre privately noted that this interference in the self-governing nature of the Academy was a novelty of which he himself did not entirely approve. Delambre, 'Lui-même'.

11. For the proposal to revive the Balearic extension, see BL, 'Procès-verbaux', 19 fructidor X [6 September 1801]. For Méchain's report, see AN F17 3712, [Méchain], 'Rapport aux Consuls sur la continuation de la mesure de la méridienne de France depuis Barcelone jusqu'aux îles Baléares' [September–October 1802]. Antonio E. Ten, 'Le problème du 45e parallèle et les origines du système métrique décimal', in *Scientifiques et sociétés pendant la Révolution et l'Empire*, 114e Congrès National des Sociétés Savantes, 1989 (Paris: CTHS, 1990), pp. 441–52.

12. Méchain to Rolland, 10 floréal VIII [11 May 1799], in Dougados, 'Lettres de Méchain', p. 114. Méchain was 'alarmingly' ill in March–April 1801; see *ASPV* 2 (6 germinal, 1 prairial VIII [27 March, 21 May 1800]), pp. 128, 169. The young savant whom Delambre had in mind was Lalande's pupil Henri, then triangulating his way through Bavaria; see Lalande, *Bibliographie astronomique*, pp. 701, 704, 791, 868; also Delambre, *Grandeur*, pp. 223–4.

13. For Humboldt's pleasure at obtaining results approximating to those of Méchain, see CUS, Humboldt to Delambre, 23 floréal VII [12 May 1799]. Humboldt also compared his data to Méchain's in Humboldt to Zach, 12 May 1799, in Humboldt, *Die jugendbriefe Alexander von Humboldts, 1787–1799*, eds Ilse Jahn and Fritz G. Lange (Berlin: Akademie-Verlag, 1973), pp. 671–2. Humboldt's latitude result, taken on 8 January 1799, was 41°23′28″ and 41°22′59″, as compared with Méchain's result of 41°22′47″, supplied in a private letter from Méchain, which Humboldt mentioned. This suggests that Méchain also knew that Humboldt planned to take a measurement at the Fontana de Oro. See Delambre's commentary on these values in SBB Autgr. J1792(3), Delambre to Humboldt, 10 November 1807. The results were published in 1810 in a two-volume work dedicated to Delambre; see Alexandre de Humboldt and Jabbo Oltmanns, *Recueil d'observations astronomiques, d'opérations trigonométriques et de mesures barométriques* (Paris: Schoell, 1810), 2, pp. 3–6.

14. Méchain to Rolland, 6 frimaire VII [26 November 1798], in Dougados, 'Lettres de Méchain', p. 91.

15. AN F17 3712, [Méchain], 'Rapport aux Consuls sur la continuation de la mesure de la méridienne de France depuis Barcelone jusqu'aux îles Baléares' [September–October 1802]. For state approval of the mission, see BL, 'Procès-verbaux', 5 vendémiaire XI [27 September 1802]. For Laplace and Napoleon's interest in the project, see Zach's views, undoubtedly based on information from Lalande; KBD NKS1304, Zach to Bugge, 19 January 1803.

16. For Méchain's recruitment of Le Chevalier, see BML 26CA6, Méchain to Le Chevalier, 25 ventôse XI [16 March 1803]. For Le Chevalier's mission, see AN F17 3712, Le Chevalier to Min. Int. [Chaptal], 8 ventôse XI [27 February 1803]. For Méchain's acquaintance with Le Chevalier, see KBD NKS1304, Méchain to Bugge, 1 vendémiaire X [23 September 1801]. For Le Chevalier's career in astronomy and his relations with Méchain, see Bigourdan, *Astronomie d'observation*, pp. 137–40. For Le Chevalier's murky dealings in Madrid in 1796–7, see Camille Pitollet, 'Comment fut accueilli en Espagne la première ambassade française en faveur du système métrique', *Archivo de investigaciones históricas* 1 (1911), pp. 457–73. For Augustin Méchain, see KBD NKS1304, Méchain to Bugge, 1 vendémiaire X [23 September 1801]. Méchain's eldest son, Jérôme-Isaac, back safely from Egypt, had renounced astronomy – despite having discovered his own comet – in favour of a career in the Levant as a diplomatic attaché in the Dardanelles; see Méchain to Rolland, 5 jour comp. VI, 12 brumaire, 26 pluviôse, 30 prairial VIII, 18 germinal X [21 September 1798, 3 November 1799, 15 February, 19 June 1800, 8 April 1802], in Dougados, 'Lettres de Méchain', pp. 86, 110–11, 115, 125.

17. Méchain to Dezauche, 6 germinal XII [27 March 1804], in [A.-M. Dezauche], 'La dernière mission de l'astronome Méchain, 1804', *Revue rétrospective* 15 (1891), pp. 145–68, quotation on pp. 155–7. On the British reflectors, see AN F17 3702, Méchain to Min. Int., 8 pluviôse XI [28 January 1803].

18. For Delambre's account of Méchain's handing over his data at the last minute before his departure, see SBB J1792(3), Delambre to Humboldt, 10 September 1807.

19. For the ten-month claim, see KBD NKS1304, Méchain to Bugge, 2 ventôse XI [21 February 1803]. For his expected departure date of 1 February, see KBD NKS1304, Zach to Bugge, 19 January 1803. For his actual departure date, see BL, 'Procès-verbaux', 6 floréal XI [26 April 1803]. Méchain had originally intended to measure the length of a pendulum at Bordeaux; but he gave up this project to save time on his way south. He sailed to Barcelona from Montpellier.

20. For the delays in Barcelona, see AN F17 3712, Le Chevalier to Min. Int. [Chaptal], 25 floréal XI [5 May 1803]. Enrile had been in Paris during the preparations for the expedition; see AN F12 3712, Méchain, 'Etat de position', 12 frimaire XI [3 December 1802].

21. Coronado to Godoy, 29 December 1796, 6 January 1797, in Pitollet, 'Comment fut accueilli en Espagne', pp. 565–70.

22. AOP B4-9, Chaix to Méchain, 14 August 1804. Chaix had also assisted

Méchain briefly during his first expedition in 1792. Gonzales also warned Méchain about Coronado; see AOP B4-10, Gonzales to Méchain, 24 September 1803. Le Chevalier, who had tangled with Coronado in 1796–7, also blamed him for the delays in Méchain's mission; see AOP MS1054, Le Chevalier to Min. Int., 20 January 1806.

23. For the savant's intercession with the British, see CUS, Lalande (on behalf of the Bureau des Longitudes) to Min. Aff. Etr. [Talleyrand], 5 messidor XI [24 June 1803]. Gonzales did not think Méchain had much to worry about from the British; see AOP B4-10, Gonzales to Méchain, 30 July 1803.

24. For Méchain's foray down the coast, see Méchain to Delambre, 30 vendémiaire XII [23 October 1803], in Guillaume Bigourdan, 'La prolongation de la méridienne de Paris, de Barcelone aux Baléares, d'après les correspondances inédites de Méchain, de Biot et d'Arago', *Bulletin astronomique* 17 (1900), pp. 348–68, 390–400, 467–80; see especially pp. 352–6. The geodetic results from this portion of Méchain's survey were published in Jean-Baptiste Biot and Dominique-François-Jean Arago, *Recueil d'observations géodésiques, astronomiques et physiques* (Paris: Courcier, 1821), pp. 1–40.

25. For Méchain's accusation of lying, see Méchain to Delambre, 15 pluviôse XII [5 February 1804], in Bigourdan, 'Prolongation', p. 364. In fact, Ibiza *is* visible from Montsia under the right weather conditions.

26. Méchain to Delambre, 15 frimaire XII [7 December 1803], in Bigourdan, 'Prolongation', p. 357.

27. For the dispersal of his team, see Méchain to Delambre, 15 frimaire XII [7 December 1803], in Bigourdan, 'Prolongation', p. 359. The Spaniards, such as Enrile's second in command, Cini, refused to guard the reflectors, and the French (Le Chevalier) were irritated at being denied a chance to observe; Méchain to Dezauche, 4 vendémiaire, 28 pluviôse XII [27 September 1803, 18 February 1804], in [Dezauche], 'La dernière mission', pp. 145–7, 151–2.

28. On the poorly packed reflectors, see Méchain to Dezauche, 4 vendémiaire XII [27 September 1803], in [Dezauche], 'La dernière mission', pp. 145–7.

29. Méchain to Delambre, 30 vendémiaire XII [23 October 1803], in Bigourdan, 'Prolongation', p. 356.

30. On Canellas, see Méchain to Dezauche, 28 pluviôse XII [18 February 1804], in [Dezauche], 'La dernière mission', pp. 151–5.

31. On the trip across to Ibiza, see Méchain to Delambre, 4 pluviôse XII [25 January 1804], in Bigourdan, 'Prolongation', p. 361; Méchain to Dezauche, 4 pluviôse XII [25 January 1804], in [Dezauche], 'La dernière mission', pp. 147–9.

32. Méchain to Dezauche, 4 pluviôse XII [25 January 1804], in [Dezauche], 'La dernière mission', pp. 147–9. For Ibiza, see André Grasset de St-Sauveur, *Voyage dans les Iles Baléares et Pithiuses, fait dans les années 1801, 1802, 1803, 1804 et 1805* (Paris: Collin, 1807), pp. 249–87; Christian Augustus Fischer, *A Picture of Valencia*, trans. Frederic Shoberl (London: Colburn [1803], 1808), pp. 290–7.

33. Méchain to Dezauche, 15 pluviôse XII [5 February 1804], in [Dezauche], 'La dernière mission', pp. 149–50.
34. Méchain to Dezauche, 15 pluviôse XII [5 February 1804], in [Dezauche], 'La dernière mission', pp. 149–50.
35. Méchain to Delambre, 4 pluviôse XII [25 January 1804], in Bigourdan, 'Prolongation', p. 363.
36. On the sun clock, see Grasset de St-Sauveur, *Baléares*, pp. 97–9; Alexandre de Laborde, *Itinéraire descriptif de l'Espagne*, 2nd edn (Paris: Nicolle, 1809), 4, p. 441; Fischer, *Valencia*, p. 272; Sand, *Winter*, pp. 69–70.
37. For the eclipse, see Méchain, 'Mémoire sur l'éclipse de soleil du 20 pluviôse XII', 10 February 1804, *CT pour l'an XV* (pub. frimaire XII [November–December 1804]), pp. 476–82. For a recapitulation of Gonzales' expedition to Mallorca in December 1792, see AOP B4-10, Gonzales to Méchain, 24 September 1803.
38. Delambre to Méchain, sent 5 ventôse XII [25 February 1804], received in Palma on 22 ventôse XII [13 March 1804], in Bigourdan, 'Prolongation', pp. 390–1.
39. Méchain to Dezauche, 6 germinal XII [27 March 1804], in [Dezauche], 'La dernière mission', pp. 155–7. For Méchain's deference to the Bureau, see Méchain to Delambre, 16 germinal XII [6 April 1804], in Bigourdan, 'Prolongation', pp. 391–3; letter read to the Bureau of Longitudes, BL, 'Procès-verbaux', 7 floréal XII [7 March 1804].
40. On the dangers of delay and Coronado's machinations, see AOP B4-10, Gonzales to Méchain, 24 September 1803. On Valencia, see Fischer, *Valencia*, pp. 46–8, 200–4; Swindburne, *Travels*, 1, pp. 153–4; Alexandre de Laborde, *Voyage pittoresque et historique de l'Espagne* ([Paris]: n.p., 1806), vol. 1; Laborde, *Itinéraire*, 1, pp. 175–250; Richard Twiss, *Travels Through Portugal and Spain in 1772 and 1773* (London: Robinson, 1775), p. 201.
41. For Méchain's reconnoitring trip, see Méchain to Delambre, 11 fructidor XII [29 August 1804], in Bigourdan, 'Prolongation', pp. 396–400, 467–70. For a description of travel at this time of year in these regions in this period, see Twiss, *Travels*, pp. 213–14.
42. For Méchain's sunburn, see Méchain to [Mme Méchain?], 4 messidor XII [23 June 1804], in Bigourdan, 'Prolongation', pp. 393–4. For the flares, see Méchain to Delambre, 11 fructidor XII [29 August 1804], in Bigourdan, 'Prolongation', pp. 396–400, 467–70. For a description of the archbishop three years later, see François Arago, 'Histoire de ma jeunesse', in *Œuvres complètes*, 2nd edn (Paris: Legrand, Pomey et Crouzet, 1865), 1, pp. 32, 37.
43. For the operations of that summer, see Méchain to Delambre, 11 fructidor XII [29 August 1804], in Bigourdan, 'Prolongation', pp. 396–400, 467–70. On Canellas' mistake, see Ten, *Medir el metro*, p. 155.
44. Méchain to Delambre, 11 fructidor XII [29 August 1804], in Bigourdan, 'Prolongation', pp. 400, 467.
45. AAS Dossier Méchain, Méchain to Jaubert, 13 messidor XII [2 July 1804].

46. For a description of Méchain on the Sierra de Espadán, see Delambre, *Notice historique de Méchain*, p. 24. For the region around the Sierra de Espadán, see Townsend, *Journey Through Spain*, 3, pp. 296–8.
47. Dezauche, 'Journal' [1794], in [Dezauche], 'La dernière mission', pp. 159–68.
48. For a copy of Méchain's death certificate from the parish register of Castellón de la Plana, see AAS Dossier Méchain, 'Dn. Pedro Méchain', 14 September 1804.
49. For a late eighteenth-century diagnosis of intermittent fever (malaria), see Vicq d'Azyr and Jeanrol, 'Rapport . . . au sujet de l'épidémie qui a régné à Villeneuve-lès-Avignon', *Histoire de la Société Royale de Médicine* (1776), pp. 213–25. For the most widely used treatise, translated into all the major languages of Europe and continuously in print since the middle of the eighteenth century, see James Lind, *Essai sur les maladies des européens dans les pays chauds*, ed. and trans. Thion de la Chaume (Paris: Barrois, 1785). See also the compendia in J.-L. Alibert, *Traité des fièvres pernicieuses*, 3rd edn (Paris: Crapelet, 1804). For the history of malaria in Spain, see Leonard Jan Bruce-Chwatt and Julian de Zulueta, *The Rise and Fall of Malaria in Europe: A Historico-Epidemological Study* (Oxford: Oxford University Press, 1980), pp. 123–8.
50. On Méchain's obsession with his papers, see Delambre, *Notice historique sur M. Méchain*, p. 28.
51. For the news of Méchain's death, see *ASPV* 3 (16 vendémiaire XIII [8 October 1804]), p. 138.
52. On the sale of Méchain's library, see BN Delta 49306, *Catalogue des livres et instruments du feu M. P.-Fr.-And. Méchain*, sale on 4 floréal XIII [24 April 1805] (Paris: Bleuet, XIII [1805]). On the penury of the Méchain family after his death, see AN F17 1541, Jaubert to Min. Int. [Champagny], 18 ventôse XIII [9 March 1805]. On Mme Méchain's efforts to get a pension, see AN F17 1541, Mme Méchain to Min. Int. Champagny, received 3 nivôse XIII [24 December 1804]; Min. Int. to Mme Méchain, 11 ventôse XIII [2 March 1805]; Min. Int. 'Rapport à Sa Majesté l'Empereur', 24 pluviôse XIII [14 February 1805]; Napoleon, 'décret ce qui suit . . . Méchain', n.d. The final pension was 1500 *francs* per year, one third of which would come to Méchain's daughter upon her mother's death.
53. AOP E2-21, Augustin Méchain, 'Notice' [1804–5].
54. For Lalande's obituary, see Lalande, 'Nécrologie', *Moniteur* 22 (7 nivôse XIII [28 December 1804]): 78. See also Barón de la Puebla, 'Nachtrag zu Méchains Biographie', *MC* (1805), pp. 367–9.
55. Delambre, *Notice historique sur M. Méchain, lue, le 5 messidor XIII* [24 June 1805] (Paris: Baudouin, January 1806), p. 19.
56. Delambre, *Notice historique sur M. Méchain*, p. 30. Delambre did publish a few of Méchain's observations found amid his papers; see *ASPV* 3 (22 pluviôse XIII [11 February 1805]), p. 180.
57. For the publication of the eulogy, see BA MS2038, Delambre to Baudouin, 21 January 1806.

58. For a contemporary account of the pressure on Delambre to complete his book manuscript, and the fact that although he had received the first batch of Méchain's papers he had yet to go through them, see UBL MS074, Delambre to Van Swinden, 10 ventôse XIII [1 March 1805]. Delambre himself says he did not have Méchain's papers in time for Volume 1, in Delambre, *Base*, 2, pp. v–x. For a retrospective account, see Delambre, *Grandeur*, p. 224. On the meaning of the metre, see Ludwig Wittgenstein, *Philosophical Investigations*, trans. G. E. M. Anscombe, 2nd edn (Oxford: Blackwell, 1963), section 50. Also Saul Kripke, *Naming and Necessity* (Cambridge: Harvard University Press, 1972).

Chapter 11 *Méchain's Mistake, Delambre's Peace*

1. William Shakespeare, *Julius Caesar*, Act I, scene ii, in *The Riverside Shakespeare* (Boston: Houghton Mifflin [1599], 1974), p. 1108.

2. Jean-Baptiste-Joseph Delambre, *Histoire de l'astronomie moderne* (Paris: Courcier, 1821), 1, p. xli. He is speaking of his treatment of Descartes; see below.

3. Delambre, *Rapport historique*, p. 68.

4. KM, Delambre, *Base*, 1, title page. As it is not clear when Delambre presented the *Base* to Napoleon, it may well have been when all three volumes were published in 1810; see BA MS2038, Delambre to Baudouin, 21 January 1806. Delambre deliberately refrained from dedicating the *Base* to the Emperor to avoid the impression of kowtowing; see Delambre, 'Lui-même'.

5. AOP E2-9, Delambre's comments at end of Méchain's notebook.

6. For Méchain's alterations, see AOP E2-9, 23 January 1793. In one instance, Delambre could tell that the data had been recopied because of the way they were laid out on the page; yet Méchain had appended the times of the observations to the top and bottom of the page, something that served no purpose except to make the page appear to be original.

7. Delambre, in KM, Delambre, *Base*, 1, p. 510. For Delambre's notes on Méchain's sections, see KM, Delambre, *Base*, 1, pp. 289–510. For Méchain's 'falsifications' at Saint-Pons, see p. 345. The many alterations at Carcassonne (sixteen series suppressed) shifted the final outcome by 1.91 seconds; see p. 374. Delambre notes a case in which Méchain presented data to the International Commission that had been altered by two full seconds without any plausible explanation; see p. 386. Delambre also made corrections for typographical errors and occasionally tweaked his own equations.

8. For the latitude of Paris, then and now, see Delambre, *Grandeur*, p. 222; Bigourdan, *Système métrique*, p. 154.

9. Delambre's comments in AOP E2-9. See also KM, Delambre, *Base*, 1, p. 484; AOP MS1033b, Anon. [Tranchot?], 'Pour Delambre seul' [c. 1807–9]. Méchain published an abbreviated version of his results for Barcelona; see *CT pour l'an XII* (Paris, X [1801–2]), pp. 242–3.

10. Delambre, *Base*, 2, p. 619; Delambre, 'Auszug aus einem Briefe', 1 February 1808, *MC* 18 (May 1806), pp. 45–9. For his promise to publish all the data, see Delambre, 'Base du système métrique', *CT pour l'an 1808* (1807), pp. 463–6.

11. For the deposit of the papers, see Burckhardt, Biot and Bouvard, 'Dépôt des manuscrits à l'Observatoire impérial', 12 August 1807; and Bouvard, Burckhardt and Arago, 'Dépôt', 19 September 1810, in *Base*, 3, pp. 698–704. Delambre made the deposit in two lots because he wanted to hold on to some of the material while writing Volume 3 of the *Base*. However, unlike the published accounts of the second deposit in 1810, the manuscript version notes that Méchain's letters had been placed under seal. See the original at AOP D5-38, 'Dépôt', 26 September 1810.

12. AOP E2-9, Delambre's final comments in Méchain's notebook.

13. Delambre (c. 1810), marginal note to AOP E2-19, Méchain to Delambre, 7 brumaire VII [28 October 1798].

14. AOP E2-19, Méchain to Lalande, 11 ventôse IV [1 March 1796]. For the claim that the double star Mizar-Alcor may have ruined Méchain's data, see Jean-Nicolas Nicollet, *Mémoire sur un nouveau calcul des latitudes de Mont-Jouy . . . lu à l'Académie des sciences le 10 mars 1828* (Paris: Huzard-Courcier [1828]); also published in *CT pour 1831*, pp. 58–77. However, Méchain had publicly noted that Mizar was a double star; see Méchain, *MC* 8 (November 1803), p. 455. Moreover, the astronomer royal of England, George Airy, later examined Mizar-Alcor through the repeating circle which Méchain had sold to the Milanese, and found that he was able to resolve the double star. George Airy, 'The Figure of the Earth', *Encyclopaedia Metropolitana* (London: Fellowes et al., 1845), 5, p. 230. On the issue of refraction, see Delambre, 'Auszug, aus einem Briefe', 1 February 1808, *MC* 18 (1808), pp. 45–9. Delambre thought that Méchain's observations of Mizar proved that the Bradley tables were in error; see Delambre, *Base*, 2, p. 595.

15. For Delambre's speculations, see Delambre, *Base*, 2, pp. 618–19. Méchain himself worried that the Pyrénées would distort his readings; see AOP E2-19, Méchain to Lalande, 3 brumaire IV [25 October 1795]. Méchain's friend, the German astronomer Baron von Zach, cited Méchain's experience at Barcelona as evidence of the gravitational pull of mountains; see Franz-Xaver Zach, *L'attraction des montagnes et ses effets sur les fils à plomb* (Avignon: Seguin, 1814), p. 19. At the time, some geodesers noted that the tug of the local geography should have created a discrepancy in the opposite direction; see Joseph Rodriguez, 'Observations on the Measurement of Three Degrees of the Meridian', *Philosophical Transactions* 102 (1812), p. 344.

16. Some have suggested that Méchain sold his circle to the Milanese astronomers to get rid of the defective apparatus. This seems unlikely. Méchain gave the Milanese astronomers free choice between his two circles, and they chose the 360° old-fashioned circle rather than the 400° decimal circle, which is the one Méchain himself had used almost exclusively; see Delambre, *Base*, 3, pp. 503–4. For his sale to the

Milanese, see AOAB Cart. 88, Méchain to Oriani, 12 February 1795. For the Catalan conspiracy theory, see Moreu-Rey, *Naixement del metre*, pp. 91–2.

17. For Delambre's defence of the Fontana de Oro data, see Delambre, *Base*, 2, p. 620. Méchain's data for Mont-Jouy appear accurate by modern standards. In 1931, a Catalan astronomer measured the latitude of the exact same location on the Mont-Jouy tower, and using eighteenth-century methods of data reduction found the latitude to be 41°21'44.62", which differs by only 0.44 seconds from the latitude found by Méchain. But without comparable measurements at the Fontana de Oro, which is no longer in business today, this information tells us little about the discrepancy between the two results. See Isadore Polit, cited in A. Ten, 'Les expéditions de Méchain et Biot-Arago', in *Figure de la terre du XVIIIe siècle à l'ère spatiale*, eds Henri Lacombe and Pierre Costabel (Paris: Gauthier-Villars, 1988), pp. 245–65, especially p. 263.

18. For Nicollet's critique, see Nicollet, *Mémoire*. See the appreciation of this work in Airy, 'Figure'. See the misguided critique of this work in Anon., 'Réflexions sur un Mémoire de M. T. [sic] N. Nicollet', *Philosophical Magazine*, new series, 5 (1829), pp. 180–8. For eighteenth-century methods of calculating latitude, see Charles Cotter, *A History of Nautical Astronomy* (New York: Elsevier, 1968), pp. 123–79; also F. Marguet, *Histoire générale de la navigation du XVe au XXe siècle* (Paris: Société d'Editions Géographiques, Maritimes et Coloniales, 1931).

19. Arriving in the United States, Nicollet got in touch with Ferdinand Rudolph Hassler, described his geodetic instruments, books and skills to him, and asked to become part of the US geodetic survey; see NYPL Ford Collection, Nicollet to Hassler, 25 July 1832.

20. As Delambre did not measure any stars which passed south of the zenith, it is not possible to verify the accuracy of his results. Interestingly, Delambre was aware that one might balance 'north' and 'south' stars in this way, but he neither measured any southern stars, nor did he manipulate Méchain's data in this way, in part because he would have needed to know the declinations with much greater accuracy than was available in contemporary tables, and in part because the two stars he picked were much easier to locate than the stars Méchain selected; see Delambre, *Base*, 2, p. 186.

21. Delambre, 'Lui-même'. Delambre, *Tables écliptiques des satellites de Jupiter* (Paris: Courcier, 1817). On the savant who discovered an error in those tables, see Anon., 'Nécrologie de M. le chevalier Delambre', *Nouvelles annales des voyages* 15 (1822), pp. 425–8.

22. For Delambre's attitude towards perfection in nature, see UBL MS1872, Delambre to Moll, 21 July 1820. An obituary in *Ami de la religion* 33 (1822), pp. 111–12, called Delambre an atheist who respected faith, principally because he did not use geodesy or astronomy to challenge publicly the age of the earth as stated in the Bible.

23. Delambre, *Base*, 3, p. 103.

24. For Delambre's revised metre, see Delambre, *Base*, 3, p. 135.

25. For Delambre's mnemonics, see Delambre, *Base*, 3, p. 299.

26. WL 65667, Delambre to Lindeman, 1 May 1811.
27. For Legendre's discovery, see A.-M. Legendre, *Nouvelles méthodes pour la détermination des orbites des comètes avec un supplément*, 6 March 1805 (Paris: Courcier, 1806), especially pp. 72–80. Stephen M. Stigler calls the least-squares method the Ford Model-T of statistics; see 'Gauss and the Invention of Least Squares' [1981], in *Statistics on the Table: The History of Statistical Concepts and Methods* (Cambridge: Harvard University Press, 1999), p. 320. V. Parisot, 'Adrien-Marie Legendre', *Biographie universelle*, ed. Michaud, new edn (Paris: Desplaces, n.d.), 23, pp. 610–15; Elie de Beaumont, 'Eloge historique de Adrien-Marie Legendre', *MA* 32 (1864), pp. xxxvii–xciv.
28. For Laplace's *ancien régime* methods, see Laplace, 'Mémoire sur la figure de la terre' (1783, pub. 1786), in *Œuvres*, 11, pp. 5–9.
29. For Delambre's use of Legendre's analysis, see Delambre, *Base*, 3, p. 92.
30. For Gauss's interpretation of least squares, see Ch.-Fr. Gauss, *Méthodes des moindres carrés*, trans. J. Bertrand (Paris: Mallet-Bachelier, 1855). For evidence of his priority, Gauss claimed to have discussed it with several colleagues – although none of them seems to have understood him. Gauss clearly did have some kind of similar method at an early date because he was able to catch a typographical error in Delambre and Méchain's geodetic data, published in the *Allgemeine Geographische Ephemeriden* of 1799–1800. Delambre, in his role as Permanent Secretary of the Academy of Sciences, offered a fair-minded adjudication of this dispute, granting Legendre priority for the publication and the clarity of his presentation, while conceding Gauss's undeniable contributions; see Delambre, 'Analyse des travaux', *MA* (1811, pub. 1814), pp. iii–xiii. See also Churchill Eisenhart, 'The Meaning of Least Squares', *Journal of the Washington Academy of Sciences* 54 (1964), pp. 24–33; R. L. Plackett, 'The Discovery of the Method of Least Squares' [1972], in *Studies in the History of Statistics and Probability*, eds Maurice Kendall and R. L. Plackett (London: Griffin, 1977), 2, pp. 239–51; Stigler, 'Gauss and the Invention of Least Squares', pp. 320–31. For Gauss's work on least squares and geodesy, see O. B. Sheynin, 'C. F. Gauss and the Theory of Errors', and 'C. F. Gauss and Geodetic Observations', *Archive for History of the Exact Sciences* 20 (1979), pp. 21–72; 46 (1994), pp. 253–82. See also Laura Tilling, 'The Interpretation of Observational Errors in the Eighteenth and Early Nineteenth Centuries' (Ph.D. diss., Imperial College of Science and Technology, University of London, 1973). Bernard Bru, 'Laplace et la critique probabilistique des mesures géodesiques', in *Figure de la terre*, eds Lacombe and Costabel, pp. 223–44; M. Armatte, 'Théorie des erreurs, moyenne et loi normale', in *Moyenne, milieu, centre: Histoires et usages*, eds Jacqueline Feldman, Gérard Lagneau and Benjamin Matalon (Paris: Editions de l'Ecole des Hautes Etudes en Sciences Sociales, 1991), pp. 63–84; Eberhard Knobloch, 'Historical Aspects of the Foundations of Error Theory', in *The Space of Mathematics: Philosophical, Epistemological, and Historical Explorations*, eds Javier Echeverria, Andoni Ibarra and Thomas Mormann (Berlin: Gruyter, 1992), pp. 253–79.

31. For the first French discussion of what later became known as the personal equation, see Delambre and Laplace, 'Rapport sur la théorie de Mars, par Lefrançois-Lalande', *ASPV* 2 (21 brumaire X [12 November 1801]), pp. 426–9. The first cited instance of the personal equation was the discrepancy between Maskelyne and his assistant in England in 1796, though this was not 'explained' for another twenty years. See Stephen M. Stigler, *History of Statistics: The Measurement of Uncertainty before 1900* (Cambridge: Harvard University Press, 1986), pp. 240–2. I am indebted for my analysis of the transformation of the astronomical discipline to Simon Schaffer, 'Astronomers Mark Time: Discipline and the Personal Equation', *Science in Context* 2 (1988), pp. 115–45. See also Lorraine Daston, 'Enlightenment Calculators', *Critical Inquiry* 21 (1994), pp. 182–202; Giora Hon, 'Towards a Typology of Experimental Errors: An Epistemological View', *Studies in the History and Philosophy of Science* 20 (1989), pp. 469–504.

32. For the role of nineteenth-century statistics in the various sciences, see Theodore M. Porter, *The Rise of Statistical Thinking, 1820–1900* (Princeton: Princeton University Press, 1986).

33. Jean-Paul Marat, *Les Pamphlets – 1792*, ed. C. Vellay (Paris: Fasquelle, 1911), p. 295. This is, however, a precocious use of the term. The term 'scientist' was introduced into English in the 1840s by William Whewell. The equivalent term in French, 'scientifique', used as a noun, did not come into general use in France until the early twentieth century, and the term 'savant' continued to be used there throughout the nineteenth century. Nevertheless, many of the elements of the new scientific 'professionalism' (and the new political world which sustained it) emerged in the early nineteenth century.

34. Napoleon to Lalande, 15 frimaire V [5 December 1796], in Napoleon, *Correspondance*, 2, pp. 175–6.

35. BVCS MS99, Lalande, 'Journal', 23 November 1805. Jérôme Lalande, *Histoire céleste française* (Paris: Imprimerie de la République, IX [1801]). A few years later, his grandson Isaac, named after Isaac Newton, was hard at work in the family astronomical workshop; see Lalande, 'History of Astronomy for the Year 1805', *Philosophical Magazine* 26 (1806–7), p. 362.

36. BNR MS Fr 12273 fol. 73, Anon. 'Air des fraises: Voyez du nain des savans / La fierté peu commune / Il voulait savoir des vents / Si l'on parle de lui dans / La lune, la lune, la lune.' On Mlle Henry, see Aulard, *Thermidorienne*, 4 (22 messidor VI [10 July 1798]), p. 533.

37. *Gazette de France*, 4 ventôse XIII [23 February 1805], cited in François-Alphonse Aulard, *Paris sous le premier Empire* (Paris, Cerf, 1912–23), 1, p. 620.

38. BN R43050, Lalande in Sylvain M[aréchal] [with Jérôme Lalande], *Dictionnaire des athées anciens et modernes* (Paris: Grabit, VIII [1799]), p. 57.

39. BN Ln27 11115, Anon., *Grand conseil tenu par les sylphes* (Paris: Sourds-Muets, n.d.). For attacks on Lalande's atheism in the press, see Aulard, *Consulat*, 1 (12 December 1799, 18 March 1800), pp. 48, 221–2.

40. NL FRC18618, Lalande, *Notice sur Sylvain Maréchal* (n.p., n.d.), pp. 48–9. 'Les hommes fous, méchants ou bêtes / Prouvent que tout est mal dans cet indigne lieu. / Un scélérat suffit pour renverser les têtes; / L'homme ne serait plus s'il existait un Dieu.'

41. Delambre, 'Lui-même'.

42. Jérôme Lalande, *Seconde supplément au Dictionnaire des Athées* (n.p., 1805), p. 76. The publisher only printed the dictionary after being told by the Ministry of Police that he might do so for Lalande's private use. Lalande then 'accidentally' left some copies in the antechamber of the French Senate, where they were discovered. For a full discussion of this episode, see Aulard, 'Napoléon et l'athée Lalande', *Révolution française*, 4th series, 9 (1904), pp. 303–16. For Lalande's views on the function of religion, see NL FC18618, Lalande, *Notice sur Sylvain Maréchal* (n.p., n.d.), pp. 36–7, and his conversation with the Pope, p. 88. See also his cheeky attack on Napoleon as a warmonger in *Journal de l'Empire*, 13 fructidor XIII [31 August 1805], in Aulard, *Empire*, 2, p. 147.

43. For Napoleon's accusation against Lalande, see Napoleon (at Schönbrunn) to Min. Int. Champagny, 23 frimaire XIV [13 December 1805], in Napoleon, *Correspondance*, 11, pp. 574–6. For Lalande's acceptance of the conditions, see AN AF IV 1050, Delambre to Min. Int., 5 nivôse XIV [26 December 1805]; and CUS, Président de l'Institut, 'Certifie que ce qui suit . . .', 5 nivôse XIV [26 December 1805]. Delambre gave a detailed account of this episode and his attempt to make Lalande conform, while preserving academic freedom; see BI MS2041 v. 2, fol. 610, Delambre, 'Lalande' [1805]. For the re-edition of the *Dictionary*, see BN 8°R13719, Jérôme Lalande, *Seconde supplément au Dictionnaire des Athées* (n.p., 1806).

44. Lalande, 'Testament moral' [21 October 1804], in Amiable, *Lalande*, p. 53, emphasis added.

45. Delambre, 'Lalande', *Biographie universelle*, p. 613; Salm-Reifferscheid-Dyck, *La Lande*, p. 34. For rumours about Lalande's will, see *Gazette de France*, 8 April 1807, and *Journal de Paris*, 11 April 1807, in Aulard, *Empire*, 3, pp. 115, 117. For Delambre's increasingly harsh assessment of Lalande after his death, see Delambre, 'Eloge historique de M. de Lalande', 4 January 1808, *MA* (1807), pp. 30–57; Delambre, 'Lalande', *Biographie universelle*, pp. 603–13; Delambre, 'Lalande', *Astronomie au dix-huitième*, pp. 547–621.

46. BMA Arch. Rev. 2K10, Delambre to [Louis-François] Janvier, 12 August 1806.

47. On Delambre's debilitating illness of 1803, see CUS, Delambre to Cagnoli, 6 August 1810.

48. For Delambre's loans to Mme de Pommard at favourable interest in 1800–1, see BI MS1041 fol. 29. Her property was located in Courcelles-sur-Viosne; see AN M.C. Etude II 797, 'Donation entrevis par Madame Delambre', 5 vendémiaire XIII [27 September 1804]. For his housing, see CUS, Delambre to Delambre (his cousin, a notary), 7 floréal [c. 1800]. For a description of Mme Delambre, see Charles Dupin, 'Notice nécrologique sur M. Delambre', *Revue encyclopédique* 48 (December

1822), pp. 22–3. For their mutual friendship with Humboldt, who may have introduced them, see Delambre to Humboldt, 22 January 1801, in Alexander von Humboldt, *Briefe aus Amerika, 1799–1804* (Berlin: Akademie, 1993), p. 121.

49. BI MS2041, in Delambre's hand: 'An Athenian Air, Translation from the Roman.' Charles de Pommard's examiner for the Ecole Polytechnique was Biot, one of Delambre's scientific protégés, who ranked him near the middle of the incoming class; see BEP II/1, Ecole Impériale Polytechnique, 'Liste des élèves admis à l'Ecole Polytechnique en l'an 9' [1801]. Delambre to Humboldt, 22 January 1801, in Humboldt, *Briefe aus Amerika*, p. 120. WL 7080/7, Humboldt to Delambre, 27 August 1807.

50. BMA Arch. Rev. 2K10, Delambre to [Louis-François] Janvier, 12 August 1806. On Delambre's change of residence, see Delambre, *Grandeur*, p. 192. See his addresses in *Almanach national*.

51. For the view of Gauss, see Gauss to Bessel, 13 November 1814, in *Briefwechsel C. F. Gauss–F. W. Bessel* (Hildesheim: Olms, 1975), 1, p. 202. For a more generous view, see Wilson, 'Perturbations', pp. 283–96. Delambre's tables were plagiarized while still in manuscript form by the German astronomer Baron von Zach, who took advantage of Lalande's generosity in conveying scientific results. This infuriated Delambre, though he restrained himself publicly; see Delambre, 'Lui-même'.

52. For Delambre's military research, see AN AF IV 1205, Delambre to Gen. Duroc, 23 vendémiaire XII [16 October 1803]; see also Fernand Beaucour, 'Un problème d'optique posé par Napoléon à l'Institut, en 1803, résolu par Delambre', *Bulletin historique de la Société de Sauvegarde du Château Impérial de Pont-de-Briques* (1972), pp. 196–206. For the background of the threatened invasion, see Edouard Desbrière, *Projets et tentatives de débarquement aux îles Britanniques* (Paris: Chapelot, 1902), vol. 3.

53. For Delambre's key role in saving the life of James Smithson, who later endowed the Smithsonian Institution, see CUS, Delambre to Clarke (Min. Guerre), 16 April 1809. Delambre also convinced the Emperor to release the English astronomer Edmond Pigott; see APS MS 76-932, 'Letter book, 1802–6.' See also Banks to Delambre, 30 January 1804; Delambre to Banks, 11 October 1806, in Gavin de Beer, *The Sciences Were Never at War* (London: Nelson, 1960), pp. 138, 177, also pp. 154–5. Delambre sent eight copies of his *Base* to England; see Delambre to Banks, 27 April 1807, 2 March 1808, 8 March 1812, in de Beer, *Sciences*, pp. 179, 181, 192. For his general wishes for peace, see Delambre to Banks, 18 [March 1813], in de Beer, *Sciences*, pp. 193–5.

54. *ASPV 4* (27 August 1810), p. 375.

55. For the complaint filed by Méchain's 'children', see *ASPV 4* (24 June 1811), p. 490. As Jérôme-Isaac was then in the Dardanelles, this presumably means Augustin (and perhaps his sister). There is no mention of Mme Méchain's role. For the committee's response, see Arago et al., 'Rapport sur la réclamation de la famille Méchain', *ASPV 4* (8 July 1811), pp. 496–7. The committee was not stacked too heavily in Delambre's

favour. Although Arago, one of Delambre's protégés, was the author of the committee's report, the committee also included Guyton-Morveau, one of the two savants who had voted against Delambre's admission to the Academy in the first place; see Delambre, 'Lui-même'. The other members were Charles and Vauquelin. The final report asserted that Delambre had a conflict of interest because he was on the jury; but he was not on the jury, he was the Permanent Secretary. Institut de France, *Rapports et discussions de toutes les classes de l'Institut de France sur les ouvrages admis au concours pour les prix décennaux* (Paris: Baudouin, 1810), 1, pp. 130–7.

56. For Lalande's repudiation of the calendar he had once taken credit for, see Aulard, *Consulat*, 1 (21 November 1801), p. 618; 4 (27 September 1803), pp. 400–1; also 6 brumaire XIII [28 October 1804], and 18 messidor XIII [7 July 1805], in Aulard, *Empire*, 1, p. 349; 2, p. 43. Also see his 'I told you so', in Jean-Etienne Montucla, *Histoire des mathématiques*, ed. Lalande (Paris: Agasse, X [1802]), 4, pp. 329–33. Delambre also reminded his colleagues of the calendar's flaws; see *ASPV* 1 (26 pluviôse V [14 February 1797]), p. 172; 2 (6 vendémiaire IX [28 September 1800]), p. 233. For a general discussion of the repudiation of the calendar, see *PVCIP*, 6, pp. 207–13; Léon de Lanzac de Laborie, *Paris sous Napoléon* (Paris: Plon-Nourrit, 1905–), 3, pp. 202–6; [Blagdon], *Paris as It Was*, 2, pp. 79–80.

57. For exhortations to continue to use the metric measures, see Chaptal (Min. Int.) to Préfets, 2 frimaire XI [23 November 1802], in *Moniteur* 111, 112, 113 (21, 23, 24 nivôse XI [11, 13, 14 January 1803]), pp. 446, 454–5, 459–60. See 'Préfecture de Police', *Moniteur* 17 (17 vendémiaire XI [9 October 1802]), p. 63. For the ongoing use of old measures, see AN F2I/106/31, Duplantier (Préfet de Landes) to Conseiller d'Etat, 26 thermidor XIII [14 August 1805]. For a denial of the rumours of the metric system's demise, see ADSe VD* 430, Bureau Central de Paris, *Avis: poids et mesures, 26 brumaire VIII* [17 November 1799] (Paris: Lottin, VIII [1799]).

58. BLUC Laplace Box 10, Laplace to Napoleon, 7 May 1811. For the lobbying of Laplace and Delambre, see BLUC Laplace Box 10, Laplace to [Min. Int. Chaptal], 13 pluviôse XII [3 February 1804]; also printed in [Arthur-Jules] Morin, 'Notice historique sur le système métrique', *Annales du Conservatoire des Arts et Métiers* 9 (1871), p. 607. For the attempt to connect the metre to Napoleon's conquests, see AN F12 1289, Anon., unaddressed letter, 2 March 1811.

59. On the 'ordinary measures', see Napoleon, 'Décret concernant l'universalité des poids et mesures', 12 February 1812, in *Moniteur* 50 (19 February 1812), p. 199; Monalivet (Min. Int.) aux Préfets, 28 March 1812, in *Moniteur* 116 (25 April 1812), pp. 454–5.

60. Benjamin Constant, *De l'esprit de conquête* (Paris: Librairie de Médicis, 1813), pp. 53–4. See also Benjamin Constant, *Cours de politique constitutionelle* (Paris: Gullaumin, 1872), 2, pp. 170–5.

61. AN F12 1290, Préfet (Bouches du Rhin) to Min. Int., 6 July 1813; see also, Min. Int. to Préfet, 20 July 1813; Préfet to local mayors and

administrators, June 1813; Préfecture du Dépt. des Bouches du Rhin, broadsheets in French and Dutch, 12 February 1813 (Bois-le-Duc: Lion [1813]).

62. Napoleon, *Mémoires*, 4, pp. 211–15. By 'forty million people' Napoleon was referring to the population of imperial France. See also his bitter views on the savants and the metric system in Napoleon, *Sainte-Hélène, Journal inédite de 1815 à 1818*, ed. Gaspard Gourgaud (Paris: Flammarion, 1899), 1, p. 95; 2, p. 28.

63. UBL MS1872, Delambre to Moll, 7 May 1814.

64. On Delambre's second change of residence, see CUS, Delambre to Mlle Delambre (his sister), 18 November 1815; Delambre, 'Lui-même'. As Permanent Secretary, Delambre had a salary of 6000 *francs*, but he lost his salary as Treasurer of the University, which was much larger – 12,000 *francs*; see *ASPV* 6 (27 March 1816), p. 43. On Delambre's defence of the savants' political neutrality, see Delambre to Min. Int., 18 April 1816, in Bigourdan, 'Bureau des Longitudes' (1928), p. A49.

65. For the Report to the Emperor, see Jean Dhombres, 'Introduction' to *Rapports à l'Empereur sur le progrès des sciences, des lettres et des arts depuis 1789, vol. 1, Sciences mathématiques*, ed. Jean Dhombres (Paris: Belin, 1989), pp. 13–37.

66. CUS, Delambre to Cagnoli, 6 August 1810.

67. For the only comparable history of science prior to Delambre's work, see the history of (applied) mathematics begun by Montucla, with volumes 3 and 4 completed by Lalande in 1802; see Montucla, *Histoire des mathématiques*. For the relationship of Delambre's *Histoire* to his *Traité*, see Delambre, *Astronomie moderne*, 1, p. lii. For an appreciation of Delambre as a historian, see I. Bernard Cohen, 'Introduction', in Delambre, *Histoire de l'astronomie moderne* (New York: Johnson Reprint, 1969), pp. ix–xx.

68. On his historical method, see Delambre, *Histoire de l'astronomie ancienne* (Paris: Courcier, 1817), 1, pp. xviii–xx, xxxvi. He especially attacked the speculative *ancien régime* histories of Bailly; see Delambre, *Histoire de l'astronomie du moyen-age* (Paris: Courcier, 1819), pp. xxxiv–xxxvii. For Delambre's interest in the Egyptian expedition, see BI MS1041, Nouet to Delambre, 21 fructidor IX [8 September 1801]; 8 floréal X [28 April 1802]. Delambre and Méchain were the Académie's examiners for the results brought back from Egypt by Nouet and Jérôme-Isaac Méchain; see *ASPV* 2 (1 floréal X [21 April 1802]), p. 493. For his doubts about the pyramid claims, see BI MS 1042 fol. 388, Delambre, '*Recherches sur les sciences de l'Egypte* par M. Fourier', n.d. Delambre, *Astronomie ancienne*, 1, pp. 89–90. Delambre, *Astronomie du moyen-age*, pp. vi, lxv.

69. Delambre, *Astronomie moderne*, 1, p. xli; 2, p. 235, emphasis in original.

70. Delambre, *Astronomie moderne*, 2, pp. 199–200.

71. On Descartes' skull, see Delambre, 'Crâne venu de Suède et que l'on dit être celui de Descartes', *ASPV 7* (14 May, 8 October 1821), pp. 193–7, 232–3.

72. On Delambre's destruction of private letters and papers, see Charles Dupin, 'Notice nécrologique sur M. Delambre', *Revue encyclopédique* 48

(December 1822), pp. 12–13. For his autobiography and biography, see Delambre, 'Lui-même'; Mathieu, 'Delambre', *Biographie universelle*, pp. 304–8.

73. For Delambre's death certificate, see AN Etude CVIII 987, Jean-Eustache Montand, 'Actes de décès: Delambre', 26 August 1822. For the sale of Delambre's large collection of fifteen hundred books, see AOP 22569, *Catalogue des livres composant la bibliothèque de feu M. le Chevalier Delambre*, 10–20 May 1824 (Paris: Gaudfroy et Bachelier, 1824). For his eulogy, see Joseph Fourier, 'Eloge de M. Delambre', 2 July 1823, *MA* 4 (1824), pp. cciv–ccxxviii. For Fourier's appointment, see *ASPV* 7 (26 August 1822), p. 362.

74. UBL MS1872, Delambre to Moll, 21 July 1820.

75. For Delambre's last biography of Méchain, see Delambre, 'Méchain', *Histoire de l'astronomie au dix-huitième siècle*, ed. Claude-Louis Mathieu (Paris: Bachelier, 1827), pp. 755–67. Lalande himself said in his eulogy for Méchain that he had met the young man through a correspondence; see Lalande, 'Nécrologie', *Moniteur* 22 (7 nivôse XIII [28 December 1804]). Delambre wrestled several times in his manuscripts with the phrase: 'funeste résolution d'en faire mystère'; see BYU folder 32, Delambre, 'Méchain'; also BI MS2041 fol. 10, Delambre, 'Méchain'. Delambre also wrote a mini-biography of Méchain, published in 1821, which takes much the same tone; see Delambre, 'Méchain', *Biographie universelle*, ed. Michaud (Paris, 1821), 28, pp. 454–8.

76. Delambre, 'Méchain', *Astronomie au dix-huitième*, pp. 766–7.

77. Delambre, *Grandeur*, pp. 231, 234.

78. The sealed manuscripts were presumably opened by Guillaume Bigourdan at the very end of the nineteenth century, although he made no use of them in his *Système métrique* of 1901.

Chapter 12 The Metred Globe

1. G. K. Chesterton, 'The Rolling English Road', *The Flying Inn* (London: Methuen, 1914).

2. Josephus, *Jewish Antiquities* (1, p. 61), in *Works*, trans. H. St. J. Thackeray (London: Heinemann, 1930), 4, p. 29. Cain was the first to lay out territorial boundaries and to build a city.

3. John Quincy Adams (Secretary of State), 'Weights and Measures', US Senate, 22 February 1821; 16th Congress, 2nd Session, no. 503, Class 10, vol. 2, pp. 656–750; see p. 672. The report was written in response to a request made by the Senate on 3 March 1817 and echoed by President James Madison. See US House of Representatives, 15th Congress, 2nd Session, no. 463, pp. 538–42.

4. Armand Machabey, 'Aspects de la métrologie au XVII siècle', *Les Conférences du Palais de la Découverte*, Series D, 14 (1955), p. 5.

5. J. Q. Adams, 'Weights and Measures', p. 699.

6. Jefferson to J. Q. Adams, 1 November 1817, in Jefferson, *Writings*, 7, p. 87.

7. Mathieu pointed out that there was no need to remeasure the meridian should the Archive Metre be damaged, since its length was also known in relation to a pendulum; see *AP2* 111 (10 May 1837), p. 29. Laplace's son claimed the measure was equal to an 'aliquot' part of the meridian; see *AP2* 112 (12 June 1837), p. 496.

8. Anon., in *AP2* 111 (20 May 1837), p. 482. The physicist was Joseph-Louis Gay-Lussac. For the history of the legislation, see *AP2*, Min. de Commerce Martin du Nord, 107 (28 February 1837), pp. 627, 690–2; Rapport de Mathieu, 111 (10 May 1837), pp. 28–36; Débats, 111 (20 May 1837), pp. 478–84; Min. de Commerce Martin du Nord, 112 (27 May 1837), pp. 19–21; Rapport de Laplace, 112 (12 June 1837), pp. 495–500; Débats, 112 (16 June 1837), pp. 637–46, 779–80; Débats, 113 (22 June 1837), pp. 151–61; Débats, 112 (24 June 1837), pp. 305–6, 347–50; Débats, 112 (27 June 1837), pp. 462–7. The vote in the House of Deputies was 224 to 9; in the House of Peers, 65 to 21.

9. Charles Gilles, 'Ma Varlope' [c. 1848], in Pierre Brochon, ed., *Le pamphlet du pauvre, du socialisme utopique à la Révolution de 1848* (Paris: Edition Sociales, 1957), p. 112. 'Bravant la routine et sa haine / Dans sa valeur puisant son droit, / La mesure républicaine / A détrôner le pied de roi.' Quoted and translated in Eugen Weber, *Peasants into Frenchmen: The Modernization of Rural France, 1870–1914* (Stanford: Stanford University Press, 1976), pp. 30, 509. See also the instructional verses in Anon., *Complainte sur les poids et mesures* (Paris: Esculier, 1840).

10. The riot in Clamecy, dépt. de Nièvre, was sparked by 'new' measures, which were a decimal version of the 1825 'usual' measures, see *Moniteur* 109, 116 (19, 26 April 1837), pp. 923, 1002. Also *Echo de la Nièvre*, 9 April 1837, in Gustave Tallent, *Histoire du système métrique* (Paris: Soudier, 1910), pp. 88–91.

11. Anon., 'Les nouveaux poids et mesures', Tallent, *Système métrique*, p. 92. 'De quoi qu'nous sert c'te loi nouvelle? / A c't'heur' nous ne pourrons plus jamais / Demander un' liv' de chandelle, / Pas même un quart'ron de beurre frais. / Faudra qu' dans l'épic'rie / On mett' de vrais sorciers, / Ou que l'Académie / Fourniss' des épiciers. / *Chorus:* / Ce n'est pas d' nos faiseurs de lois / L' système / Décimal que j'aime. / Viv' les mesur' d'autrefois! / Au diabl' les nouveaux poids.'

12. For the continued use of the old measures, see Weber, *Peasants*, p. 36; Gaudefroy, *Mesures anciennes en usage à Amiens*, p. 30; Arthur Edwin Kennelly, *Vestiges of Pre-metric Weights and Measures Persisting in Metric-System Europe, 1926–1927* (New York: Macmillan, 1928), p. 30.

13. The Dutch law of 21 August 1816 made the metric system (minus the nomenclature) obligatory throughout the Low Countries on 1 January 1820 (delayed until 1821). The Belgian law of 18 June 1836 reinstated the Classical nomenclature. On Belgium, see J. Mertens, 'L'introduction du système métrique dans les Pays-Bas Meridionaux', *Janus* 60 (1973), pp. 1–12. On Holland, see Van Swinden to Delambre, 28 June 1802, in Bigourdan, *Système métrique*, pp. 242–4. On Luxembourg (and the rest of the Low Countries), see Henri Thill, 'Esquisse de l'histoire du système métrique dans notre pays', *Institut grand-ducal de Luxembourg: Section*

des sciences naturelles, physiques et mathématiques, Archives, new series, 20 (1951–3), pp. 95–130.

14. On Italy, see Kula, *Measures and Men,* pp. 268–75.

15. For the international movement for the metric system, see Edward Franklin Cox, 'The Metric System: A Quarter-Century of Acceptance, 1851–1876', *Osiris* 13 (1959), pp. 358–79. On the statistics conferences, see M. Engel, ed., *Compte-rendu général des travaux du congrès International de Statistique dans ses séances tenues à Bruxelles 1853, Paris 1855, Vienne 1857, et Londres 1860* (Berlin: Imprimerie Royale, 1863), pp. xx, 56, 192–3.

16. For the impact of the 1851 World's Fair, see Leone Levi, *Theory and Practice of the Metric System of Weights and Measures* (London: Griffith, 1871), pp. 2–3. For 1867, see Michel Chevalier, ed., *Exposition Universelle de 1867 à Paris: Rapports du Jury International* (Paris: Dupont, 1868), 2, pp. 485–500.

17. For the development of German laws, see 'No. 28, Maas- und Gewichtsordnung für den Norddeutschen Bund', 27 August 1868, in *Bundes-Gesetzblatt des Norddeutschen Bundes* (1868), pp. 473–80. International Statistical Congress (5th), *Programm der fünften Sitzungsperiode* (Berlin: Königliche Geheime Ober-Hofbuchdruckerei, 1863), pp. 79–87, 201–6.

18. For the improved knowledge of geodesy, see L.-B. Francoeur, *Géodesie, ou traité de la figure de la terre* (Paris: Bachelier, 1835), pp. 189–93; Louis Puissant (vols 6 and 7) and E. Peytier (vol. 9), *Nouvelle description géométrique de la France,* part of *Mémorial du Dépôt Général de la Guerre* 6 (Paris: Piquet, 1832), pp. 42, 126–9; 7 (Paris: Maulde, 1840), pp. 601–44; 9 (Paris: Maulde, 1853).

19. Baeyer to Min. War of Prussia, 'Entwurf zu einer mitteleuropäischen Gradmessung', April 1861, in Levallois, *Mesurer la terre,* p. 152.

20. For histories of geodesy in the nineteenth century, see Georges Perrier, *Petit histoire de la géodesie* (Paris: Alcan, 1939); Marie-Françoise Jozeau, 'Mesure de la terre au XIXe siècle', in *La mesure, instruments et philosophies,* ed. Jean-Claude Beaune (Paris: Callon, 1994), pp. 95–106; Levallois, *Mesurer la terre,* pp. 141–56.

21. Le Verrier, *CR* 57 (1863), p. 36. See the debate between Faye (speaking for the Bureau des Longitudes and pro-cooperation) versus Le Verrier (speaking for the Paris Observatory and anti-cooperation), in *CR* 56 (1863), pp. 28–37.

22. On the proposal for a new expedition, see Pontécoulant, *CR* 69 (27 September 1869), pp. 728–30. For the response, see Faye, 'Observations sur la lettre de M. de Pontécoulant,' *CR* 69 (4 October 1869), pp. 737–43.

23. Jacobi, Struve and Wild, 'Confection des étalons prototypes des poids et mesures métriques: Rapport de la Commission nommée par la Classe Physio-Mathématique de l'Académie des Sciences de Saint-Pétersbourg', in *CR* 69 (16 August 1869), pp. 425–8. These Russians were present at the Berlin meeting and reflect its views.

24. C. Bruhns, W. Foerster and A. Hirsch, eds, *Bericht über die Verhandlungen . . . der Europäischen Gradmessung* (Berlin: Reimer, 1868),

p. 126. Countries which sent delegates to this meeting included Holland, Belgium, Italy, Russia, Switzerland and all the German states. The damage to the bar was due to the fact that Lenoir had constructed the 1799 bar *à bouts* (an end-standard defined by two projecting flanges at the ends of the bar), not *à traits* (a line-standard defined by two marks precisely placed on the bar). Steinheil, *Abhandlungen der Baierischen Academie* 4 (1837), p. 251. Morin et al., 'Procès-verbal de comparaison entre étalons prototypes', 5 March 1864, in *Annales du Conservatoire Impériale des Arts et Métiers* 5 (1864), p. 6. On the impurities in platinum, see McDonald and Hunt, *Platinum*, pp. 147–77.

25. For the French debate over the challenge to the metre, see Mathieu, Laugier and Faye, 'Rapport du Commission', 24 December 1867, in Bigourdan, *Système métrique*, pp. 253–4. See also Morin et al., 'Rapport . . . sur la révision des étalons des bureaux de vérification des poids et mesures de l'Empire français en 1867 et 1868', *Annales du Conservatoire Impérial des Arts et Métiers* 9 (1871), pp. 5–63. AN F17 3715, Min. Instruction Publique, 'Note sur la construction d'un étalon métrique', 24 July 1869; Dumas et al., 'Rapport sur les prototypes du système métrique', 23 August 1869, *CR* 69 (1869), pp. 514–19.

26. Min. Aff. Etr. to French diplomatic agents, 16 November 1869, in Bigourdan, *Système métrique*, pp. 272–3. Min. Commerce, 'Rapport à S. M. l'Empereur', 1 September 1869, in Bigourdan, *Système métrique*, pp. 265–72.

27. Adolph Hirsch, in Commission Internationale du Mètre, *Session de 1870, Procès-verbaux des séances* (Paris: Baudry, 1871), p. 29. For the run-up to the meeting, see Morin, 'Notice historique sur le système métrique,' *Annales du Conservatoire des Arts et Métiers* 9 (1871), pp. 573–640.

28. For the meeting of 1872, see Commission Internationale du Mètre, *Procès-verbaux des séances du Comité des Recherches Préparatoires*, April 1872 (Paris: Viéville, 1872). Commission Internationale du Mètre, 'Procès-verbaux', September–October 1872, in *Annales du Conservatoire des Arts et Métiers* 10 (1872), p. 3–229.

29. For the convention of 1875, see Charles-Edmond Guillaume, *La création du Bureau International des Poids et Mesures* (Paris: Gauthier-Villars, 1927); US Department of Commerce, National Bureau of Standards, *The International Bureau of Weights and Measures, 1875–1975*, NBS Special Bulletin 420 (Washington, DC: USGPO, 1975). For internal French debates, see AN F17 3715, Bureau des Longitudes, 'Rapport sur la proposition relative à l'adoption d'une annèxe géodesique', 15 October 1875. The British and Dutch were also against the creation of a permanent international bureau.

30. For the making of the new metres, see Bigourdan, *Système métrique*, pp. 338–52.

31. For India, see Lal C. Verman and Jainath Kaul, eds, *Metric Change in India* (New Delhi: Indian Standards Institution, 1970).

32. [John Playfair], 'Base du Système Métrique', *Edinburgh Review* 9 (January 1807), pp. 373–91.

33. For Delambre's appreciation of Playfair, see Delambre, *Base*, 3, p. 308.

34. For measurement in nineteenth-century Britain, see Simon Schaffer, 'Metrology, Metrication and Victorian Values', *Victorian Science in Context*, ed. Bernard Lightman (Chicago: University of Chicago Press, 1997), pp. 438–74. A. D. C. Simpson, 'The Pendulum as the British Length Standard: A Nineteenth-Century Legal Aberration', in *Making Instruments Count: Essays on Historical Scientific Instruments*, eds R. G. W. Anderson, J. A. Bennett and W. F. Ryan (Cambridge: Variorum, 1993), pp. 174–90.

35. International Association for Obtaining a Uniform Decimal System of Measures, Weights and Coins, *Sixth Annual Report* (July 1862), p. 15.

36. William G. Armstrong, at BAAS, quoted in International Association for Obtaining a Uniform Decimal System of Measures, Weights and Coins, *Seventh Annual Report* (December 1863), p. 9.

37. For the most complete discussion of the nineteenth-century campaigns for metric conversion in Britain, see Edward Franklin Cox, 'A History of the Metric System of Weights and Measures, With Emphasis on Campaigns for its Adoption in Great Britain and in the United States Prior to 1914' (Ph.D. diss., Indiana University, 1956).

38. *The Times* was anti-metric, though it denied being anti-French. See Ewart, *The Times* (16 June 1864), p. 11.

39. James Yates, *What Is the Best Unit of Length?* (London: Bell and Daldy, 1858), pp. 20, 24–46.

40. For Herschel's axis inch, see John F. W. Herschel, 'The Yard, the Pendulum and the Metre' [1863], *Familiar Lectures on Scientific Subjects* (New York: Routledge, 1869), pp. 419–51. Also, John F. W. Herschel, 'The Battle of the Standards', *The Times* (21 June 1864), p. 7; (4 July 1864), p. 11; (2 May 1864), p. 12.

41. William John Macquon Rankine, 'The Three-Foot Rule', *Songs and Fables* (Glasgow: Maclehouse, 1874).

42. Anon., 'Chains of Habit', *Decimal Educator* 1 (September 1918), p. 19.

43. Editorial in *Toronto Star*, quoted in Grace Ellen Watkins, 'Metrication in the United States: A Social Constructivist Approach' (Ph.D. diss., Southern Illinois University, 1998), p. 305. See also Minister of Trade and Commerce, Government of Canada, *White Paper on Metric Conversion in Canada*, January 1970; Gerald Black, *Canada Goes Metric* (Toronto: Doubleday, 1974).

44. For American diversity, see J. Q. Adams, 'Weights and Measures', pp. 741–3. In the 1820s the number of troy grains in an avoirdupois pound still differed by as much as 5 per cent in customs houses up and down the Atlantic coast, and the capacity of a bushel basket by as much as 20 per cent. Louisiana had a different colonial experience and so different measures.

45. The best history of the metric system in America is Charles F. Treat, *A History of the Metric System Controversy in the United States*, National Bureau of Standards, Pub. 345-10 (Washington, DC: USGPO, 1971). See also, Cox, 'A History of the Metric System'. For the arrival of the 1799 iron metre in the US, see APS, Hassler, 'Confrontation des toises faite par le Comité des poids et mesures à Paris', November 1806. For

the sale of the measures, see APS, Hassler, 'Livres concernant les mesures de degrés' [1808]. Hassler also visited Lalande and Delambre in Paris in the 1790s to collect metrical standards and geodetic instruments, including a Borda repeating circle; see Hassler, *Memoirs*, trans. Emil Zschokke (Nice: Gauthier [1877], 1882), pp. 36–8, 53–7. Florian Cajori, *The Chequered Career of Ferdinand Rudolph Hassler* (Boston: Christopher, 1929), pp. 20–3, 35–6. For Hassler's report, see 'Weights and Measures', 29 June 1832, 22nd Congress, 1st Session, *Register of Debates* (Washington, DC: Duff Green, 1832), pp. 1–123. For general information on weights and measures at this time, see *North American Review* 97 (October 1837), pp. 269–92. National Academy of Sciences, *A History of the First Half-Century of the National Academy of Sciences, 1863–1913* ([Washington, DC]: n.p., 1913), pp. 206–13.

46. Josh Billings, in Aubrey Drury, *World Metric Standardization: An Urgent Issue* (San Francisco: World Metric Standardization Council, 1922), p. 157.

47. Charles Latimer, *Proceedings of the Ohio Auxiliary Society of the International Institute*, January 1887, in Treat, *History of the Metric System*, p. 89. Charles Latimer, *The French Metric System, or, The Battle of the Standards* (Chicago: Wilson, 1880), pp. 28–9. Edward F. Cox, 'The International Institute: First Organized Opposition to the Metric System', *Ohio Historical Quarterly* 68 (1959), pp. 54–83.

48. President Gerald Ford quoted in Susan Fraker Holt, revised by Gretchen Borges, *The United States and the Metric System* (Federal Reserve Bank of Minneapolis, December 1976), p. 36, emphasis added. Daniel V. De Simone, ed., *A Metric America: A Decision Whose Time Has Come*, National Bureau of Standards, Pub. 345 (Washington, DC: USGPO, 1971).

49. Bob Greene, 'Man from WAM! Puts His Foot Down', *Chicago Tribune* (11 April 1978), 2, p. 1.

50. For the US in the 1990s, see Gary P. Carver, *A Metric America: A Decision Whose Time Has Come – for Real* (Gaithersburg, MD: National Institute of Standards and Technology, June 1992). George Gallup, Jr, *The Gallup Poll* (Wilmington, DE: Scholarly Resources, 1991), p. 210.

51. Arthur G. Stephenson et al., 'Mars Climate Orbiter Mishap Investigation Board: Phase I Report', NASA, 10 November 1999. *Science* 286 (1999), pp. 18, 207. *New York Times* (21, 24 September, 1 October 1999).

Epilogue The Shape of Our World

1. Jean-Paul Sartre, *La nausée* (Paris: Gallimard, [1938]), pp. 179–80.

2. On the Melun baseline in the nineteenth century, see ADSM MDZ333, Bassot et al., 'Vérification faite en 1882 des travaux géodesiques des astronomes Delambre, Laplace et de Prony', 12 August 1882. For the traffic accident, see Levallois, *Mesurer la terre*, p. 134.

Time-Line of the French Revolution and the Adoption of the Metric System

Date	The French Revolution	The Metric System
1788	Louis XVI convokes the Estates-General	
1789		
April		Lalande urges the Estates-General to adopt a uniform physical standard of measurement
June	The Estates-General becomes the National Assembly	
July	Fall of the Bastille	
August	Declaration of the Rights of Man, end of feudal privilege	
1790		
February		The National Assembly hears a proposal for uniform measures based on Paris standards
March		The National Assembly adopts Talleyrand's proposal for measures based on a pendulum
1791		
March		The National Assembly approves a measurement standard based on the meridian
June	The 'Flight to Varennes' disgraces Louis XVI	
1792		
June	War between France and Prussia	Delambre and Méchain set out on the meridian expedition

DATE	THE FRENCH REVOLUTION	THE METRIC SYSTEM
1792 *continued*		
July		Delambre scouts out stations near Paris
		Méchain arrives in Barcelona to scout out stations in Catalonia
August	10 August uprising ends the monarchy	
September	The fall of Verdun fortress opens Paris to Prussians	Delambre arrested at Lagny and Saint-Denis
	The Republic is declared	Méchain triangulates in Catalonia
	The battle of Valmy begins the Prussian retreat	
October		Delambre triangulates in the Paris region
December		Méchain measures the latitude at Mont-Jouy
1793		
January	Execution of Louis XVI	
February	War between France and Britain	
March	War between France and Spain	
April	Creation of the Committee of Public Safety	The Spanish oblige Méchain to relocate to the Fontana de Oro inn in Barcelona
May		Méchain's near-fatal accident at the pumping station
June		Delambre begins to triangulate south from Dunkerque towards Paris
August	Academy of Sciences is abolished	The metric law establishes a provisional metre and a set of systematic prefixes
September	The Terror begins	Méchain triangulates from the peaks of the Pyrénées
October	Revolutionary calendar begins with the year II	
December		Delambre is officially purged from the expedition
		Méchain measures the latitude at the Fontana de Oro
1794		
March		Méchain discovers a discrepancy in his two values for the latitude of Barcelona

DATE	THE FRENCH REVOLUTION	THE METRIC SYSTEM
May	French conquest of the Low Countries	Méchain leaves Barcelona for Italy, stays in Genoa for a year
July	Robespierre is executed; the Terror ends	
1795		
April	Peace treaty between France and Prussia	Law of 18 germinal III relaunches the meridian expedition, finalizes the metric system
		Méchain leaves Genoa for Marseille; stays there for six months
May	Hyperinflation gathers momentum	Delambre resumes his triangulations between Orléans and Bourges
July	Peace treaty between France and Spain	
August		Méchain resumes his triangulations between Perpignan and Carcassonne
1796		
May	Napoleon's invasion of Italy	Delambre triangulates south of Bourges towards Evaux
1797		
February	Abolition of paper money	
April		Delambre triangulates south of Evaux towards Rodez
August		Delambre completes triangles at Rodez and returns to Paris
October		Méchain triangulates at Pic de Nore, then returns to Carcassonne
1798		
April		Delambre begins to measure baseline in Melun, near Paris
May	Napoleon departs on Egyptian expedition	
July		Delambre arrives in Perpignan to measure baseline
		Madame Méchain meets her husband in Rodez
August		Méchain resumes his triangulations from Rodez towards Carcassonne

DATE	THE FRENCH REVOLUTION	THE METRIC SYSTEM
1798 *continued*		
September		Official start date for International Commission to determine length of the metre
October		Delambre waits in Carcassonne while Méchain completes final stations
November		Delambre and Méchain return to Paris
		First meeting of International Commission to determine metre
1799		
February		Delambre's geodetic data accepted by International Commission
March		Méchain's geodetic data accepted by International Commission
June		Definitive platinum metre bar presented to legislature
September		Metre becomes obligatory in region surrounding Paris
October	Napoleon returns from Egypt	
November	Napoleon's coup d'état of 18 brumaire	
1800		
November		Abolition of systematic prefixes for metric units, return to 'ordinary names'
1801		
July	Napoleon's Concordat with Catholic Church	
September		Metric system obligatory throughout France
1802		
March	Peace of Amiens between France and Britain	
August	Napoleon made Consul for Life	
1803		
January		Delambre made Permanent Secretary of the Academy
April		Méchain leaves for second expedition to Spain

DATE	THE FRENCH REVOLUTION	THE METRIC SYSTEM
June	France and Britain edge towards war	Méchain triangulates down the Catalan coast
1804		
January		Méchain crosses to Ibiza, then Mallorca
April		Méchain triangulates in Valencia
September		Death of Méchain
October		Méchain's son returns to Paris with his father's papers
December	Napoleon crowned Emperor	
1805		
November		Delambre becomes aware of Méchain's error
December	Austria, Prussia and Russia defeated at Austerlitz	
1806		
January	Revolutionary calendar ends	Delambre publishes first volume of *Base*
1807		Delambre publishes second volume of *Base*
1812	Napoleon invades Russia	Napoleon rescinds metric system
1815	Fall of Napoleon, restoration of Bourbons	
1820		Metric system adopted in the Low Countries
1822		Death of Delambre
1830	Fall of the Bourbons, constitutional monarchy	
1837		French legislature revives the metric system
1840		Metric system becomes obligatory throughout France

SELECTED BIBLIOGRAPHY

Selected Primary Sources

Adams, John Quincy. 'Weights and Measures.' US Senate, Sixteenth Congress, Second Session, 22 February 1821. In *American State Papers: Documents*, no. 503, class 10, vol. 2, pp. 656–750. Eds Walter Lowrie and Walter S. Franklin. Washington, DC: Gales and Seaton, 1834.

Aulard, François-Alphonse, ed. *Paris pendant la réaction thermidorienne et sous le Directoire*. Paris: Cerf, 1898–1902.

Aulard, François-Alphonse, ed. *Paris sous le Consulat*. Paris: Cerf, 1903.

Aulard, François-Alphonse, ed. *Paris sous le premier Empire*. Paris: Cerf, 1912–23.

Bigourdan, Guillaume, ed. 'La prolongation de la méridienne de Paris, de Barcelone aux Baléares, d'après les correspondances inédites de Méchain, de Biot et d'Arago.' *Bulletin astronomique* 17 (1900), pp. 348–68, 390–400, 467–80.

Biot, Jean-Baptiste and Dominique-François-Jean Arago. *Recueil d'observations géodesiques, astronomiques et physiques*. Paris: Courcier, 1821.

Borda, Jean-Charles. *Description et usage du cercle de reflection*. Paris: Didot, 1787.

Bugge, Thomas. *Reise nach Paris in den Jahren 1798 und 1799*. Trans. Johann Nicolaus Tilemann. Copenhagen: Brummer, 1801. English version: *Travels in the French Republic*. Trans. John Jones. London: Phillips, 1801. Reissue in English: *Science in France in the Revolutionary Era*. Ed. Maurice Crosland. Cambridge, MA: MIT Press, 1969.

Cassini IV, Jean-Dominique, Pierre-François-André Méchain and Adrien-Marie Legendre. *Exposé des opérations faites en France en 1787, pour la jonction des observatoires de Paris et de Greenwich*. Paris: Institution des Sourds-Muets, 1790.

Delambre, Jean-Baptiste-Joseph. *Grandeur et figure de la terre*. Ed. Guillaume Bigourdan. Paris: Gauthier-Villars, 1912.

Delambre, Jean-Baptiste-Joseph. *Histoire de l'astronomie au dix-huitième siècle*. Ed. Claude-Louis Mathieu. Paris: Bachelier, 1827.

Delambre, Jean-Baptiste-Joseph. *Histoire de l'astronomie moderne*. Paris: Courcier, 1821.

Delambre, Jean-Baptiste-Joseph. *Méthodes analytiques pour la détermination d'un arc du méridien*. Dated: 11 germinal VII [31 March 1799]. Bound with Adrien-Marie Legendre. *Méthode pour déterminer la longueur exacte du quart du méridien*. Dated: 9 nivôse VII [29 December 1798]. Paris: Crapelet, VII [1799].

Delambre, Jean-Baptiste-Joseph. *Notice historique sur M. Méchain, lue le 5 messidor XIII* [24 June 1805]. Paris: Baudouin, January 1806.

Delambre, Jean-Baptiste-Joseph. *Rapport historique sur les progrès des sciences mathématiques depuis 1789*. Paris: Imprimerie Impériale, 1810.

Delambre, Jean-Baptiste-Joseph, ed. *Base du système métrique décimal, ou mesure de l'arc du méridien compris entre les parallèles de Dunkerque et Barcelone, exécutée en 1792 et années suivantes, par MM. Méchain et Delambre*. Paris: Baudouin, 1806, 1807, 1810.

[Dezauche, A.-M.]. 'La dernière mission de l'astronome Méchain, 1804.' *Revue retrospective* 15 (1891), pp. 145–68.

Dougados, [Isidore]. 'Lettres de l'astronome Méchain à M. Rolland.' *Mémoires de la Société des Arts et des Sciences de Carcassonne* 2 (1856), pp. 74–130.

Gales, Joseph, ed. *Debates and Proceedings in the Congress of the United States*. Washington, DC: Gales and Seaton, 1834–56.

[Haüy, René-Just]. *Instructions sur les mesures déduites de la grandeur de la terre*. First edn. Paris: Imprimerie Nationale, II [1794].

Jefferson, Thomas. *The Papers of Thomas Jefferson*. Ed. Julian P. Boyd. Princeton: Princeton University Press, 1950–.

Jefferson, Thomas. *The Writings of Thomas Jefferson*. Ed. H. A. Washington. New York: Derby and Jackson, 1859.

Lalande, Joseph-Jérôme Le Français de. *Bibliographie astronomique; avec l'histoire de l'astronomie depuis 1781 jusqu'à 1802*. Paris: Imprimerie de la République, XI [1803].

Laplace, Pierre-Simon. *Oeuvres complètes de Laplace*. Paris: Gauthier-Villars, 1878–1912.

Lavoisier, Antoine-Laurent. *Oeuvres de Lavoisier*. Paris: Imprimerie Impériale: 1862–93.

Leblond, Auguste-Savinien. *Sur la fixation d'une mesure et d'un poid – lu à l'Académie des Sciences le 12 mai 1790*. Paris: Demonville, 1791.

Legendre, Adrien-Marie. *Nouvelles méthodes pour la détermination des orbites des comètes avec un supplément*. Dated: 6 March 1805. Paris: Courcier, 1806.

Méchain, Pierre-François-André. 'Recherches sur les comètes de 1532 et 1661.' *Mémoires de mathématique et de physique* 10 (1782), pp. 330–96.

Miller, John Riggs. *Speeches in the House of Commons upon the Equalization of the Weights and Measures of Great Britain*. London: Debrett, 1790.

Nicollet, Jean-Nicolas. *Mémoire sur un nouveau calcul des latitudes de Mont-Jouy . . . lu à l'Académie des sciences le 10 mars 1828*. Paris: Huzard-Courcier, [1828]. Also published in *Connaissance des temps . . . pour 1831*, pp. 58–77.

Paucton, Alexis-Jean-Pierre. *Métrologie, ou traité des mesures, poids et monnoies des anciens peuples et des moderns*. Paris: Veuve Desaint, 1780.

Prieur-Duvernois, Claude-Antoine. *Mémoire sur la nécessité et les moyens de rendre uniformes dans le royaume toutes les mesures*. Dijon: Causse, 1790.

Raspail, Julien. 'Papiers de Lalande.' *La Révolution française* 74 (1921), pp. 236–54.

Young, Arthur. *Travels during the Years 1787, 1788, and 1789*. Dublin: Gross, 1793.

Selected Secondary Sources

Alder, Ken. *Engineering the Revolution: Arms and Enlightenment in France, 1763–1815.* Princeton: Princeton University Press, 1997.

Baker, Keith Michael. 'Science and Politics at the End of the Old Regime.' In *Inventing the French Revolution: Essays on French Political Culture in the Eighteenth Century,* pp. 153–66. Cambridge: Cambridge University Press, 1990.

Bigourdan, Guillaume. 'Le Bureau des Longitudes, son histoire et ses travaux, de l'origine (1795) à ce jour.' *Annuaire du Bureau des Longitudes* (1928), pp. A1–72; (1929), pp. C1–92; (1930), pp. A1–110; (1931), pp. A1–151; (1932), pp. A1–117.

Bigourdan, Guillaume. *Histoire de l'astronomie d'observation et des observatoires en France, seconde partie.* Paris: Gauthier, 1930.

Bigourdan, Guillaume. *Le système métrique des poids et mesures.* Paris: Gauthier-Villars, 1901.

Bouchard, Georges. *Un organisateur de la victoire: Prieur de la Côte-d'Or.* Paris: Clavreuil, 1946.

Cox, Edward Franklin. 'A History of the Metric System of Weights and Measures, with Emphasis on Campaigns for its Adoption in Great Britain and in the United States prior to 1914.' Ph.D. diss., Indiana University, 1956.

Cox, Edward Franklin. 'The Metric System: A Quarter-Century of Acceptance, 1851–1876.' *Osiris* 13 (1959), pp. 358–79.

Crosland, Maurice. '"Nature" and Measurement in Eighteenth-Century France.' *Studies on Voltaire and the Eighteenth Century* 87 (1972), pp. 277–309.

Crosland, Maurice. 'The Congress on Definitive Metric Standards, 1798–1799: The First International Scientific Conference?' *Isis* 60 (1969), pp. 226–31.

Débarbat, Suzanne. 'Coopération géodesique.' *Echanges d'influences scientifiques et techniques entre pays européens de 1780 à 1830,* pp. 47–76. Actes du 114e Congrès National des Sociétés Savantes, Paris, 1989. Paris: Comité des Travaux Historiques et Scientifiques, 1990.

Devic, J.-F.-S. *Histoire de la vie et des travaux scientifiques et littéraires de J.-D. Cassini IV.* Clermont: Daix, 1851.

Favre, Adrien. *Les origines du système métrique.* Paris: Presses Universitaires de France, 1931.

Fischer, Joachim. *Napoleon und die Naturwissenschaften.* Stuttgart: Steiner, 1988.

Garnier, Bernard and Jean-Claude Hocquet, eds. *Genèse et diffusion du système métrique.* Caen: Editions du Lys, 1990.

Gillispie, Charles Coulson. *Pierre-Simon Laplace, 1749–1827: A Life in Exact Science.* Princeton: Princeton University Press, 1997.

Greenberg, John L. *The Problem of the Earth's Shape from Newton to Clairaut: The Rise of Mathematical Science in Eighteenth-Century Paris and the Fall of 'Normal' Science.* Cambridge: Cambridge University Press, 1995.

Hahn, Roger. *The Anatomy of a Scientific Institution: The Paris Academy of Sciences, 1666–1803.* Berkeley: University of California Press, 1971.

Heilbron, John L. 'The Measure of Enlightenment.' In *The Quantifying Spirit in the Eighteenth Century*, pp. 207–42. Eds Tore Frängsmyr, John L. Heilbron and Robin E. Rider. Berkeley: University of California Press, 1990.

Heilbron, John L. *Weighing Imponderables and Other Quantitative Science around 1800*. Berkeley: University of California Press, 1993.

Kula, Witold. *Measures and Men*. Trans. R. Szreter. Princeton: Princeton University Press, 1986.

Lacombe, Henri and Pierre Costabel, eds. *La figure de la terre du XVIIIe siècle à l'ère spatiale*. Paris: Gauthier-Villars, 1988.

Laissus, Joseph. 'Un astronome français en Espagne: Pierre-François-André Méchain (1744–1804).' In *Comptes rendues du 94e Congrès National des Sociétés Savantes*, pp. 36–59. Paris: Bibliothèque Nationale, 1970.

Levallois, Jean-Jacques. *Mesurer la terre: 300 ans de géodesie française*. Paris: Presses de l'Ecole Nationale des Ponts et Chaussées, 1988.

Mascart, Jean. *La vie et les travaux du chevalier Jean-Charles de Borda, 1733–1799*. Lyon: Rey, 1919.

Miller, Judith. *Mastering the Market: The State and Grain Trade in Northern France, 1700–1860*. Cambridge: Cambridge University Press, 1999.

Moreu-Rey, Enric. *El naixement del metre*. Palma de Mallorca: Moll, 1956.

Morin, [Arthur-Jules]. 'Notice historique sur le système métrique.' *Annales du Conservatoire des Arts et Métiers* 9 (1871), pp. 573–640.

Noël, Yves and René Taton. 'La réforme des poids et mesures, origines et premières étapes (1789–91).' In *Oeuvres de Lavoisier*. Volume 7: *Correspondance*, 6, pp. 339–65. Ed. Patrice Bret. Paris: Académie des Sciences, 1997.

Porter, Theodore M. *Trust in Numbers: Objectivity in Science and Public Life*. Princeton: Princeton University Press, 1995.

Roncin, Désiré. 'Mis en application du système métrique (7 avril 1795–4 juillet 1837).' *Cahiers de métrologie* 2 (1984), pp. 3–86; 3 (1985), pp. 19–130.

Rothschild, Emma. *Economic Sentiments: Adam Smith, Condorcet, and the Enlightenment*. Cambridge, MA: Harvard University Press, 2001.

Sahlins, Peter. *Boundaries: The Making of France and Spain in the Pyrénées*. Berkeley: University of California Press, 1989.

Schaffer, Simon. 'Astronomers Mark Time: Discipline and the Personal Equation.' *Science in Context* 2 (1988), pp. 115–45.

Schaffer, Simon. 'Metrology, Metrication, and Victorian Values.' In *Victorian Science in Context*, pp. 438–74. Ed. Bernard Lightman. Chicago: University of Chicago Press, 1997.

Stigler, Stephen M. *History of Statistics: The Measurement of Uncertainty before 1900*. Cambridge, MA: Harvard University Press, 1986.

Ten, Antonio E. *Medir el metro: La historia de la prolongatión del arco de meridiano Dunkerque-Barcelona, base del Sistemo Métrico Decimal*. Valencia: Universitat de València, 1996.

Treat, Charles F. *A History of the Metric Controversy in the United States*. National Bureau of Standards, Pub. 345–10. Washington, DC: GPO, 1971.

Turner, Anthony John. *From Pleasure and Profit to Science and Security: Etienne Lenoir and the Transformation of Precision Instrument-Making in France, 1760–1830*. Cambridge: Whipple Museum, 1989.

Wolf, Charles. 'Recherches historiques sur les étalons de l'Observatoire.' *Annales de chimie et de physique* 5th series, 25 (1882), pp. 5–112.

Zupko, Ronald. *French Weights and Measures before the Revolution: A Dictionary of Provincial and Local Units.* Bloomington: Indiana University Press, 1978.

Zupko, Ronald. *Revolution in Measurement: Western European Weights and Measures since the Age of Science.* Philadelphia: American Philosophical Society, 1990.

INDEX

Page number in *italics* denotes illustration
n denotes note